Christopher B. Johnson
Chlodwig Franz
Editors

Breeding Research on Aromatic and Medicinal Plants

Breeding Research on Aromatic and Medicinal Plants has been co-published simultaneously as *Journal of Herbs, Spices & Medicinal Plants*, Volume 9, Numbers 2/3 and 4 2002.

Pre-publication
REVIEWS,
COMMENTARIES,
EVALUATIONS . . .

"The 20th Century has been the century of agricultural development. Tremendous advances in this applied field have resulted in marked improvements in the fight against famine and in enhancing the quality of life. The immense accumulation of knowledge in agricultural practices has also helped developments in the field of aromatic and medicinal plants. Better understanding of natural floras has initiated selection and adaptation for cultivation of wild plants. Success in agricultural production encouraged agrobased industries to flourish. The boom experienced in the last three decades in the successful marketing of phytopharmaceuticals, health food and natural products heralded in coming of a new age in the breeding of medicinal and aromatic plants.

Fifty-four well written papers covering a wide range of topics from conventional breeding to modern techniques, through biodiversity and conservation to economic, ethical and legal aspects provide valuable scientific and technical information for agronomists and plant breeders on over thirty-five aromatic and medicinal plants.

This book, in this context, can be seen as a *'millennial'* commemorating the status of medicinal and aromatic plant breeding entering into the new millennium. It will be a good legacy for new generations to take the developments into new heights."

K. Husnu Can Baser, PhD, FLS
Professor of Pharmacognosy
Faculty of Pharmacy
Director
Medicinal and Aromatic Plant
and Drug Research Centre (TBAM)
Anadolu University, Turkey
Secretary-General
International Council for Medicinal
and Aromatic Plants (ICMAP)

"These conference proceedings show that research into aromatic and medicinal plant breeding has come of age. Many papers deal with the latest techniques of biochemistry, biotechnology and molecular genetics applied to this area, alongside reports on more conventional breeding research. For example, there are papers on the manipulation of natural product accumulation in plants through genetic engineering, the application of protoplast fusion technology to tansy, DNA fingerprinting by RAPD of various species and breeding for resistance to biotic and abiotic factors in medicinal and aromatic plants. It is also good to see that a fair proportion of the papers are devoted to biodiversity and conservation issues, as many medicinal plants harvested in the wild are now under threat of extinction, and controlled cultivation is an important issue for their conservation. Economic, ethical and legal aspects also have not been forgotten, as the keynote papers on the economic value of plants for medicinal application and on 'European plant intellectual property' show.

I can warmly recommend this book, not only to those involved in research on aromatic and medicinal plant breeding, but also to natural product chemists, pharmacognosists and the users of these economically increasingly important plants."

Renée J. Grayer, PhD
Plant Chemist
Royal Botanic Gardens
Kew, UK

"Medicinal and aromatic plants continue to attract growing interest both from scientists and the general public. This volume deals with the somewhat neglected area of breeding these plants so as to enhance their quality and productivity. It covers not only traditional methods of breeding, with a series of papers on well known species such as *Origanum vulgare* (oregano), *Ocimum basilicum* (basil) and *Coriandrum sativum* (coriander), but more modern approaches such as DNA fingerprinting, propoplast fusion technology and genetic engineering. The editors recognize that medicinal and aromatic plants are natural biological resources that have to be husbanded and sustainably used if they are to continue to be available for present and future generations. Accordingly some of the papers address the issues of conservation and genetic resources, highlighting the need for better knowledge of the taxonomy, distribution, and ecology of these species so that their wild populations can be properly monitored and appropriately managed in reserves. The need for involvement by the primary stakeholders such as farmers and rural communities is stressed by some of the contributors while others address the complex issues of intellectual property rights, economic valuation of these plants and the controversial question of access to their genetic resources. This is a valuable and important contribution to our knowledge of breeding research on medicinal and aromatic plants and to their role in human societies."

Vernon Heywood, PhD, DSc
Professor
School of Plant Sciences
The University of Reading, UK
President
International Council for Medicinal and Aromatic Plants (ICMAP)

Breeding Research on Aromatic and Medicinal Plants

Breeding Research on Aromatic and Medicinal Plants has been co-published simultaneously as *Journal of Herbs, Spices & Medicinal Plants*, Volume 9, Numbers 2/3 and 4 2002.

Breeding Research
on Aromatic
and Medicinal Plants

Christopher B. Johnson
Chlodwig Franz
Editors

Breeding Research on Aromatic and Medicinal Plants has been co-published simultaneously as *Journal of Herbs, Spices & Medicinal Plants*, Volume 9, Numbers 2/3 and 4 2002.

CRC Press
Taylor & Francis Group
Boca Raton London New York

CRC Press is an imprint of the
Taylor & Francis Group, an informa business

Breeding Research on Aromatic and Medicinal Plants has been co-published simultaneously as *Journal of Herbs, Spices & Medicinal Plants*, Volume 9, Numbers 2/3 and 4 2002.

Reprinted 2009 by CRC Press

Cover design by Jennifer M. Gaska

Library of Congress Cataloging-in-Publication Data

Breeding research on aromatic and medicinal plants / Christopher B. Johnson, Chlodwig Franz, editors.
     p. cm.
   "Breeding research on aromatic and medicinal plants has been co-published simultaneously as Journal of herbs, spices & medicinal plants, volume 9, numbers 2/3 & 4 2002."
   Includes bibliographical references.
   ISBN 0-7890-1972-8 (hard : alk. paper)–ISBN 0-7890-1973-6 (pbk. : alk. paper)
   1. Aromatic plants–Breeding. 2. Medicinal plants–Breeding. 3. Aromatic plants–Breeding–Research. 4. Medicinal plants–Breeding–Research. I. Johnson, Christopher B. II. Franz, Chlodwig.
SB301.B7 2002
633.8′1–dc21   2002005197

Breeding Research on Aromatic and Medicinal Plants

CONTENTS

Breeding Research
on Aromatic and Medicinal Plants

CONTENTS

ABOUT THE EDITORS

Christopher B. Johnson is Senior Research Associate at the Mediterranean Agronomic Institute of Chania, Greece, and Lecturer in plant physiology and biochemistry at the University of Reading, UK. Dr. Johnson has been a member of the Department of Natural Products and Biotechnology at MAICh since 1997. His main research interests include plant photoreceptors and their modes of action, photocontrol of gene expression, control of nitrate reduction in higher plants, and environmental control of secondary product information in aromatic plants. Dr. Johnson's current research projects include a study on the potential for extracting antioxidants from rosemary and other aromatic plants for the food and pharmaceutical industry.

Chlodwig Franz is Full Professor, Head of the Department of Applied Botany and Vice Chancellor (Research) of the University of Veterinary Medicine in Vienna, Austria. For more than 30 years, he has devoted his professional life to research and teaching on medicinal and aromatic plants. He established the Division of Medicinal and Aromatic Plants Research at the Technical University of Munich in 1978. Professor Franz's main research interests include biodiversity, plant breeding, and genetics with a special emphasis on secondary substances; the domestication of wild crafted plants from Europe, tropical, and subtropical countries; the development of quality assurance systems (GAP-guidelines); and the use of herbs in animal husbandry. He holds leading positions in several scientific socieities and has been the organizer of a number of international meetings, including the 1st ISHS Symposium on Medicinal and Aromatic Plants, Munich 1977.

Preface:
Breeding Research
on Aromatic and Medicinal Plants

The papers collected in this volume represent a significant portion of those presented at the second international conference on "Breeding Research on Medicinal and Aromatic Plants" held at the Mediterranean Agronomic Institute of Chania, Crete, in July, 2000. The first conference had taken place in Quedlinburg, Germany in 1996.

In the period since the previous symposium public interest in medicinal and aromatic plants has increased considerably and with this interest has come increasing exploitation of many of the plants, with a concomitant present or future threat to the survival of many of the species involved.

It is widely recognized that a strategy of "conservation through use," by which plant collection from wild harvesting is replaced by controlled cultivation, is the best way forward if we are to balance human demands with the necessary conservation of the biodiversity represented by these species. This provides one major driving force for research on breeding of aromatic and medicinal plants. The other concerns the very real need for improving the quality control of materials made available on the market, both to satisfy consumer demand and to conform with (justifiably) increasing requirements for standardization and precise identification of the composition of the plant materials being sold for human use.

The meeting was arranged under four umbrella headings: *Conventional Breeding*; *Biochemistry, Biotechnology, Molecular Genetics and Physiology*; *Biodiversity and Conservation of Medicinal and Aromatic Plant Resources*; and *Economic, Ethical and Legal Aspects*, a grouping that has been maintained in this publication. Within each heading, the topic was introduced by one or two keynote lectures and these appear at the start of each section.

[Haworth co-indexing entry note]: "Preface: Breeding Research on Aromatic and Medicinal Plants." Johnson, Christopher B. Co-published simultaneously in *Journal of Herbs, Spices & Medicinal Plants* (The Haworth Herbal Press, an imprint of The Haworth Press, Inc.) Vol. 9, No. 2/3, 2002, pp. xxix-xxx; and: *Breeding Research on Aromatic and Medicinal Plants* (ed: Christopher B. Johnson, and Chlodwig Franz) The Haworth Herbal Press, an imprint of The Haworth Press, Inc., 2002, pp. xix-xx. Single or multiple copies of this article are available for a fee from The Haworth Document Delivery Service [1-800-HAWORTH 9:00 a.m. - 5:00 p.m. (EST). E-mail address: getinfo@haworthpressinc.com].

We hope that this volume will give a flavor of the exciting developments in the field of aromatic and medicinal plants: the reader can only imagine the enthusiastic and lively discussion that followed especially the more controversial issues, discussion that often continued well into the evenings in the many bars and taverns of Chania.

Christopher B. Johnson

CONVENTIONAL BREEDING

Breeding for Resistance
to Biotic and Abiotic Factors
in Medicinal and Aromatic Plants:
General Situation and Current Results
in Annual Caraway
(*Carum carvi* L. var. *annum*)

Jutta Gabler

SUMMARY. An analysis of 302 scientific publications on resistance research and breeding in medicinal and aromatic plants (MAP) during

Jutta Gabler is affiliated with the Federal Centre for Breeding Research on Cultivated Plants, Institute for Resistance Research and Pathogen Diagnostics, Theodor-Roemer-Weg 4, 06449 Aschersleben, Germany.

The author thanks Marion Urban and Edith Brenner for the reliable performance of the experiments, Karin Klingebeil and Herta Gulde for seasonal assistance, and Gabriele Kulessa and Silvia Ballhause for the technical preparation of the manuscript.

[Haworth co-indexing entry note]: "Breeding for Resistance to Biotic and Abiotic Factors in Medicinal and Aromatic Plants: General Situation and Current Results in Annual Caraway (*Carum carvi* L. var. *annum*)." Gabler, Jutta. Co-published simultaneously in *Journal of Herbs, Spices & Medicinal Plants* (The Haworth Herbal Press, an imprint of The Haworth Press, Inc.) Vol. 9, No. 2/3, 2002, pp. 1-11; and: *Breeding Research on Aromatic and Medicinal Plants* (ed: Christopher B. Johnson, and Chlodwig Franz) The Haworth Herbal Press, an imprint of The Haworth Press, Inc., 2002, pp. 1-11. Single or multiple copies of this article are available for a fee from The Haworth Document Delivery Service [1-800-HAWORTH 9:00 a.m. - 5:00 p.m. (EST). E-mail address: getinfo@haworthpressinc.com].

the last 3 decades shows that the activities have been intensified in this field, mainly in Europe and America. This reflects the increasing importance of MAP. The current state of resistance research and breeding of poppy, coriander and fennel, investigated very intensely, is presented in detail. Our own studies in annual caraway were used as an example of which approaches are necessary before resistance breeding can begin.

Within the last 30 years, mainly Europe and America have experienced an increased trend towards healthy diet and natural products. This trend led to a growing demand for medicinal and aromatic plants (MAP), which is partly satisfied by collections of wild-growing plants, but to an increasing extent by cultivation. A negative consequence of the growing concentration in cultivation is an increase of pathogens and pests. They can cause economically considerable losses in yield and quality unless appropriate control measures are introduced. Approved pesticides are usually not available. In addition, there are reservations against the application of pesticides. Therefore, the breeding of resistant cultivars is gaining more importance, as is reflected by the high number of topic-related scientific publications. *[Article copies available for a fee from The Haworth Document Delivery Service: 1-800-HAWORTH. E-mail address: <getinfo@haworthpressinc.com> Website: <http://www.HaworthPress.com> © 2002 by The Haworth Press, Inc. All rights reserved.]*

KEYWORDS. Medicinal and aromatic plants, resistance research and breeding, poppy, diseases, pests, coriander, fennel, caraway, *Phomopsis diachenii*

ANALYSIS OF SCIENTIFIC PUBLICATIONS ON RESISTANCE RESEARCH AND BREEDING IN MEDICINAL AND AROMATIC PLANTS DURING THE LAST THREE DECADES

The literature research results presented here encompass 302 contributions published worldwide in relevant journals from 1970 to 1999. This analysis raises no claim to completeness. Its aim is to show general tendencies in the field of resistance research and breeding in MAP with special regard to Europe.

Number of Topic-Related Publications per Continent During the Period 1970-1999

The majority of the registered publications (149) was attributed to Asia, a good third (104) to Europe, 35 to America, 12 to Africa and 4 to Australia.

Number of Topic-Related Publications per Continent and Decade

In Europe and America, the number of topic-related contributions increased continuously from decade to decade. In Asia, a similar evolution took place, in particular from the 1970s to the 1980s. At this period, the number of contributions approximately tripled. For Africa an opposite trend can be suggested (under reservation), whereas a slight increase was recorded in Australia from 1980s-1990s.

Number of MAP-Genera Investigated During the Period 1970 to 1999

In the course of the last 30 years, MAP of 62 genera in total were investigated in Europe. In Asia, it was 73, in America, 46, in Africa, 26 and in Australia, 4 genera. In Asia, the activities, above all, concentrated on MAP genera used traditionally which represent a sustainable and important economic factor. In such a way, almost a third of all contributions was allotted to *Piper*, followed by *Coriandrum* (11%) and *Zingiber* (8%). In Europe, the concentration on specific MAP was less than in Asia. In the period under review, 9% of the contributions dealt with *Papaver, Mentha* and *Foeniculum*, 7% with *Coriandrum, Brassica, Datura* and *Sinapis*, and 6% with *Solanum* and *Valeriana*. The share of the other genera was smaller. I will later refer to the current state of resistance research and breeding in *Papaver, Foeniculum* and *Coriandrum*. In the 1970s, *Capsicum* was in the focus of interest in Europe (26% of contributions). In the 1980s, the emphasis of the investigations shifted to *Papaver* (19% of the contributions) as well as *Coriandrum* and *Mentha* (each 14% of the contributions). In the 1990s, the majority of contributions was concerned with *Brassica* (16%), followed by *Foeniculum* (11%). The increasing interest in the use of MAP is also recognizable regarding the increasing diversity of genera processed in Europe. Compared with the 1970s, 15 further MAP genera were investigated in the 1980s, and in the 1990s there were an additional 28 genera which had not yet been investigated in the 1980s.

Frequency of Different Biotic and Abiotic Damaging Factors Described in Topic-Related Publications per Decade in Europe from 1970 to 1999

The majority of the contributions was concerned with fungal diseases, reflecting the exceptional economic importance of this pathogen group in Europe as a damaging factor. In the 1980s, this dominance was especially pronounced, since 70% of the publications were on fungal diseases. In the 1990s, the number of contributions about pests (nematodes and arthropodes) increased significantly compared to the previous years; in addition, the first contributions appeared for the topic of herbicide resistance. On the other hand, the contributions about viral and bacterial diseases as well as abiotic stress factors showed a slightly regressive tendency compared to the 1970s.

FUNDAMENTALS OF RESISTANCE BREEDING

Resistance to Biotic Damaging Factors

The capacity of plants to resist biotic factors (viruses, bacteria, fungi and pests) is called resistance. There are interactions between the genotypes of the host and the genotypes of the pathogen. Two types of resistance are known–monogenic (MoR) and polygenic resistance (PoR)–that differ fundamentally in their inheritance, effect and breeding procedures. Both types can be applied in the resistance breeding of MAP. MoR is race-specific, prevents the disease attack by specific races completely, is easier to apply than PoR, but is not durable. PoR is not race-specific and does not prevent the attack; however, it reduces its impacts. PoR is widespread and largely durable. There are many variants to assess resistance; however, selection in the field is an indispensable part of every breeding program. There are different strategies for application of resistances. Each of them is characterized by specific advantages and disadvantages regarding duration and degree of the prevention as well as the breeding extension. Therefore, an individual decision on what is the most suitable breeding strategy is necessary in all cases (7,18).

Tolerance to Abiotic Factors

The robustness to abiotic stress factors (cold, heat, drought, wetness, nutrition deficiency or oversupply, salinization and others) should be called tolerance because interactions between the systems are missing here (7). Improvements in stress tolerance are therefore durable but the procedure of achieving it is more difficult. Factors like cold, heat, dryness and wetness are complex, more or less diffuse terms and can rarely be selected for in the greenhouse or lab.

CURRENT STATE OF RESISTANCE RESEARCH AND BREEDING IN SOME SELECTED MAP (POPPY, CORIANDER AND FENNEL)

Poppy (Papaver somniferum L.)

In Bulgaria, studies dealt with the improvement in opium poppy cultivars in the early 1980s. Two new cultivars were bred. One of them (Sadovo 1235) showed higher winter hardiness and other improved properties compared with the original population. The second cultivar (Sadovo 1242) was characterized by higher winter hardiness and higher resistance against diseases as well as other improved properties (35). There were similar efforts in the Ukraine. Between 1965 and 1985, poppy and other MAP were the basis for cultivars with winter hardiness, resistance against diseases and pests and other improved properties (22). In Romania, too, cultivars of poppy and other MAP were improved by breeding (31). There, the first cultivar (Extaz) with resistance to *Botrytis cinerea* (gray mould), which belongs to the most important pathogens of poppy, was bred by single plant selection from Olanda 245 × Cluj A (4). In the Soviet Union, in the middle of the 1980s, 13 poppy, cultivars were tested for resistance to the seedborne fungal pathogen *Helminthosporium papaveris* (parasitic leaf drying) in a field test. Not one of the tested cultivars proved to be resistant (27). In Hungary, *Fusarium oxysporum* (dry rot disease) was identified as the cause of a new and economically important poppy disease in 1988 that has long been confused with the bacterial wetrot disease. None of the fungicides approved in Hungary at that time was effective. A resistance screening was carried out. It turned out that only one (L 1241) among 60 breeding lines showed resistance to this pathogen (19). In India in the 1990s, a cultivar with resistance against *Peronospora arborescens* (downy mildew) was selected from 36 breeding lines under field conditions (2).

Coriander (Coriandrum sativum)

Intensive efforts were made in the Soviet Union in the 1980s to find resistance against *Ramularia coriandri* (ramulariosis). The fungus caused economically important losses there.

Researchers succeeded in proving that the resistance level of a cultivar can be recognized by specific membrane changes occurring after infection but before first symptoms are visible. Consequently, the method allowed for a selection for resistance in a very early stage of infection (17). Ukrainian breeders began in the early 1980s to evaluate 509 forms, 20 inbred lines, 15 progenies of regional cultivars, 37 intervarietal hybrids and 118 chemically induced M4-mutants for resistance to *R. coriandri*. The extensive material showed differences in resistance level and other useful characteristics (37). It was also observed that the time of fruit formation in early cultivars coincided with the stage of maximum infection. Selected hybrids were used for the breeding of resistant and early cultivars by intervarietal hybridization (38). Two further genotypes with relative resistance to *Ramularia* were found by evaluation of additional material (39). In the middle of the 1980s in the Ukraine, 100 additional coriander forms were subject to a screening for resistance to *Ramularia* and other valuable properties in order to get promising basic material for breeding (8). It was mainly India which studied the fungus *Protomyces macrosporus* (stem gall disease). At the beginning of the 1970s, 7 cultivars with moderate resistance to the pathogen were found among 18 tested (16). In the 1990s, 8 other cultivars were subject to a screening for resistance to stem galls and for other useful properties. Among these, 'Pant Haritma' showed the lowest and 'Veldurthy' the highest incidence of stem galls (41). In addition, India investigated 83 cultivars to elucidate the host-parasite interactions in the pathosystem coriander/*P. macrosporus*. This led to the finding that phenolics and reducing sugars probably play a role in stem gall disease (29). *Erysiphe polygoni* (powdery mildew) was also investigated in India where the disease regionally caused considerable yield losses at the beginning of the 1990s. It was suggested to control the disease by a fungicide measure in late maturing cultivars (20). Among several Indian cultivars, three had a low and four a moderate level of susceptibility to powdery mildew (21). A bacterial disease caused by *Pseudomonas syringae* pv. *coriandricola* (bacterial blight) was intensely studied in Germany. In a screening of 267 coriander genotypes, five of them showed high resistance to this pathogen. In the leaves of the resistant genotypes, an acting mechanism could be proved (26). In addition, two different ways of inheritance of resistance were available (cytoplasmatic inheritance and polygenic resistance) (25). Three nematode species were studied in India. Coriander forms with moderate resistance to *Meloidogyne javanica* were found in pot experiments (30). Among 16 genotypes, 2 were highly resistant and 4 moderately resistant to *Meloidogyne incognita*. However, the seed yield was influenced more by genotype than by nematode attack (9,10). In addition, two cultivars with high resistance and four with moderate resistance against *Meloidogyne hapla* were selected from among 16 (9).

Fennel (Foeniculum vulgare)

At the beginning of the 1970s, investigations were carried out in Switzerland with regard to several cultivation-related properties of fennel cultivars (3). Similar activities were also started in Indonesia in the 1990s (6). At the beginning of the 1990s, the survey of Italian breeding suggested the improvement of genotypes in fennel and other MAP (24). Experiments with fennel oil for control of *Leptosphaeria maculans* in rape were carried out in Germany in the middle of the 1990s. A control effect was only obtained under laboratory conditions but not in the field (42). At the end of the 1970s, an outbreak of *Phomopsis foeniculi* (umbel browning and stem necrosis) occurred in France and was described in 1981 as a new species (11). In 1990, the presence of the pathogen was detected for the first time in Italy (28). The fungus, regarded as highly aggressive, caused considerable losses in yield. Since then, Italian scientists are in particular occupied with this fennel disease. In Germany, *P. foeniculi* was detected for the first time in 1991; however, there have been no appreciable problems up to now (34). Meanwhile, it was possible to identify several phytotoxins produced by *P. foeniculi* (1,12). In addition, the first results about the *in vitro* selection for resistance were presented. A significant difference in cell growth and cell vitality as well as in callus growth appeared in the presence of cultural filtrate (70%) of the pathogen. This effect showed quantitative grading corresponding to the different susceptibilities of the genotypes. The potential suitability of cultural filtrate for in vitro selection for resistance to *P. foeniculi* could be shown (5). On the basis of these results, the breeding of resistant fennel cultivars was started in Italy in 1996 (23).

In Bulgaria, the leaf and stem anthracnosis pathogen *Cercospora depressa* Vass. caused economically important losses in crops of annual fennel in the 1970s (32,36). The fungus was later renamed in *Passalora puncta* Delacr. S. Petzold with the perfect stage *Mycosphaerella anethi* Petri (33). Two cultivars ('Phenix' and 'Gigant') with field resistance to this fungus were bred by recurrent individual selection (40). The same pathogen recently caused considerable problems in Germany, too. Therefore, there is now an attempt to establish a scientific basis for the breeding of resistant cultivars within the framework of a joint project (Institute of Resistance Research and Pathogen Diagnostics Aschersleben of the Federal Centre for Breeding Research on Cultivated Plants). The possibilities of a chemical control of the disease are being tested by the cooperative partners.

CURRENT RESULTS IN ANNUAL CARAWAY

Our own investigations in annual caraway shall serve as an example to show which phytopathological questions must be answered before the actual breeding of disease-resistant cultivars can begin. The caraway is originally a biennial plant under our climate conditions. The breeder (F. Pank, Institute of Horticultural Crops of the Federal Centre for Breeding Research on Cultivated Plants in Quedlinburg) selected populations of annual caraway with the aim of allowing the producers to have annual yields. The breeder repeatedly observed an umbel browning of unknown cause with negative implications on yield and quality. For a selection for resistance, it had to be clarified whether the umbel browning is: (1) a physiological disturbance, (2) a genetic defect, or (3) a parasitic disease. In three years of investigations into the causes of umbel browning of annual caraway, the following results were obtained.

Results

Twelve caraway populations showed partially significant differences in susceptibility to the umbel browning under natural conditions. Umbel browning was mainly caused by the fungi *Phomopsis diachenii* (Pdi), *Alternaria* spp., *Botrytis cinerea, Synchytrium aureum* and *Cladosporium* sp. They mostly occurred as a complex, but could also cause symptoms separately (installation of an isolate collection for pathogenicity tests and the development of resistance screening methods). Occasionally, umbel browning was also caused by bacterial pathogens, in particular *Pseudomonas* sp. and *Erwinia* sp.; however, they played a subordinate role compared to fungi (13). A fungicide field experiment also confirmed that fungal pathogens are the dominant causal agents of umbel browning (attack level of maximal 40% compared to the untreated control). Pdi and *Alternaria* proved in pathogenicity tests to be the most important fungal pathogens. In case of Pdi, it was also the first detection of the pathogen in caraway in Germany (15). Pdi occurred for the first time the Czech Republic in 1996 and caused considerable yield losses (more than 50%) locally (Ondrej, unpublished). Outdoor observations and pathogenicity tests indicate that *Lygus* bugs probably contribute to the distribution of the disease. Pdi also showed a very high aggressivity in pathogenicity tests. The fungus did not only cause umbel browning but also the dying of whole plants within a short time by progressive stem necrosis. Its temperature optimum was between 25°C and 30°C; however, mycel growth also occurred at temperatures of 14°C, 18°C and 20°C, but stopped at 37°C. There was a positive correlation between the disease progress in the field and some climate data, above all the mean as well as the maximum air temperature (which could be a possible explanation for high disease occurrence in hot years). Pdi grows well on potato dextrose agar and V8 vegetable juice agar. Addition of caraway seeds to the agar stimulates the pycnidia formation with α- and β-pycnospores. A scale was developed for the visual score of symptom severity. As an adjunct to the visual score, a serological method for pathogen detection (PTA-ELISA) using polyclonal antisera was developed (14). The susceptibility differences between 12 caraway populations determined in the field were also confirmed in their tendencies in an ELISA test (using artificial infection).

CONCLUSIONS

The facts and results presented have shown that progress was achieved worldwide in the field of MAP-resistance research and breeding. Meanwhile, there are cultivars with resistance against a range of diseases and pests as well as tolerance to abiotic damaging factors in several MAP. Taking the increasing demand and the higher quality standards into account (homogenicity, purity, minimal content of distinct constituents as well as harmlessness regarding pesticides, microbial contamination and other environmental residues), it is foreseeable that not only the area under cultivation of the present MAP will be expanded, but also the diversity of cultivated MAP will increase. Therefore, resistance breeding researchers will continue to investigate diseases as well as pathosystems. At the same time, the research data will contribute to an overall expansion of knowledge in the field of phytopathology.

REFERENCES

1. Abouzeid, M.A., L. Mugnai, A. Evidente, R. Lanzetta, and G. Surico. 1994. Metabolit tossici prodotti da *Phomopsis foeniculi*. *In* Atti del Convegno, Aspetti molecolari e fisiologici delle interazioni pianta-patogeno. 30 maggio-1 giugno, Alghero. *Petria* 4: 265.

2. Anila, D. and B.B.L. Thakore. 1995. Sources of resistance to downy mildew disease of opium poppy. *Indian Phytopathology* 48(3):339-341.

3. Anonymous. 1972. Report for 1970/1971 of the activities of the Federal Research Institute for Fruit Growing, Viticulture and Horticulture. *Landwirtschaftliches Jahrbuch der Schweiz* 86(3-4):393-490.

4. Anonymous. 1987. Soiuri si hibrizi de mare productivitate. (High-yielding varieties and hybrids.) [Poppy]. *Productia vegetala: cereale si plante tehnice.* 39 (8): 43.

5. Anzidei, M., L. Mugnai, S. Schiff, A. Sfalanga, A. Bennici, and G. Surico. 1996. Saggi preliminari per la selezione in vitro di piante di *Foeniculum vulgare* Mill. da seme resistenti a *Phomopsis foeniculi* Du Manoir et Vegh (Preliminary tests for in vitro selection of *Foeniculum vulgare* Mill. plants resistant to *Phomopsis foeniculi* Du Manoir et (Vegh)). *In* Atti del convegno internazionale: Coltivazione e miglioramento di piante officinali, 2-3 giugno 1994, Trento, Italy, pp. 527-529.

6. Auzay, H. and A. Djisbar. 1989. Current work on essential oils and spices in Indonesia. *Industrial Crops Research Journal* 2(1):16-21.

7. Becker, H. 1993. *Pflanzenzüchtung.* Verlag Eugen Ulmer, Stuttgart, Germany. 75-92.

8. Chislova, L.S. 1988. Useful forms of coriander. *Selekcija i semenovodstvo, Moskva* (6):33.

9. Das, N. 1998. Reaction of coriander varieties to root-knot nematode (*Meloidogyne incognita* Chitwood). *Annals of Agricultural Research* 19(1):94-95.

10. Das, N., A.K. Pattnaik, A.K. Senapati, and D.K. Dash, 1997. Screening of coriander varieties against root-knot nematode (*Meloidogyne incognita*). *Environment and Ecology* 15(3):562-563.

11. Du Manoir, J. and I. Vegh. 1981. *Phomopsis foeniculi* spec. nov. sur Fenouil (*Foeniculum vulgare* Mill.). *Phytopathologische Zeitschrift* 100(4):319-330.

12. Evidente, A., R. Lanzetta, M.A. Abouzeid, M. M. Corsaro, L. Mugnai, and G. Surico. 1994. Foeniculoxin, a new phytotoxic geranylhydroquinone from *Phomopsis foeniculi*. *Tetrahedron* 50:10371-10378.

13. Gabler, J. and R. Zielke. 1998. Analyse der Ursachen der Doldenbräune beim einjährigen Kümmel. *Jahresbericht Bundesanstalt für Züchtungsforschung an Kulturpflanzen*, 65-66.

14. Gabler, J. 2000. Entwicklung eines PTA-ELISA zum Nachweis von *Phomopsis diachenii* Sacc. an Kümmel. *In* BDGL-Schriftenreihe 18. Kurzfassungen Vorträge und

Poster der 37. Gartenbauwissenschaftlichen Tagung, 8-10. März 2000, Zürich, Switzerland. p. 98.

15. Gabler, J. and F. Ehrig. 1999. *Phomopsis diachenii* Sacc., ein gefährlicher Doldenbräune Erreger an Kümmel-Erstnachweis für Deutschland. *In* Beiträge 2. Symposium Phytomedizin und Pflanzenschutz im Gartenbau, 27-30 September 1999, Wien, Austria. pp. 149-151.

16. Gupta, R.N. and S. Sinha. 1973. Varietal field trials in the control of stem-gall disease of coriander. *Indian Phytopathology* 26(2):337-340.

17. Guzhova, N.V., L.A. Zavjalova, L.G. Veselova, V.I. Tjutjunnik, and N.V. Glumova. 1984. Permeability of the coriander leaf cell membrane as related to resistance to *Ramularia* leaf spot. *Sel' skochozajstvennaja Biologija* 9:93-95.

18. Hartleb, H., R. Heitefuss, and H.-H. Hoppe. 1997. *Resistance of Crop Plants against Fungi*. Gustav Fischer Verlag, Germany. 544 p.

19. Hörömpöli, T. 1988. A mak (*Papaver somniferum* L.) uj betegsege Magyarorszagon, a fusariumos *tokorhadas* (*Fusarium oxysporum* Schl.). *Fusarium* dry rot (*Fusarium oxysporum* Schl.), a new disease of poppy (*Papaver somniferum* L.) in Hungary. *Növényvédelem* 24(5):201-207.

20. Kalra, A., T.N. Parameswaran, N.S. Ravindra, and B.P. Dimri. 1995. Effect of powdery mildew (*Erysiphe polygoni*) on yields and yield components of early and late maturing coriander (*Coriandrum sativum*). *Journal of Agricultural Science* 125(3): 395-398.

21. Keshwal, R.L. and R.K. Khatri. 1998. Reaction of some high yielding varieties of coriander to powdery mildew. *Journal of Mycology and Plant Pathology* 28(1): 58-59.

22. Kondratenko, L.M., ko-L.A. Shelud, and A.P.-Taranich. 1985. Breeding and seed production of medicinal crop plants. *Lekarstvennoe rastenievodstvo v usloviayakh Ukrainy*. 77-89.

23. Landi, R. 1994. Il miglioramento genetico delle specie aromatiche e medicinali (Breeding aromatic and medicinal species). *Georgofili* 40(1-4):161-189.

24. Landi, R. 1996. Problemi colturali del finocchio da seme (*Foeniculum vulgare* Mill.) in Italia e possibilita di diffonderne la coltivazione (Cultural problems with seed fennel (*Foeniculum vulgare* Mill.) in Italy and possibilities for cultural diffusion). *In* Atti del convegno internazionale: Coltivazione e miglioramento di piante officinali, Trento, Italy, 2-3 giugno 1994, pp. 165-188.

25. Liehe, A. and K. Rudolph. 1998. Zur Vererbung der Resistenz gegenüber dem bakteriellen Doldenbrand am Koriander verursacht durch *Pseudomonas syringae* pv. *coriandricola*. (Inheritance of resistance to bacterial umbel blight of coriander caused by *Pseudomonas syringae* pv. *coriandricola*.) *Gesunde Pflanzen* 50(7):189-195.

26. Liehe, A., T. Al-Shinawi, H. Toben, K. Rudolph, and F. Pank. 1996. Characterization of coriander genotypes with resistance to *Pseudomonas syringae* pv. *coriandricola*. *In* Proceedings International Symposium. Breeding research on medicinal and aromatic plants, June 30-July 4, 1996, Quedlinburg, Germany. Beitrage zur Züchtungsforschung. Bundesanstalt für Züchtungsforschung an Kulturpflanzen. 2(1):49-52.

27. Martynovskaya, N.M. 1986. Producing a *Helminthosporium* infection regime and evaluating opium poppy varieties under artificial infection. *In* Tr. 7 Konf. mol.

Uchenykh VNII lekarstv. rast., Moskva, apr., 1985. Ch. 1., Dep. 3339V, 225-231; MS deposited in VINITI, Moscow. (Ref.: Referativnyi Zhurnal (1986) 10.65.469.)

28. Mugnai L. and M. Anzidei. 1994. Casi di necrosi corticale da *Phomopsis foeniculi* del finocchio da seme in Italia. *Petria* 4:237-244.

29. Naqvi-SAMV. 1987. Possible role of phenolics and reducing sugars in stem gall of coriander. *Plant Disease Research* 2:2, 73-76.

30. Paruthi, I.J., R K. Jain, and D.C. Gupta. 1987. A note on reaction of some spices to root knot nematode (*Meloidogyne javanica*). *Haryana Journal of Horticultural Sciences* 16(1-2):154-155.

31. Paun, E., A. Dumitrescu, A. Mihalea, and M. Verzea. 1985. Noile soiuri romanesti de plante medicinale si aromatice, mijloc principal de asigurare a cantitatii si calitatii materiei prime vegetale. (New Romanian varieties of medicinal and essential-oil crops, the chief means of assuring the quality and quantity of the crude products of the plants.) *Productia vegetala: Cereale si plante tehnice* 37(7):6-10.

32. Peneva, P.T. 1986. The fennel variety Pomorie. *Rasteniev'dni nauki* 23(9): 33-38.

33. Petzold, S. 1989. Zur Biologie, Epidemiologie und Schadwirkung des Erregers der Blatt-und Stengelanthraknose (*Mycosphaerella anethi* Petr.) am Fenchel (*Foeniculum vulgare* Mill.)-1. *Drogenreport; Mitt. über Arznei- und Gewürzpflanzen*, 3, 49-65.

34. Plescher, A. 1992. Bericht über das Auftreten wichtiger Schadpathogene im Arznei- and Gewürzpflanzenanbau im Jahre 1991 in den ostdeutschen Bundesländern. *Drogenreport, Mitt. über Arznei- und Gewürzpflanzen*; 5, 7: 9-12.

35. Popov, P., I. Dimitrov, T. Deneva, S. Georgiev, and L. Iliev. 1981. New poppy varieties (*Papaver somniferum* L.). *Rasteniev'dni Nauki* 18(6):67-72.

36. Raev, R.T., B. Bojadzieva, P. Tsalbukov, and V. Topalova. 1982. Results of breeding annual fennel (*Foeniculum officinale* All. var. *dulce*). *Rasteniev'dni Nauki.* 19(4):60-64.

37. Romanenko, L.G., N.V. Nevkrytaja, and L.N. Serkov. 1987. Evaluation of coriander accessions for resistance to *Ramularia. Zašita rastenij introdutsentov ot vred. organizmov.* Kiev. pp. 69-72.

38. Romanenko, L.G, N.V. Nevkrytaja, and L.N. Serkov. 1985. Ways of producing initial material in coriander with resistance to *Ramularia. Selekcija i semenovodstvo,* USSR (3):31-33.

39. Romanenko, L.G., I.E. Omelchenko, N.V. Nevkrytaja, and L.N. Serkov. 1986. Breeding coriander for earliness and resistance to *Ramularia. Selekcija i semenovodstvo,* Ukrainian SSR (60):37-39.

40. Shulga, E. B. and M.M. Urbenko. 1984. A study of *Cercospora* resistance in fennel breeding material under artificial infection. *Trudy vsesojuznogo nauchno issledovatelskogo instituta efiromaslichnyk kultur* 16:72-80.

41. Singh, D.K., N.P. Singh, and R.S. Tewari. 1995. Evaluation of coriander cultivars in Himalayan foots of U.P. *Annals of Agricultural Research* 16(4):481-482.

42. Winter, W., M. Zoller, J.P. Burdet, M. Gygax, I. Banziger, and H. Krebs. 1996. *Phoma* an Raps: Samenbefall, Aggressivitat und Bekämpfung. *Agrarforschung* 3(5): 234-237.

Combining Ability
of *Origanum majorana* L. Hybrids:
Sensorial Quality

Johannes Novak
Wolfram Junghanns
Wolf-Dieter Blüthner
Rudolf Marchart
Carla Vender
Leon van Niekerk
Friedrich Pank
Jan Langbehn
Chlodwig Franz

Johannes Novak, Rudolf Marchart, and Chlodwig Franz are affiliated with the Institute for Applied Botany, University of Veterinary Medicine, Veterinärplatz 1, A-1210 Wien, Austria.

Wolfram Junghanns is affiliated with Dr. Junghanns GmbH, Untere Dorfstr. 8, 06449 Groß Schierstedt, Germany.

Wolf-Dieter Blüthner is affiliated with N.L. Chrestensen Samenzucht und Produktion GmbH, 99092 Erfurt, Germany.

Carla Vender is affiliated with ISAFA, Piazza Nicolini 6, 38050 Villazzano di Trento, Italy.

Leon van Niekerk is affiliated with SC Darbonne, 6, Bd. Joffre, 91490 Milly-la-Foret, France.

Friedrich Pank, and Jan Langbehn are affiliated with the Institute for Horticultural Crops, Federal Centre for Breeding Research on Cultivated Plants, Neuer Weg 22/23, 06484 Quedlinburg, Germany.

[Haworth co-indexing entry note]: "Combining Ability of *Origanum majorana* L. Hybrids: Sensorial Quality." Novak, Johannes et al. Co-published simultaneously in *Journal of Herbs, Spices & Medicinal Plants* (The Haworth Herbal Press, an imprint of The Haworth Press, Inc.) Vol. 9, No. 2/3, 2002, pp. 13-19; and: *Breeding Research on Aromatic and Medicinal Plants* (ed: Christopher B. Johnson, and Chlodwig Franz) The Haworth Herbal Press, an imprint of The Haworth Press, Inc., 2002, pp. 13-19. Single or multiple copies of this article are available for a fee from The Haworth Document Delivery Service [1-800-HAWORTH 9:00 a.m. - 5:00 p.m. (EST). E-mail address: getinfo@haworthpressinc.com].

13

SUMMARY. Combining ability tests of *cms* and pollinator marjoram lines were carried out to explore the feasibility of a marjoram hybrid variety system. The performance of colour and taste of the hybrids was influenced mainly in the first year by pollinated as well as pollinator compounds of the hybrids. Smell was only influenced by the pollinators.

Bitter tasting hybrid combinations were found in the first test and bitterness was clearly determined by the genetic background of the combinations. These bitter tasting hybrid combinations were strictly eliminated from the breeding process and bitterness hardly occurred in the second test. The correlations between the sensorial parameters were medium to high, but only within each experimental year. The medium correlation between colour and smell or taste indicated a possible influence of the colour impression on consecutive sensorial estimations of a test panel. This correlation furthermore indicated the importance of sensorial colour impression on consumer behaviour not only perceived separately but in correlation to consumers' impressions of smell and taste. *[Article copies available for a fee from The Haworth Document Delivery Service: 1-800-HAWORTH. E-mail address: <getinfo@haworthpressinc.com> Website: <http://www.HaworthPress.com> © 2002 by The Haworth Press, Inc. All rights reserved.]*

KEYWORDS. *Origanum majorana,* marjoram, Lamiaceae, smell, taste, bitterness, colour

INTRODUCTION

The advantages of hybrid variety systems and the novelty of this system for medicinal and aromatic plants have been outlined by Pank et al. (1,2). Sensorial characteristics like colour, smell and taste are without doubt very important quality characteristics for aromatic plants. Most companies producing aromatic herbs do accompany their production by sensorial panel testing of production lots to assure quality. This is especially important in a market where different qualities can lead to a significant market differentiation allowing higher prices for higher quality as it is the case for marjoram. This communication completes the agronomical characteristics and the essential oil content of combining ability tests (2) by adding the panel-assessed evaluation of the sensorial characteristics colour, smell and taste and by comparing these characteristics in a two-years experiment.

MATERIALS AND METHODS

Male sterile marjoram lines and their maintainers as well as pollinator lines were developed during the preceding years (3). Test crossings of four cytoplasmic male sterile lines (cms) and 8 pollinator lines were performed in the greenhouse in 1997 and 1998. A breeding scheme is shown in Langbehn et al. (3).

The precultivated F_1 plantlets were planted into the experimental field at one location in 1998 (Erfurt, Germany) and at three locations in 1999 (Gross Schierstedt, Germany; Milly-la-Foret, France; Villazzano di Trento, Italy). A uniform experimental design was used in both years using a randomized block design. Combinations not performing well in 1998 were eliminated from the breeding process and were therefore not present in the 1999 field trial, narrowing the variability in 1999. On each plot three rows with 27 plants were planted, the distance being 40 cm between the rows and 15 cm within a row. Plot samples were taken when 10% of the plants per plot were blooming. The herb was dried in the shade and the leaf-flower-fraction thereafter separated from the stems.

The samples were scored by a sensorial panel according to a company standard based on the standards no. TGL 37785/03 (appearance of marjoram, for categories see Table 1) and TGL 45611 (smell and taste of marjoram, for categories see Table 2) of the former GDR.

The statistical evaluation was done by calculating the non-parametric Kruskal-Wallis test basing on rank orders for the group of male sterile genotypes and the group of pollinators that computed the respective experimental hybrids separately. So the influence of each parental side was clearly made visible. Experimental years were also handled separately.

TABLE 1. Scores and their description for appearance of marjoram (based on TGL 37785/03).

category	description
0	spoilt
1	colour deviating, very high content of small stalks and stems, very high content of tiny parts
2	greyish green to greyish brown round leaves, parts of leaves and flowers too high, high content of small stalks and stems, very high content of tiny parts
3	green to brownish green round leaves, parts of leaves and flowers slightly augmented, small stalks augmented, stems in a less degree, high content of tiny parts
4	green to greyish green round leaves and parts of leaves and flowers, small stalks only sporadic, low content of tiny parts
5	green to greyish green round leaves and parts of flowers, small stalks only sporadic, very low content of tiny parts

TABLE 2. Scores and their description of smell and taste of marjoram (based on TGL 45611).

category	description
SMELL	
0	spoilt
1	not typical, strange
2	hardly typical, impure, hardly aromatic, not spicy
3	still typical, not completely pure, slightly aromatic, not so spicy
4	typical, pure, aromatic, spicy
5	typical, pure, fully aromatic, spicy
TASTE	
0	spoilt
1	not typical, strange
2	hardly typical, impure, hardly aromatic, strong bitter note, not spicy, haylike
3	still typical, not completely pure, slightly aromatic, bitter note somewhat dominating, not so spicy, slightly haylike
4	typical, pure, aromatic, mild bitter note, spicy
5	typical, pure, fully aromatic, mild bitter note, spicy

RESULTS AND DISCUSSION

Sensorial characteristics colour, smell and taste. The sensorial impression of colour was the character most influenced by the genotypes composing the experimental hybrids (Table 3). The influence of male sterile as well as pollinator genotypes was extremely high in 1998, whereas in 1999 only the pollinators contributed significantly to a colour differentiation.

"Smell" was only significantly influenced by the pollinators in 1998. Male sterile genotypes as well as pollinators contributed to a different evaluation regarding "taste" in 1998. In 1999, however, none of the hybrids components was responsible for marked differences with the male sterile component being much closer to the significance level.

Comparing the two years, the reason for the lower degree of genotypical influences in 1999 could be due to two reasons: (1) Three locations in 1999 compared to only one location in 1998. Genotype $\times$ environment interaction could here lead to non observable differences between the genotypes. This argument can be rejected, however, since a Kruskal-Wallis test calculated for the genotypes for their locations separately did not show any significant differences either (data not shown). (2) The selection step between 1998 and 1999. The

TABLE 3. Non-parametrical Kruskal-Wallis test of the genetic influence of pollinated and pollinator genotypes in experimental hybrid combinations of marjoram ($df_{POLLINATED}$ = 3; $df_{POLLINATOR}$ = 7).

character	hybrid component	1998		1999	
		chi-square	sig.	chi-square	sig.
colour	male sterile	12.9	0.005	2.9	0.408
	pollinator	21.0	0.004	15.6	0.030
smell	male sterile	4.6	0.207	2.6	0.461
	pollinator	20.2	0.004	8.4	0.297
taste	male sterile	10.7	0.013	6.7	0.082
	pollinator	16.0	0.025	8.8	0.325

constitution of the experimental hybrids in 1998 was not equal compared to 1999. The material in 1999 presented an improved set of hybrids, since they were computed from strongly preselected material. Between the experimental hybrids of the two years there was a pronounced single plant based selection step with a high selection intensity especially on pollinator side.

The pollinated components (cytoplasmic sterile plants as well as their maintainers), however, consisted of material already pre-selected also regarding their sensorial properties before project start. This narrowed range (sensorially as well as regards the number of maintainers) may be the reason for the non-significant differences in "smell."

Bitterness. Some combinations were found to be clearly bitter in 1998 (Table 4). It was interesting to note that in most cases all replications of a combination were bitter. The sensorial panel acted of course "blindfolded," the samples being coded throughout the evaluation process. All bitter combinations except "200 × 4" where eliminated from the breeding process. So no information is available on the possible influence of selection regarding "bitterness." However, the clear genotypical determination of "bitterness" could be shown. A further interesting aspect can be seen in the combinations with the pollinator "18." In 1998, all combinations except "400 × 18" were bitter. This exceptional composition "400 × 18" remained "bitter-free" in 1999. This "bitter-free" combination within a set of bitter combinations maybe due to "combinational effects."

Correlations between the sensorial characteristics. No connection was visible comparing the two years for each sensorial characteristic (Table 5). The clear correlation between smell and taste in the respective years did not astonish very much. Additionally, there was a clear significant correlation between

TABLE 4. Frequencies of samples (%) declared as 'bitter' (n = 3 in 1998 and n = 9 in 1999; '-' no combination tested in 1999).

male sterile → pollinator ↓	100		200		300		400	
	'98	'99	'98	'99	'98	'99	'98	'99
1	0	0	100	-	0	0	0	0
2	0	0	0	0	0	0	0	0
4	0	0	100	0	0	0	0	0
7	0	0	0	0	0	0	0	0
9	100	-	33	0	0	0	0	0
15	0	0	0	0	0	0	0	0
17	0	0	33	33	33	0	0	33
18	100	-	100	-	66	-	0	0

TABLE 5. Correlations between the sensorial characteristics in both years (Spearman rho–shaded correlations are significant at the level of 0.01).

	SMELL99	TASTE98	TASTE99	COLOUR98	COLOUR99
SMELL98	.40	.73	.32	.40	0.28
SMELL99		.34	.89	.34	.55
TASTE98			.19	.59	.19
TASTE99				.26	.68
COLOUR98					.30

colour and smell as well as taste within the two years indicating a possible influence of the colour (evaluated first) on the other steps of sensorial testing. This fact indicates the importance of the sensorial colour evaluation as being an integrated part of sensorial herb testing and the importance of the sensorial impression of colour on consumers behaviour.

REFERENCES

1. Pank, F., J. Langbehn, J. Novak, W. Junghanns, J. Franke, F. Scartezzini, C. Franz, and A. Schröder. 1999. Eignung verschiedener Merkmale des Majorans (*Origanum majorana* L.) zur Differenzierung von Populationen und für die indirekte Selektion. 1. Mitteilung: Genetische Variabilität. *Zeitschrift für Arznei- und Gewürzpflanzen.* 4: 8-18.

2. Pank, F., C. Vender, L. van Niekerk, W. Junghanns, J. Langbehn, W.-D. Blüthner, J. Novak, and C. Franz. 2001. Combining ability of *Origanum majorana* L. Strains–Agronomical traits and essential oil content: Results of the field experiment series in 1999. *Journal of Herbs, Spices and Medicinal Plants* 9(2/3):31-37.

3. Langbehn J., F. Pank, J. Novak, and C. Franz. 2001. Influence of selection and inbreeding on *Origanum majorana* L. *Journal of Herbs, Spices and Medicinal Plants* 9(2/3):21-29.

Influence of Selection and Inbreeding on *Origanum majorana* L.

Jan Langbehn
Friedrich Pank
Johannes Novak
Chlodwig Franz

SUMMARY. The comparison of two generations of marjoram single plants grown in Quedlinburg, Germany in 1997 and 1999 (non-inbred initial accessions versus I_2 lines derived from elite plants) is given for important traits. The impact of selection was indicated for resistance to lodging and anther status. Severe inbreeding depression of yield of leaf-flower fraction could be avoided by selection. The superiority of all I_2 lines over the standard and all initial accessions with respect to the content of essential oil was most distinct; in some numbers there was also higher homogeneity. The percentage of *cis*-sabinene hydrate in the essential oil showed a diversification in the I_2 due to the segregation of chemotypes. *[Article copies available for a fee from The Haworth Document Delivery Service: 1-800-HAWORTH. E-mail address: <getinfo@haworthpressinc. com> Website: <http://www.HaworthPress.com> © 2002 by The Haworth Press, Inc. All rights reserved.]*

KEYWORDS. *Origanum majorana*, breeding, inbreeding, selection

Jan Langbehn and Friedrich Pank are affiliated with the Institute for Horticultural Crops, Federal Centre for Breeding Research on Cultivated Plants, Neuer Weg 22/23, 06484 Quedlinburg, Germany.

Johannes Novak and Chlodwig Franz are affiliated with the Institute of Applied Botany, University of Veterinary Medicine, Veterinärplatz 1, A-1210 Wien, Austria.

[Haworth co-indexing entry note]: "Influence of Selection and Inbreeding on *Origanum majorana* L." Langbehn, Jan et al. Co-published simultaneously in *Journal of Herbs, Spices & Medicinal Plants* (The Haworth Herbal Press, an imprint of The Haworth Press, Inc.) Vol. 9, No. 2/3, 2002, pp. 21-29; and: *Breeding Research on Aromatic and Medicinal Plants* (ed: Christopher B. Johnson, and Chlodwig Franz) The Haworth Herbal Press, an imprint of The Haworth Press, Inc., 2002, pp. 21-29. Single or multiple copies of this article are available for a fee from The Haworth Document Delivery Service [1-800-HAWORTH 9:00 a.m. - 5:00 p.m. (EST). E-mail address: getinfo@haworthpressinc.com].

INTRODUCTION

Marjoram (*Origanum majorana* L., Lamiaceae) is one of the economically most important aromatic plants and is mainly used as a condiment in meat products (1). In Europe, cultivation of marjoram is mostly located in Germany, France and Hungary. For the improvement of the competitiveness of this minor crop, a research project supported by the European Commission (FAIR3-CT96-1914) was launched in 1997. One aim of this project was the improvement of yield, quality and homogeneity of the marjoram drug by a hybrid variety system. The development of the required pollinator inbred lines was to be started by a proper sub-task, some results of which are presented here by comparing non-inbred material from the first project year with I_2-lines of the third year.

MATERIALS AND METHODS

Twenty marjoram accessions that were cultivated annually as initial material in Quedlinburg, Germany in 1997 have been listed in Pank et al. (3). The methods of single plant trait evaluation were visual assessment (anther status), rating for resistance to lodging (5 classes from 1 = very low to 9 = very high), manual separation (yield of leaf-flower-fraction), hydrodistillation and GC/MS (content and composition of essential oil). Details can be found in Pank et al. (3,4). Elite plants resulting from mass selection among the initial material were selfed in the greenhouse over winter. Nineteen I_1 lines were cultivated annually in 1998, and a new cycle of mass selection and selfing led to 19 I_2 lines evaluated in 1999. The criteria for selection are listed in Table 1.

TABLE 1. Criteria for the selection of single plants.

year of sowing the plants	time of selection	anther status	resistance to lodging	content of essential oil [ml/100 g dry matter of leaf-flower-fraction]	yield of leaf-flower-fraction [g]
1997	spring 1998	01 or 11	≥ 5	≥ 1.11	≥ 17.8
1998	spring 1999	11	≥ 5	≥ 1.24	≥ 17.2

In order to make the results of the different years comparable, the variety 'Marcelka' (represented by the number 01) was always chosen as the standard.

RESULTS

Resistance to lodging. The experimental mean in 1997 was 4.6 (78% of the standard). The average in the I_2 of 6.0 (105% of the standard) is indicative of the selection steps.

Anther status. The anthers were screened as a marker for male fertility and the possible seed production over winter by selfing. In 1997 this need was not known before August, therefore this trait was recorded during the flowering of the regrowth in October. The partition of the material is shown in Figure 1. The numbers 02 to 20 do not identify the same strains in 1997 and 1999.

All the plants showing normal anthers were regarded as fertile (classes x1, 01, 11). The diminuation of class 00 in 1999 is evident.

Yield of leaf-flower fraction. This variable is the amount of the trade product and is listed in Tables 2 and 3. No I_2 line significantly surpassed the standard, but some had almost the same mean.

Content of essential oil. This variable is an important quality parameter and is treated in Figure 2 and Tables 4 and 5. The differences were not significant in 1997. In 1999 the average of every I_2 line was higher than in the standard (in 5 cases significantly higher). The average advantage was approximately 70%. As can be seen in Figure 2 and by the lower c.v.s in the Tables, some numbers also showed higher homogeneity.

Cis-sabinene hydrate. This compound of the essential oil was no subject of selection. It is reported in Tables 6 and 7 as it is said to be responsible for the aromatic properties (2). Table 7 shows a diversification of the material (means up to 194% of the standard) and reduced variation within some I_2 lines by smaller c.v.

CONCLUSIONS

The investigations reveal that the development of marjoram lines can lead to improved plant material. However, further breeding work seems to be advisable.

FIGURE 1. Mosaic plots of the anther status of one marjoram standard (no. 01, always 'Marcelka'), 19 accessions in 1997 (no. 02-20, left hand), and 19 I_2 lines in 1999 (no. 02-20, right hand); 1997: flowering of the regrowth; 1999: flowering of first growth, n ≈ 48 each. xx = no anthers at all; 00 = only degenerated anthers; 11 = only normal anthers; 01 = both flowers of class 00 and 11; x1 = both flowers of class xx and 11.

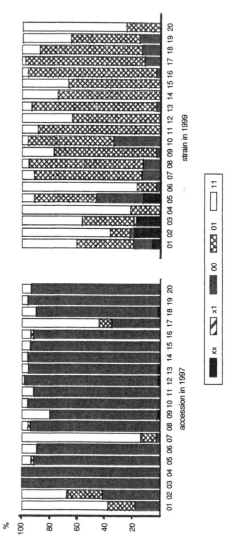

TABLE 2. Mean yield of leaf-flower-fraction in 1997 (g/single plant, 14% water cont.), c.v. = coefficient of variation.

accession	g	[% Marc.]	c.v.	statistical significance (Tukey) + = significant difference α = 5% accession																		
				01	02	03	04	05	06	07	08	09	10	11	12	13	14	15	16	17	18	19	20
01	27.3	100.0	28.4																				
02	24.7	90.5	38.1																				
03	18.6	68.1	42.8	+																			
04	19.7	72.2	48.1	+																			
05	19.9	72.9	44.8	+																			
06	19.2	70.3	36.2	+																			
07	22.0	80.6	29.6																				
08	20.8	76.2	26.6	+																			
09	22.8	83.5	38.4																				
10	22.0	80.6	43.3																				
11	20.9	76.6	34.7	+																			
12	20.6	75.5	37.8	+																			
13	22.1	81.0	39.5																				
14	20.3	74.4	33.6	+																			
15	20.2	74.0	37.2	+																			
16	22.6	82.8	35.7																				
17	20.2	74.0	43.9	+																			
18	19.1	70.0	33.9	+																			
19	17.8	65.2	39.8	+	+																		
20	18.2	66.7	35.6	+	+																		
total	20.9	76.6	38.7																				

TABLE 3. Mean yield of leaf-flower-fraction in 1999 (g/single plant, 14% water cont.).

number	g	[% Marc.]	c.v.	statistical significance (Tukey) + = significant difference α = 5% accession																		
				01	02	03	04	05	06	07	08	09	10	11	12	13	14	15	16	17	18	19	20
01	26.9	100.0	24.5																				
02	22.8	84.8	29.6																				
03	23.8	88.5	15.5																				
04	22.9	85.1	22.3																				
05	26.5	98.5	18.8																				
06	21.1	78.4	22.3																				
07	26.1	97.0	40.2																				
08	13.6	50.6	36.3	+				+		+													
09	17.2	63.9	23.1	+																			
10	19.0	70.6	42.3																				
11	19.2	71.4	44.3																				
12	20.0	74.3	40.2																				
13	19.0	70.6	53.5																				
14	17.8	66.2	41.3																				
15	17.5	65.1	43.5																				
16	18.5	68.8	54.7																				
17	22.4	83.3	24.1																				
18	23.8	88.5	18.7																				
19	23.0	85.5	22.2																				
20	18.1	67.3	23.6																				
02-20	20.7	77.0	36.3																				
total	21.0	78.1	36.1																				

FIGURE 2. Box plots of the content of essential oil (%, v/w) of one marjoram standard (no. 01, always 'Marcelka'), 19 accessions in 1997 (no. 02-20, left hand), and 19 I$_2$ lines in 1999 (no. 02-20, right hand), n ≈ 48 each.

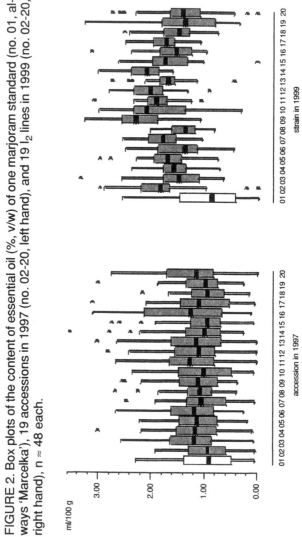

TABLE 4. Mean content of essential oil in 1997 (ml/100 g dry matter of leaf-flower-fraction).

accession	ml	[% Marc.]	c.v.
01	0.98	100.0	54.5
02	0.99	101.0	46.7
03	1.28	130.6	43.9
04	1.19	121.4	50.0
05	1.15	117.3	44.1
06	1.26	128.6	47.7
07	1.06	108.2	50.1
08	1.13	115.3	39.0
09	1.18	120.4	44.6
10	1.07	109.2	55.1
11	1.22	124.5	52.4
12	1.20	122.4	52.3
13	1.24	126.5	59.6
14	1.15	117.3	61.4
15	1.07	109.2	55.4
16	1.42	144.9	57.3
17	1.06	108.2	58.2
18	0.99	101.0	43.3
19	1.07	109.2	45.7
20	1.26	128.6	54.2
total	1.15	117.3	51.9

TABLE 5. Mean content of essential oil in 1999 (ml/100 g dry matter of leaf-flower-fraction).

number	ml	[% Marc.]	c.v.	statistical significance (Tukey) + = significant difference α = 5% number																			
				01	02	03	04	05	06	07	08	09	10	11	12	13	14	15	16	17	18	19	20
01	1.01	100.0	67.4	■																			
02	1.87	185.1	29.3		■																		
03	1.56	154.5	36.7			■																	
04	1.57	155.4	23.8				■																
05	1.72	170.3	24.7					■															
06	1.50	148.5	30.9						■														
07	1.80	178.2	20.1							■													
08	1.40	138.6	21.0								■												
09	2.21	218.8	22.8	+						+		■											
10	1.90	188.1	35.0	+									■										
11	1.90	188.1	16.6	+										■									
12	2.06	204.0	19.9	+											■								
13	1.71	169.3	20.7													■							
14	2.15	212.9	15.8	+					+								■						
15	1.69	167.3	25.3															■					
16	1.59	157.4	25.0																■				
17	1.78	176.2	17.3																	■			
18	1.53	151.5	26.8																		■		
19	1.43	141.6	50.6								+											■	
20	1.43	141.6	41.6								+						+						■
02-20	1.73	171.3	29.9																				
total	1.69	167.3	32.5																				

TABLE 6. *Cis*-sabinene hydrate/linalool in the essential oil in 1997 (%).

accession	%	[% Marc.]	c.v.	01	02	03	04	05	06	07	08	09	10	11	12	13	14	15	16	17	18	19	20
01	31.56	100.0	46.0	■																			
02	26.16	82.9	55.8		■																		
03	20.68	65.5	68.4	+		■																	
04	20.24	64.2	72.4	+			■																
05	18.65	59.1	52.3	+				■															
06	18.62	59.0	55.2	+					■														
07	36.19	114.7	28.6			+	+	+	+	■													
08	25.39	80.4	52.6							+	■												
09	20.44	64.8	73.0	+						+		■											
10	24.48	77.6	62.8							+			■										
11	22.12	70.1	57.5							+				■									
12	19.43	61.6	60.9	+						+					■								
13	19.31	61.2	67.3	+						+						■							
14	9.54	61.9	51.2	+						+							■						
15	19.52	61.9	55.3	+						+								■					
16	23.77	75.3	64.1							+									■				
17	27.12	85.9	57.9																	■			
18	18.12	57.4	42.8	+						+											■		
19	20.19	64.0	62.9	+						+												■	
20	20.88	66.2	52.3	+						+													■
total	22.65	71.8	59.7																				

TABLE 7. *Cis*-sabinene hydrate/linalool in the essential oil in 1999 (%).

number	%	[% Marc.]	c.v.	01	02	03	04	05	06	07	08	09	10	11	12	13	14	15	16	17	18	19	20
01	29.13	100.0	68.0	■																			
02	46.01	157.9	39.5		■																		
03	56.47	193.9	5.8			■																	
04	17.75	60.9	30.4				■																
05	16.11	55.3	37.0					■															
06	16.13	55.4	49.9			+			■														
07	49.58	170.2	32.2							■													
08	25.42	87.3	54.2								■												
09	16.13	55.4	28.5									■											
10	15.07	51.7	24.6			+				+			■										
11	50.28	172.6	19.1											■									
12	26.58	91.2	35.1												■								
13	32.03	110.0	54.7													■							
14	23.97	82.3	14.6												.		■						
15	15.55	53.4	30.1			+												■					
16	16.76	57.5	33.8																■				
17	46.42	159.4	5.0																	■			
18	46.03	158.0	4.3																		■		
19	27.93	95.9	38.7																			■	
20	17.27	59.3	37.3																				■
02-20	24.69	84.8	59.1																				
total	24.92	85.5	59.8																				

REFERENCES

1. Burdock, G.A. 1994. *Fenaroli's handbook of flavor ingredients.* Boca Raton, Florida, USA: CRC Press.

2. Fischer, N., S. Nitz, and F. Drawert. 1988. Original composition of marjoram flavor and its changes during processing. Journal of Agricultural and Food Chemistry, 36: 996-1003.

3. Pank, F., J. Langbehn, J. Novak, J. Junghanns, W. Franke, F. Scartezzini, C. Franz, and A. Schröder. 1999a. Eignung verschiedener Merkmale des Majorans (*Origanum majorana* L.) zur Differenzierung von Populationen und für die indirekte Selektion. 1. Mitteilung: Genetische Variabilität. (Suitability of different traits of marjoram (*Origanum majorana* L.) for population differentiation and indirect selection. Part 1: Genetical variation.). Zeitschrift für Arznei- und Gewürzpflanzen, 4: 8-18.

4. Pank, F., J. Langbehn, J. Novak, W. Junghanns, J. Franke, C. Bitsch, F. Scartezzini, C. Franz, and A. Schröder. 1999b. Eignung verschiedener Merkmale des Majorans (*Origanum majorana* L.) zur Differenzierung von Populationen und für die indirekte Selektion. 2. Mitteilung: Korrelation der Merkmale. (Suitability of different traits of marjoram (*Origanum majorana* L.) for population differentiation and indirect selection. Part 2: Correlation of the traits). Zeitschrift für Arznei- und Gewürzpflanzen, 4: 141-150.

Combining Ability
of *Origanum majorana* L. Strains–
Agronomical Traits and Essential Oil Content:
Results of the Field Experiment Series
in 1999

Friedrich Pank
Carla Vender
Leon van Niekerk
Wolfram Junghanns
Jan Langbehn
Wolf-Dieter Blüthner
Johannes Novak
Chlodwig Franz

Friedrich Pank and Jan Langbehn are affiliated with the Institute for Horticultural Crops, Federal Centre for Breeding Research on Cultivated Plants, Neuer Weg 22/23, 06484 Quedlinburg, Germany.

Carla Vender is affiliated with the Istituto Sperimentale Assestamento Forestale e Alpicoltura, Piazza Nicolini 6, 38050 Villazzano di Trento, Italy.

Leon van Niekerk is affiliated with SC Darbonne, 6, bd Joffre, 91490 Milly-la-Foret, France.

Wolfram Junghanns is affiliated with Dr. Junghanns GmbH, Untere Dorfstr. 8, 06449 Groß Schierstedt, Germany.

Wolf-Dieter Blüthner is affiliated with N.L. Chrestensen Samenzucht und Produktion GmbH, 99092 Erfurt, Germany.

Johannes Novak and Chlodwig Franz are affiliated with the Institute of Applied Botany, University of Veterinary Medicine, Veterinärplatz 1, A-1210 Wien, Austria.

[Haworth co-indexing entry note]: "Combining Ability of *Origanum majorana* L. Strains–Agronomical Traits and Essential Oil Content: Results of the Field Experiment Series in 1999." Pank, Friedrich et al. Co-published simultaneously in *Journal of Herbs, Spices & Medicinal Plants* (The Haworth Herbal Press, an imprint of The Haworth Press, Inc.) Vol. 9, No. 2/3, 2002, pp. 31-37; and: *Breeding Research on Aromatic and Medicinal Plants* (ed: Christopher B. Johnson, and Chlodwig Franz) The Haworth Herbal Press, an imprint of The Haworth Press, Inc., 2002, pp. 31-37. Single or multiple copies of this article are available for a fee from The Haworth Document Delivery Service [1-800-HAWORTH 9:00 a.m. - 5:00 p.m. (EST). E-mail address: getinfo@haworthpressinc.com].

31

SUMMARY. A combining ability test of 4 cms and 8 pollinator marjoram lines was carried out by a field experiment series on 3 locations to explore the feasibility of a marjoram hybrid variety system. Most traits revealed both–negative and positive–general and specific combining effects depending on the parent components and on the traits. Due to good combining ability high performance hybrids occurred in particular for the essential oil content. In further experiments improved hybrid effects can be expected using more different and homogeneous genotypes. *[Article copies available for a fee from The Haworth Document Delivery Service: 1-800-HAWORTH. E-mail address: <getinfo@haworthpressinc.com> Website: <http://www.HaworthPress.com> © 2002 by The Haworth Press, Inc. All rights reserved.]*

KEYWORDS. *Origanum majorana*, breeding, combining ability

INTRODUCTION

The advantages of hybrid varieties are improved performance by hybrid vigour (heterosis), the homogeneity caused by the uniformity of the F_1 population and the natural protection of the plant breeders rights because unlicensed seed production from the second generation would result in segregating progenies of minor value. Hybrid variety seed production needs the controlled fertilization of selected parent components. Components of a hybrid variety system for herbal drugs are cytoplasmatic male sterile lines with their maintainers to ease controlled pollination and pollinator lines with high per se performance and combining ability. Hybrid varieties are already common in vegetables and the main agricultural crops. The breeding of hybrid varieties is just beginning in medicinal and aromatic plants (3). The following report contains the results of the combining ability tests of different marjoram strains carried out in 1999 in the frame of the EU funded project FAIR3-CT96-1914.

MATERIALS AND METHODS

Male sterile marjoram lines and their maintainers were developed during the preceding years by the Institute of Applied Botany of the University of Veterinary Medicine in Vienna and the pollinator lines by selection between populations followed by one cycle of single plant selection and inbreeding by the Federal Centre of Breeding Research on Cultivated Plants in Quedlinburg, Germany in 1997 (1). Test crossings of four cytoplasmatic male sterile lines (cms) and 8 pollinator lines were performed in the glasshouse in 1998 by the seed company NLC in Erfurt, Germany. The scheme (Figure 1) shows the breeding procedure. ·

The precultivated F_1 plantlets were planted into the experimental field on three locations: Gross Schierstedt in Germany (Dr. Junghanns GmbH), Milly-la-Foret (SC Darbonne) in France and Villazzano di Trento (ISAFA) in Northern Italy. A uniform experimental design was used with three blocks containing all replications of each hybrid, the four cms-lines and three varieties. On each plot three rows with 27 plants were planted in a distance of 40 cm between and

FIGURE 1. Development of the parent components for the combining ability tests.

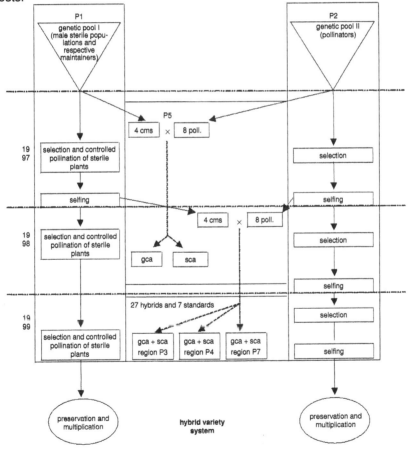

Hybrid breeding scheme

15 cm within the rows (= 4.86 m^2). The following traits were evaluated (2): anther status, plant height (cm), resistance to lodging (scores 9-1), parasitic attack (number of attacked plants/plot), precocity (days between planting and beginning of flowering), yield of leaf-flower-fraction (dt/ha with 84% dry matter content), proportion of leaf-flower fraction (percentage of the herb), content of essential oil (% v/w), and regrowth after the first harvest (scoring 9-1).

The traits (with exception of the scores) were submitted to an analysis of variance with the three factors: location (fix), variety (fix) and block within location (random). The proportion of leaf-flower-fraction was transformed with arcsin √. The glm (general linear models) procedure of the statistical analysis system (4), release 6.12 for Windows, was used. In case of the interaction variety × location, the constancy of the ranks of all varieties within locations was checked. As this procedure only showed moderate changes, main effects were then compared instead of simple effects. All variables resulting from scoring were not submitted to statistical evaluation procedures. The components of the performance of a hybrid were calculated as follows. General combining ability—for the seed parent (gca$_s$): average of all hybrids of a seed parent minus the overall average of hybrids (av); for the pollen parent (gca$_p$): average of all hybrids of a pollen parent minus av; specific combining ability of the parents of a hybrid (sca): trait expression minus (gca$_s$ + gca$_p$ + av) (5).

RESULTS

Influence of the locations and interaction location × genotype. Significant differences between the location means and significant genotype × location interaction occurred for all traits listed in Table 1 except for the leaf-flower-portion.

TABLE 1. Means of the three locations. N = 102 for each location, T* = Tukey grouping.

trait	precocity (d)		yield (dt/ha)		leaf-flower portion (%)		essential oil content (%)	
	$\bar{x}$	T*	$\bar{x}$	T*	$\bar{x}$	T*	$\bar{x}$	T*
location								
Milly La Foret	72.40	A	12.48	B	51.5	A	2.09	B
Groß Schierstedt	67.53	B	18.67	A	50.8	A	1.93	C
Villazano diTrento	64.10	C	11.94	B	51.6	A	2.18	A
LSD α = 5%	0.88		0.65		n.s.		0.087	

General combining ability. All traits revealed both-negative and positive-general combining effects depending on the parent components as demonstrated in Table 2.

The performance of the hybrids with respect to the combining ability of the parents. Tables 3 and 4 demonstrate the extremes of the performance of the hybrids and the specific combining ability (sca) for the most important traits yield and essential oil content. There are listed the absolute values of the mean, gca and sca, Tukey grouping (groups of different genotypes identified by the same letter did not differ significantly) and the performance of the genotypes related to the standard variety 'Marcelka' and to the high performance seed parent cms4 and the sca related to the mean of the concerning hybrid. The extremes of the hybrid performances are written in bold letters and the extremes of the absolute sca values in bold italic letters.

The sca has positive and negative yield effects as well. The range is between 0.88 and −1.7. The performance of the hybrids can be explained in most cases by the combining ability of the parents. The hybrid cms1 × pollen parent 01 has the highest yield of 16.40 g due to positive combining effects of both parents and sca in addition (0.28, 1.05, 0.51).

The sca has positive and negative effects also for the essential oil content. The extremes are 0.15 and −0.36. The combination cms1 × pollinator 07 has the highest content with 2.59 ml/100 g. All gca effects of the parents of this hybrid are positive with 0.05 and 0.36 and the sca-effect of the specific hybrid 0.04. The hybrid exceeds the seed parent cms4 by 27% and the standard 'Marcelka' by 108.9 per cent.

TABLE 2. General combining ability. Extremes in bold numbers.

genotype	precocity (d)	yield (dt/ha)	leaf-flower portion (%)	essential oil content (%)
parent components: cytoplasmatic male sterile lines				
1	0.5	0.28	−0.2	0.05
2	−0.4	−0.51	−0.5	0.03
3	0.2	0.34	0.7	−0.05
4	−0.2	−0.12	−0.1	−0.02
parent components: pollinator lines				
01	−0.6	**1.05**	1.1	0.19
02	−0.3	−0.15	−1.6	−0.08
04	**1.1**	−0.20	−0.5	0.22
07	−0.2	−0.07	0.6	0.36
09	**−1.7**	0.27	1.5	−0.06
15	−0.1	−0.38	0.8	−0.27
17	1.2	−0.02	−1.7	**−0.30**
18	0.2	**−0.68**	**2.1**	−0.12

TABLE 3. Yield of the leaf-flower-fraction (dt/ha, 86% dry matter).

nr	genotype	mean	gcas	gcap	sca	Tukey grouping	% from Marcelka	% from cms4	sca % from mean
4	cms4	13.61			-	AB	97.4	100.0	-
5	'Marcelka'	13.98			-	AB	100.0	102.7	-
8	$1_s \times 01_p$	**16.40**	0.28	1.05	0.51	A	117.3	120.5	3.1
25	$3_s \times 15_p$	15.38	0.34	−0.38	*0.88*	AB	110.0	113.0	5.7
32	$4_s \times 15_p$	**12.35**	−0.12	−0.38	**−1.70**	B	88.3	90.7	−13.7
	Total	14.36			-		102.7	105.5	-
LSD	$\alpha = 5\%$	3.30							

TABLE 4. Essential oil content (ml/100 g dry matter of leaf-flower-fraction).

nr	genotype	mean	gca_s	gca_p	sca	Tukey grouping	% from Marcelka	% from cms4	sca % from mean
4	cms4	2.04			-	EFGH	164.5	100.0	-
5	'Marcelka'	1.24			-	K	100.0	60.8	-
10	$1_s \times 04_p$	2.06	0.05	0.22	*−0.36*	EFGH	166.1	101.0	−17.5
11	$1_s \times 07_p$	**2.59**	0.05	0.36	0.04	A	208.9	127.0	1.5
32	$4_s \times 15_p$	1.97	−0.02	−0.27	0.12	FGH I	158.9	96.6	6.1
33	$4_s \times 17_p$	**1.71**	−0.02	−0.30	−0.12	JI	137.9	83.8	−7.0
	Total	2.07			-		166.9	101.5	-
LSD	$\alpha = 5\%$	0.31							

Expression of other traits. Resistance to lodging (res), parasitic attack (par), plant height (plh), proportion of the leaf-flower fraction of the herb (lff), regrowth (reg) were evaluated in addition. Substantial differences between the 3 locations were observed only on plh. The statistical evaluation revealed the significance of the interaction genotype × location for plh. The gca of the parent components and the sca of the specific hybrids had the following extremes for the quantitative traits: plh −1.3 to 1.7 and −1.2 to 1.6, lff −1.7 to 2.1 and −1.4 to 1.7. The extremes of the of the qualitative traits (scores) were for the 4 cms lines and for the hybrids in case of res: 5.6 to 8.0 and 4.7 to 8.8; in case of par 0.5 to 2.7 and 1.5 to 0.2 and in case of reg 7.4 to 8.7 and 7.4 to 9.

REFERENCES

1. Langbehn, J., F. Pank, J. Novak, and C. Franz. 2000. Influence of selection and inbreeding on *Origanum majorana* L. *In* Ch. Franz, F. Pank, M. Skoula and C.B. Johnson, eds. Abstracts of the Second International Symposium Breeding Research on Medicinal and Aromatic Plants, July 11th-16th, 2000 in Chania, Greece. p. A3.

2. Pank, F., J. Langbehn, J. Novak, W. Junghanns, J. Franke, F. Scartezzini, C. Franz, and A. Schröder. 1999. Eignung verschiedener Merkmale des Majorans (*Origanum majorana* L.) zur Differenzierung von Populationen und für die indirekte Selektion. 1. Mitteilung: Genetische Variabilität. (Suitability of different traits of marjoram (*Origanum majorana* L.) for population differentiation and indirect selection. Part 1: Genetical variation). *Zeitschrift für Arznei- und Gewürzpflanzen* 4: 8-18.

3. Rey, C. 1994. Une variété de thym vulgaire: "Varico" (A thyme variety: "Varico"). *Revue Suisse Viticulture, Arboriculture, Horticulture* 26: 249-250.

4. SAS Institute. 1989. SAS/STAT User's Guide, version 6, fourth edition, volume 2. SAS Institute Inc., Cary, NC.

5. Sprague, G.F. and L.A. Tatum. 1942. General vs. specific combining ability in single crosses of corn. *Journal of the American Association of Agronomy* 34: 923-932.

Distribution of Pyrrolizidine Alkaloids in Crossing Progenies of *Petasites hybridus*

R. Chizzola
Th. Langer

SUMMARY. The toxic pyrrolizidine alkaloids (PAs) senecionine and integerrimine are minor compounds in the rhizomes of the medicinal plant *Petasites hybridus*. Plants used in phytopharmacy should be as low in PAs as possible. The present work investigates whether the PA content of the rhizomes can be lowered in the progenies obtained from selected plants through experimental crossings. Starting from plants differing in the PA content, two consecutive crossing generations (F1 and F2) were established. The PAs were recorded in the young runners of the rhizome. The results showed that various progenies differed clearly in their PA content and even within a progeny the PA content may vary considerably. Plants poor in alkaloids generally generate offspring lower in PA than parent plants with a high alkaloid level. However plants totally free of alkaloids could not be generated through the crossing experiments. *[Article copies available for a fee from The Haworth Document Delivery Service: 1-800-HAWORTH. E-mail address: <getinfo@haworthpressinc.com> Website: <http://www.HaworthPress.com> © 2002 by The Haworth Press, Inc. All rights reserved.]*

R. Chizzola and Th. Langer are affiliated with the Institute for Applied Botany, University of Veterinary Medicine, A-1210 Wien, Veterinärplatz 1, Vienna, Austria.

The authors thank Mrs. H. Michitsch, Mrs. R. Pfosser, Mr. G. Sandtner and Mr. P. Weinzerl for technical assistance.

[Haworth co-indexing entry note]: "Distribution of Pyrrolizidine Alkaloids in Crossing Progenies of *Petasites hybridus*." Chizzola, R., and Th. Langer. Co-published simultaneously in *Journal of Herbs, Spices & Medicinal Plants* (The Haworth Herbal Press, an imprint of The Haworth Press, Inc.) Vol. 9, No. 2/3, 2002, pp. 39-44; and: *Breeding Research on Aromatic and Medicinal Plants* (ed: Christopher B. Johnson, and Chlodwig Franz) The Haworth Herbal Press, an imprint of The Haworth Press, Inc., 2002, pp. 39-44. Single or multiple copies of this article are available for a fee from The Haworth Document Delivery Service [1-800-HAWORTH 9:00 a.m. - 5:00 p.m. (EST). E-mail address: getinfo@haworthpressinc.com].

39

KEYWORDS. *Petasites hybridus*, pyrrolizidine alkaloids, medicinal plant quality

INTRODUCTION

Toxic pyrrolizidine alkaloids (PAs) are undesirable natural compounds occurring besides the active principles in some medicinal plants. In Germany the use of such medicinal plants had been restricted, the daily dose administered internally must not exceed 1µg of the toxic PA and their N-oxides (1).

In the case of *Petasites hybridus* GM. et Sch. (Asteraceae), butter bur, the useful substances are sesquiterpene esters (petasines) with spasmolytic properties (6) and become increasingly used not only in relieving pains in the respiratory or gastrointestinal tract but also in the case of migraine or allergic reactions as hay fever (4). However, toxic PAs, mainly senecionine and its isomer integerrimine impair severely this use and must be eliminated by various techniques from the plant extracts. These alkaloids are concentrated in the rhizomes, mainly in the young part just below the leaves and in young runners (2) and may vary from less than 2 to 500 mg·kg^{-1}.

For pharmaceutical purposes plant material low in PAs is needed. The present work investigates the alkaloid content in the young subterranean runners of two consecutive generations (F1 and F2) of offsprings derived from plants showing different PA contents collected in the wild through controlled experimental crossings. The aim is to test whether it will be possible to get plants poor in PAs through sexual reproduction of selected individuals low in alkaloids.

MATERIALS AND METHODS

The plant material originated from several locations in Austria and France and was transplanted to the experimental garden of the institute. *Petasites* builds up either male or female inflorescences which appear early in spring and the crossing experiments were carried out in this season according to the biological rhythm.

Entire inflorescences of female individuals of the dioecious plant were isolated in impregnated paper bags before the opening of the flower heads. As they reached maturity pollen was transferred directly by rubbing them with fully developed male capitules teared off from the selected crossing partner. Then they were again isolated until seed maturity, when the seeds begin to detach from the capitules. This isolation technique was efficient since isolated female inflorescences which were not experimentally pollinated did not develop any seeds.

To grow the progenies the collected seeds were sown in a standard soil and sand mixture and kept for 6 week in a growth chamber at 80% relative humidity and a 13 h, 23°C light and 11 h, 17°C dark period. Then the plants were kept in pots.

F1 crossings 1, 4 and 6 were performed in 1993, the other crossings in 1994. All F2 crossings were carried out in 1997. This time interval was necessary because it lasts 2-3 years to get flowering plants from the seeds. The plants of F1 were harvested in July 1995, those of F2 in October 1999. Additional 8 crossings yielded less than 10 offsprings each and were therefore not considered further.

From most of the plants of the progenies it was possible to get 10-20 cm long and 3-5 mm thick pieces of the runners. Then the PAs were analyzed quantitatively using an immunoassay (5) or capillary GC (3) from methanolic extracts after the reduction of the N-oxides. Both analytical methods gave comparable results (5). For a statistical analysis SPSS for Windows, version 9.0 was used.

RESULTS

The main statistical parameters of the progenies from two consecutive generations are presented in Table 1. Particularly in the F1 great differences in the PA content between the various progenies could be observed and also a great variability within a given progeny, the coefficient of variation varied between 40 and 114%. Some progenies (2, 3 and 5) showed in the F1 a higher mean than median PA content, indicating that there were several individuals with an excessive high PA level in the rhizome part tested.

From crossing population No. 1 four different progenies were obtained by crossing different individuals with plants from the original parent population. Sister crossings could not be made since all plants from this progeny produced only female inflorescences. They differed slightly in their PA content, but because of the large variability within each progeny these differences were not statistically significant. In two further crossings (No. 3 and 4) the PA content of the respective F1 and F2 generation did not differ significantly, whereas in two other crossings (No. 5 and 6) the F2 plants were lower in PAs than the F1 plants. These four F2 crossings were sister crossing where a plant from the F1 was pollinated with another one from the same progeny. As they had lower coefficients of variation, these F2 descendants were more homogenous than those in the F1.

When plants poor in alkaloids are the parents, in most of the cases the offsprings were also low in alkaloids (nr. 1, F1 and F2). Two progenies (nr. 2 and

TABLE 1. Variability of the PAs (senecionine + integerrimine) in various progenies of *Petasites hybridus*. Mean, median, minimum and maximum PA content in mg/kg in dry rhizome pieces.

Crossing		1	1	1	1	2	3	4	5	6	7
PA content of the parent plants		low				low	med.	high	low	med.	high
		low				low	low	med.	low	high	high
F1	N	64				27	53	52	27	36	73
	Mean	**25.3**				**44.0**	**40.0**	**116**	**139**	**171**	**255**
	Median	20.1				22.8	26.7	109	91.1	138	252
	C.V.%	75.5				114	106	72.9	104	63.6	40.4
	Minimum	2.4				2.7	1.0	22.0	16.2	48.1	28.2
	Maximum	84.3				199	174	414	693	416	529
PA content of the parent plants (from F1)		low	low	low	low	high	low		med.	high	
		low	low	low	low	med.	high		high	high	
F2	N	25	27	27	42	27	35		30	40	
		B	B	B	B	S	S		S	S	
	Mean	**36.5**	**25.2**	**18.0**	**24.7**	**52.9**	**60.1**		**51.3**	**64.6**	
	Median	28.2	21.6	17.6	20.9	42.3	58.0		37.0	52.3	
	C.V.%	65.2	67.9	49.4	63.2	61.4	51.9		73.8	60.2	
	Minimum	4.1	0.7	1.3	1.5	10	11.3		8.8	17.0	
	Maximum	101	63.5	37.6	88.2	119	143		160	154	

N: Number of offsprings tested in the crossing.
B: Back crossing, S: Sister crossing.
PA content of the respective parent plants: low < 20, med. < 40 and high > 40 mg/kg.

5) derived from parents low in PA exhibit higher alkaloid levels, but the PA varied also largely within them.

DISCUSSION

The accumulation of PAs in the rhizomes of *P. hybridus* is a quantitative trait, and because the alkaloids are complex substances their formation is governed certainly by a series of genes. Furthermore the PAs exhibit within the plant a very dynamic behaviour: Extreme differences in the PA contents may arise in different parts of the rhizome with maximum values in the young segments just below the leaves and in young runners (2), whereas the leaves of *Petasites* ordinary contain less than 1 mg/kg PA (4). Also in young leaves of *Cynoglossum officinale,* another PA producing plant, very high alkaloid levels could be recorded in young leaves in contrast to old leaves (7). In this context it is claimed that the ecological role of PA as toxic substances is to protect important growing zones in the plant.

Additionally in various species the PA content was dependent upon environmental influences, especially mineral nutrition and water supply (8). Therefore it is essential to test comparable plant parts, in our case ontogenetically uniform rhizome pieces, for a comparison of the PA contents in different individuals.

P. hybridus in the wild ordinary displays an intensive vegetative growth of the rhizome. Therefore the plants had to be kept in pots to avoid that the long runners intermingle. These limitations may also influence the PA content.

Although the result may suggest that in general the plants from F2 were lower in PA than those from F1 one has to keep in mind that for generating the F2 more individuals low in alkaloids were taken than for the generation of F1. The progeny 7 high in PA has not been used to breed a F2. Additionally these two generations were raised in different years, they were exposed to different climatic situations and therefore display different PA contents. On the other hand the F2 has been produced through back crossings and sister crossings and in consequence be more homogenous than the F1.

Comparing the various crossing types within the F1 or within the F2 where the plants were exposed to the same environmental conditions respectively and where comparable rhizome pieces were analyzed showed that there are clear differences in the progenies originating from different genotypes. And there is at least the genotype in crossing 1 which produces individuals low in alkaloids. This suggests that there is a genetic component determining the amounts of PA accumulated in the plant despite the great variability in the PA in natural populations and within the complex rhizome. In the case of *Senecio jacobea* at least 50% of the variability in the PAs was due to genetic differences (8).

Several individuals were very low in PAs but none completely free of alkaloids could be detected. These individuals low in PA should be monitored further under field conditions.

REFERENCES

1. Bundesgesundheitsamt: Abwehr von Arzneimittelrisiken. Stufe II. Bundesanzeiger Nr. 111 vom 17.06.1992, S. 4805. Zitiert: Pharm. Ztg. 1992; 137: 2088 und Dt. Apoth. Ztg. 1992; 132:1406.

2. Chizzola R., B. Ozlsberger, and T. Langer. 2000. Variability in chemical constituents in *Petasites hybridus* from Austria. *Biochem. Syst. Ecol.* 28, 421-432.

3. Chizzola R. 1994. Rapid sample preparation technique for the determination of pyrrolizidine alakloids in plant extracts. *J. Chromatogr. A* 668, 427-433.

4. Grossmann W. 1996. Migräneprophylaxe mit einem Phytotherapeutikum; Ergebnisse einer randomisierten, placebokontrollierten klinischen Doppelblindstudie mit Petadolex®. *Der Freie Arzt.* 37, 74-76.

5. Langer T., E. Möstl, R. Chizzola, and R. Gutleb. 1996. A competitive enzyme immunoassay for the pyrrolizidine alkaloids of the senecionine type. *Planta Medica.* 62, 267-271.

6. Meier B. and M. Meier-Liebi. 1994. *Petasites.* In: Hänsel, R., Keller, K., Rimpler, H. and Schneider G. Hagers Handbuch der pharmazeutischen Praxis. Bd. 6: Drogen P-Z, 5. Aufl. Springer, Berlin; 81-105.

7. Van Dam N.M., R. Verpoorte, and E. van der Meijden. 1994. Extreme differences in pyrrolizidine alkaloid levels between leaves of *Cynoglossum officinale. Phytochemistry.* 37, 1013-1016.

8. Vrieling K., H. de Vos, and C.A.M. van Wijk. 1993. Genetic analysis of the concentration of pyrrolizidine alkaloids in *Senecio jacobea. Phytochemistry.* 32, 1141-1144.

Breeding of Sweet Basil (*Ocimum basilicum*) Resistant to Fusarium Wilt Caused by *Fusarium oxysporum* f.sp. *basilicum*

Nativ Dudai
David Chaimovitsh
Reuven Reuveni
Uzi Ravid
Olga Larkov
Eli Putievsky

SUMMARY. Sweet basil (*Ocimum basilicum*) is one of the leading herb crops, used fresh or dry. Fusarium wilt, caused by *Fusarium oxyporum*, is a severe problem in Israel, as well as in Europe and the USA. It causes stunting, browning of vascular tissues, severe wilting without chlorosis and defoliation of the plants. In the summer of 1992 we identified several isolates of *Fusarium oxyporum* f.sp. *basilicum* originating from the stems of diseased plants of greenhouse-grown sweet basil (*O. basilicum* L.). Plants of a commercial variety in an infected field, with no symptoms, were selected for self-breeding as a source of seeds with resistant germplasm. Further selection tests were conducted in the greenhouse on infested soil in order to improve resistance up to five generations. This work has resulted in a registered variety resistant to Fusarium wilt ('Nufar').

Nativ Dudai, David Chaimovitsh, Reuven Reuveni, Uzi Ravid, Olga Larkov, and Eli Putievsky are affiliated with the Division of Aromatic Plants, Agricultural Research Organization, Newe Ya'ar Research Center, P.O. Box 1021, Ramt Yishay, Israel.

[Haworth co-indexing entry note]: "Breeding of Sweet Basil (*Ocimum basilicum*) Resistant to Fusarium Wilt Caused by *Fusarium oxysporum* f.sp. *basilicum*." Dudai, Nativ et al. Co-published simultaneously in *Journal of Herbs, Spices & Medicinal Plants* (The Haworth Herbal Press, an imprint of The Haworth Press, Inc.) Vol. 9, No. 2/3, 2002, pp. 45-51; and: *Breeding Research on Aromatic and Medicinal Plants* (ed: Christopher B. Johnson, and Chlodwig Franz) The Haworth Herbal Press, an imprint of The Haworth Press, Inc., 2002, pp. 45-51. Single or multiple copies of this article are available for a fee from The Haworth Document Delivery Service [1-800-HAWORTH 9:00 a.m. - 5:00 p.m. (EST). E-mail address: getinfo@haworthpressinc.com].

All 'Nufar' individuals and F_1 individuals obtained from the cross of 'Nufar' × susceptible source plants were resistant to Fusarium wilt. In further studies we found that this resistance is a monogenic dominant trait. At present, fresh market varieties resistant to Fusarium, with good yields and performance, as well as acceptable leaf-size, shape and shelf life, are being developed and examined for their aroma. The variability in the essential oil composition of the culinary genetic lines are shown and discussed. *[Article copies available for a fee from The Haworth Document Delivery Service: 1-800-HAWORTH. E-mail address: <getinfo@ haworthpressinc. com> Website: <http://www.HaworthPress.com> © 2002 by The Haworth Press, Inc. All rights reserved.]*

KEYWORDS. Sweet basil, *Ocimum basilicum*, *Fusarium oxysporum*, breeding, resistance, essential oils, methyl chavicol, eugenol, linalool, fresh herbs

INTRODUCTION

Sweet basil (*Ocimum basilicum* L.) is cultivated on a commercial scale in the USA and in the Mediterranean region, and is growing in popularity, as both a fresh and dried herb for use in food products. In Israel, sweet basil is the leading fresh herb crop grown in greenhouses for export, all year long, but mainly in wintertime. One of the limiting problems of this crop is Fusarium wilt, caused by *Fusarium oxysporum* f.sp. *basilicum* specific to basil ('FOB'). The disease has been known since 1956 (1) and occurs in Russia, Italy (2), France (3), USA (4) and in Israel (5,6,7,8).

The problem of Fusarium wilt became acute mainly due to the intensive mono-culture in the greenhouses and the prohibition of methyl bromide application. In 1992 we first observed in Israel the symptoms and identified the disease on the commercial variety 'Hen' which was grown in an experimental greenhouse at Newe Ya'ar, as part of agro-technical experiments (8). In that greenhouse, eventually most of the plants died and only a few plants in the area of 300 m² survived.

MATERIALS AND METHODS

Growing of the source population was in a greenhouse in Tuff (a volcanic scoria) medium, as customary in most of the commercial fields in Israel (9). In order to produce pure lines, self crossing is done by covering each plant with insect-proof nets during flowering and seed production.

Isolation and identification of the pathogen were described previously (8). Inoculum was prepared by gently brushing plates with sterile distilled water, which was then passed through cheesecloth to remove mycelial fragments.The resulting suspension was then diluted to obtain 5×10^5 conidia/ml. Seedlings at five-leaf stage were used for artificial inoculation. Each seedling was dipped in a conidial suspension to enable 10 ml of the inoculum to soak into the media (peat:vermiculite 1:1) of the developed roots, and were immediately planted in pots filled with 500 g of nonifested soil.

The specificity of the pathogen to *Ocimum basilicum* was tested by artificial inoculation of nine different species belonged to several botanical families: *Mentha piperita, Micromeria fruticosa, Lavandula officinalis* (Labiatae), *Lycopersicum esculantum, Capsicum annum* (Solanacea), *Cucumis sativus* (Cucurbitacea), *Artemisia dracunculus* (Compositae). The pathogen was very specific to basil and did not affect any other plant. Even another species of *Ocimum* like lemon basil (*O. citriodorum*) was not affected (8).

Examination of cultivars in an infected commercial field. Seedlings of the cultivars were planted in a commercial greenhouse in Gush Katif, Israel, in local sand soil, on November 23, 1999 in plots of 2 m^2 in randomized blocks, 4 replicates. The yield was determined by harvesting the plants by the commercial method, cutting 20 cm above the ground, three times: December 26, 1999; February 8, 2000; March 1, 2000.

Essential oil determination. Essential oil was obtained from plants of the various cultivars in February 2000, grown in experimental plots at the Newe ya'ar Research Center, Israel. The growing methods, hydrodistillation by clevenger apparatus and GC-MS analyses were as described previously (9).

RESULTS AND DISCUSSION

To begin the selection process, seeds were collected from the surviving plants in an infected field. This first generation seedlings were further artificially inoculated with the isolated FOB. About two thirds of the plants survived and the rest had shown typical Fusarium wilt symptoms. We collected the seeds again and the process was repeated up to the fourth generation, which was completely resistant, and so were all of the generations which followed. Resistance selection by this method yielded a similar phenotype and chemotype as the source susceptible parent variety.

By studying the inheritance mechanism we found that the resistance is a chromosomal dominant monogenic trait: F1 hybrid of the resistant variety and the susceptible source variety (R × S) yielded 100% resistant progeny. The same results observed by reciprocal crosses. The resistance of the F2 generation of the hybrids behaves according to Mendel's laws as a monogenic dominant trait. There was a segregation ratio of 3:1 between resistance to susceptible. According to these results seed companies can "protect" the variety by producing a "hybrid F1" of R × S. The first variety that we developed and released called 'Nufar' (10). Using the natural variability of the source population, new genetic lines are selected, while they tested to be homozygoic to the resistance trait by examination of each progeny.

The main criteria of breeding basil for the fresh market besides resistance to disease are: performance (mainly size and color of leaves), fresh yield, shelf life and aroma. In the last 5 years, 52 resistant lines were selected, differing mainly in their yield and morphology. Results of the examination of selected Fusarium resistant cultivars in a highly disease susceptible commercial greenhouse during winter are shown in Figure 1. 'CN' (origin: CN Co., UK) and '4' (our selection from 1995) are the main commercial cultivars in Israel. The '4/7' was selected as a very susceptible cultivar. The mortality of these cultivars during the first 5 weeks (by the first harvest) was 35-40%, and 60-100% by the second harvest (10 weeks after planting). Nevertheless, the mortality (mainly caused by infection of Botrytis disease) of the new resistant cultivars was in these conditions only 10-30% (Figure 1B). The varieties differ in total yield, due to their survival–3-6 Kg/m^2 for the resistant cultivar, while the susceptible cultivars yield was only 1-2 Kg/m^2 (Figure 1A).

The main problem in the performance of the classic varieties was the giant leaf size that was obtained under the intensive growing conditions in the greenhouses during the winter. Three new Fusarium resistant varieties that yielded an acceptable size of leaves were registered and grown commercially in Israel: Perrie, Nirit and Hagar. These varieties were selected also for their good shelf-life. The main effort of our breeding program now is to reduce the storage temperature of the basil (now 12°C), which is very sensitive compared with those that are optimal for other herbs (3-4°C).

The varieties also differ in their aromatic components (Table 1). The major components are linalool, methyl chavicol, eugenol and methyl eugenol. All of our culinary chemotypes contain (−) linalool (11), and distinguished by their main chemotypes: methyl chavicol or eugenol, as reported previously (12,13).

Now we continue the breeding also for ornamental basil. Phenotypes of compact inflorescence are being selected. The main goal is to add some good aroma back to the floral bouquet. A first variety, bearing dark red inflorescences, called 'Cardinal' has been registered (14).

FIGURE 1. Yield (A) and mortality (B) of sweet basil cultivars during winter 1999-2000 in an infected commercial field (into a greenhouse).

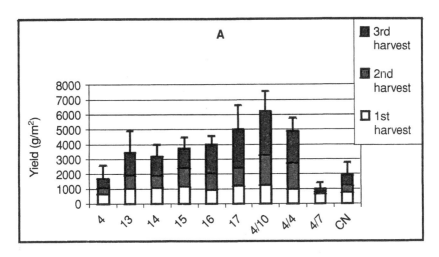

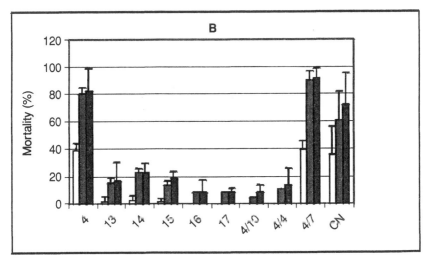

TABLE 1. Variability in the essential oil composition of some culinary sweet basil cultivars.

Relative composition, %	Cultivar														
	13	14	15	16	17	4/10	4/4	4/7	4/9	6/8	Hen	Nufar	CN	4	Geno vese
(−)-Linalool	57.6	48.1	46.0	45.0	41.0	46.3	48.4	42.2	49.3	47.3	56.4	55.9	54.1	43.7	55.4
Methyl chavicol	0.4	0.0	0.0	0.0	0.0	0.0	0.0	0.0	21.0	26.1	19.1	20.0	9.8	0.0	0.0
Eugenol	16.0	21.4	21.6	27.7	24.9	28.1	21.1	31.2	3.8	8.9	4.3	3.1	13.3	28.4	19.6
Methyl eugenol	3.0	2.4	2.4	1.4	7.1	3.0	3.0	5.0	0.1	0.3	0.0	0.0	0.0	3.4	1.0
1,8-Cineole	5.3	11.5	7.1	3.5	8.5	7.3	6.1	7.5	7.1	2.8	4.6	3.5	5.0	7.1	7.5
α-Bergamotene	4.4	4.7	0.1	9.9	8.0	1.1	0.2	6.0	6.3	t	0.5	t	0.5	3.5	4.1
Germacrene D	4.0	3.3	6.0	3.7	3.1	3.9	6.5	1.6	2.5	4.8	3.4	4.1	3.7	3.4	2.8
α-Cadinol	3.0	2.6	4.1	3.4	3.1	3.7	3.8	1.1	2.1	3.5	4.8	5.4	5.2	3.2	3.5
γ-Cadinene	2.1	1.2	3.3	1.5	0.0	2.3	3.7	0.3	1.9	2.7	1.5	1.8	1.6	1.9	1.8
Bornyl acetate	1.0	1.1	2.0	1.7	2.1	0.6	0.7	1.3	1.1	0.5	0.5	0.5	0.5	1.3	1.7
(+)-α-Terpineol	0.7	1.2	1.2	0.0	0.5	0.8	0.8	0.7	1.3	0.7	0.7	0.9	0.8	0.9	1.1
β-Elemene	0.5	0.2	1.7	0.0	0.0	0.5	1.3	0.0	0.0	1.4	1.6	1.2	1.7	0.5	0.0
(+)-Borneol	0.3	0.2	1.4	0.0	0.0	0.0	0.0	0.4	1.3	0.0	0.0	0.0	0.0	0.0	0.0
β-Ocimene (E)	0.3	0.7	0.8	0.3	0.6	1.3	0.6	0.6	0.6	0.1	0.6	0.1	0.6	1.1	0.1
α-Guaiene	0.2	0.0	1.0	0.0	0.0	0.1	0.8	0.0	0.0	0.6	0.5	0.5	0.5	0.0	0.0
Myrcene	0.1	0.2	0.1	0.1	0.1	0.2	0.1	0.2	0.1	0.1	0.2	0.1	0.2	0.2	0.1
(+)-Linalool	t	t	t	t	t	t	t	t	t	t	t	t	t	t	t
(+)-Camphor	0.2	1.0	0.1	0.6	0.4	0.2	1.1	0.1	0.1	0.1	0.3	0.8	0.3	0.6	0.8
Geraniol	0.0	0.0	0.0	0.0	0.0	0.0	0.4	1.7	0.6	0.0	0.0	0.0	0.0	0.0	0.0
β-Caryophyllene	t	t	0.1	t	t	0.0	0.1	0.0	0.7	0.0	0.0	0.3	0.2	0.0	0.0
β-Farnesene	0.0	0.1	0.4	0.6	0.0	0.2	0.8	0.0	0.0	0.0	0.0	0.0	0.0	0.6	0.4
α-Humulene	0.0	0.0	0.3	0.5	0.0	0.3	0.5	0.0	0.0	0.0	0.7	1.0	0.8	0.0	0.0
Total included	99.1	99.9	99.9	99.9	99.2	99.9	99.9	99.9	99.9	99.9	99.9	99.2	98.9	99.9	99.9

REFERENCES

1. Vergovskii, V.I. 1956. Some peculiarities in the presence of fusariosis in basil. Kratk. Otchet. Nauchno. Issled. Vseoyus. Inst. Mailichn. Efiromaslichn. Kult. 7: 195-197.

2. Grasso, D.S. 1975. Un avvizzimento del basilico de *Fusarium oxysporum*. *Inf. Fitopatol.* 25: 5-7.

3. Mercier, S. and J.C. Pionnat. 1982. Presence en France de la fusariose vasculare du basilic. *C.R. Seances Acad. Agric. Fr.* 68: 416-419.

4. Keinath, A.P. 1994. Pathogenicity and host range of *Fusarium oxysporum* from sweet basil and evaluation of disease control methods. *Plant Disease* 78: 1211-1215.

5. Dudai, N., R. Reuveni, and E. Putievsky. 1995b. Identification of resistant germplasm of *Ocimum basilicum* against *Fusarium oxysporum*. *Hassdeh* 76 (1): 47-52 (in Hebrew).

6. Gamliel, A., T. Katan, H. Yunis, and J. Katan. 1996. Fuzarium wilt and crown rot of sweet basil: Involvement of soilborne and airborne inoculum. *Phytopathology* 86: 56-62.

7. Reuveni, R., N. Dudai, and E. Putievsky. 1997. Evaluation and identification of basil germplasm for resistance to *Fusarium oxysporum* f.sp. *basilicum*. *Plant Disease* 81:1077-1081.

8. Reuveni, R., N. Dudai, and E. Putievsky. 1995. Germplasm of *Ocimum basilicum* resistant to *Fusarium oxysporum*. *Phytopathology* 85: 1178.

9. Dudai, N., E. Lewinsohn, O. Larkov, I. Katzir, U. Ravid, D. Chaimovitsh, D. Sa'adi, and E. Putievsky. 1999. Dynamics of yield components and essential oil production in a commercial hybrid sage (*Salvia officinalis* × *salvia fruticosa* cv. Newe ya'ar No. 4). *J. Food Chem.* 47: 4341-4345.

10. Dudai, N., R. Reuveni, and E. Putievsky. 1995a. Nufar–A sweet basil cultivar, resistant to Fusarium wilt. Application for registration No. 2376/95, Ministry of Agriculture, Israel Plabt Breeders' Rights Council. *Israel Plant Breeder Gazette* 42: 2.

11. Ravid, U., E. Putievsky, I, Katzir, and E. Lewinsohn. 1997. Enantiomeric composition of linalol in essential oils of *Ocimum basilicum* and in commercial basil oils. *Flvour Fragr. J.* 12: 293-296.

12. Paton, A. and E. Putievsky. 1996. Taxonomic problems and cytotaxonomic relationships between and within varieties of *Ocimum basilicum* and related species (Labiatae). *Kew Bull.* 51:509-524.

13. Putievsky, E., A. Paton, E. Lewinsohn, U. Ravid, D. Chaimovitsh, I. Katzir, D. Sa'adi, and N. Dudai. 1999. Crossability and relationship between morphological and chemical varieties of *Ocimum Basilicum* L. *J. Herbs, Spices and Med. Plants* 6: 11-24.

14. Dudai, N., E. Putievsky, U. Ravid, R. Reuveni, D. Sa'adi, O. Larkov, E. Lewinsohn, and D. Chaimovitsh. 2000. Cardinal–A sweet basil cultivar, resistant to Fusarium wilt. Application for registration No. 3035/99, Ministry of Agriculture, Israel Plabt Breeders' Rights Council. *Israel Plant Breeder Gazette* 55: 1.

The Impact of Drought Stress and/or Nitrogen Fertilization in Some Medicinal Plants

Dea Baričevič
Alenka Zupančič

SUMMARY. Several pot experiments were set up to follow the impact of drought stress and/or nitrogen fertilization on the yield and secondary metabolites content of medicinal and aromatic plants (MAP) accessions held within Genebank for MAP in Slovenia. The results of pot trials with fenugreek and with deadly nightshade are presented. In fenugreek (*Trigonella foenum-graecum* L., Fabaceae) the maximal yield of diosgenin was achieved in plants grown under an optimal irrigation regime (35% depletion of available soil water). Drought stressed seeds of cv. Margaret contained less diosgenin (p = 0.0021) than irrigated plants (0.123%, dw). In deadly nightshade (*Atropa belladonna* L., Solanaceae), the content of tropane alkaloids was determined. The results of the plant treatment responses showed that the maximal yield of tropane alkaloids (hyoscyamine: 54 mg/plant; scopolamine: 7 mg/plant) was achieved in plants grown under an optimal irrigation regime (35% depletion of available soil water) accompanied with a total nitrogen supply of 0.37 g/pot. By contrast, the maximal content of alkaloids was achieved with 95% depletion of available soil water and a nitrogen supply of 1.60 g/pot. *[Article copies available for a fee from The Haworth Document Delivery Service: 1-800-HAWORTH. E-mail address: <getinfo@haworthpressinc.com> Website: <http://www.HaworthPress.com> © 2002 by The Haworth Press, Inc. All rights reserved.]*

Dea Baričevič and Alenka Zupančič are affiliated with the University of Ljubljana, Biotechnical Faculty, Agronomy Department, Jamnikarjeva 101, 1000 Ljubljana, Slovenia.

[Haworth co-indexing entry note]: "The Impact of Drought Stress and/or Nitrogen Fertilization in Some Medicinal Plants." Baričevič, Dea, and Alenka Zupančič. Co-published simultaneously in *Journal of Herbs, Spices & Medicinal Plants* (The Haworth Herbal Press, an imprint of The Haworth Press, Inc.) Vol. 9, No. 2/3, 2002, pp. 53-64; and: *Breeding Research on Aromatic and Medicinal Plants* (ed: Christopher B. Johnson, and Chlodwig Franz) The Haworth Herbal Press, an imprint of The Haworth Press, Inc., 2002, pp. 53-64. Single or multiple copies of this article are available for a fee from The Haworth Document Delivery Service [1-800-HAWORTH 9:00 a.m. - 5:00 p.m. (EST). E-mail address: getinfo@haworthpressinc.com].

KEYWORDS. *Trigonella foenum-graccum*, fenugreek, *Atropa belladona*, deadly nightshade, tropane alkaloids, nitrogen fertilization, drought stress

INTRODUCTION

Slovenia is a small Central European country, but despite of its small area and population (20.256 km², 2 million inhabitants), it is very rich in plant diversity. There are 3261 plant species known to be autochthonous or well adapted to the Slovenian climate for centuries (1). The territory of Slovenia covers 3 centres of diversity (Mediterranean, European-Siberian, Near Eastern). There are over 100 medicinal and aromatic plants (MAP) with potential medicinal properties. Massive exploitation of wild MAP from natural habitats for market supply and natural degradation of vegetation/flora due to anthropogenic factors are the main reasons why MAP are often considered as endangered and/or threatened species in many European regions, including Slovenia. International conventions that were ratified in Slovenia in the last 4 years (in 1996, the Rio Convention; in 1999, CITES) and also national legislation (Nature Conservation Law, adopted in 1999) are considered as general documents that list more than 35,000 plant or animal species considered as endangered. Only about 50 plant species covered by this legislation belong to the group of medicinal and aromatic plants (MAP). Like in other European countries, the awareness of the need for conservation of MAP through cultivation became topical in Slovenia in the last decade. As a result of the national developmental programs in some European countries, the acreage covered by MAP is rapidly increasing. Although a few years ago the cultivation area of MAP in Europe was estimated at 70,000 ha (2), current data reveal that MAP are cultivated on an estimated 100,000 ha, the main countries being Hungary (about 40,000 ha) (3), France (25,000 ha), Spain (19,000), Italy (2,300 ha) (4), Germany (7,500 ha), Austria (4,300 ha) (5), and Yugoslavia (3,000 ha) (6).

At present in Slovenia, MAP are cultivated on only 20-25 ha (12 ha of purple cornflower, *Echinacea purpurea* Moench, with the average yield of 13 t fw/ha; other species are cultivated on smaller acreages of 0.5-1 ha). Common species like garden marigold, *Calendula officinalis* L., narrow-leaved plantain, *Plantago lanceolata* L., fenugreek, *Trigonella foenum graecum* L., and hoary willowherb, *Epilobium parviflorum* L., are cultivated mainly on fragmented areas with minor possibility of usage of modern machinery. In spite of neglected acreage under MAP cultivation and relative low demand on raw materials in target markets in Slovenia (7), MAP were identified as minor crops that could be of national interest with both realistic prospects for cultivation and a market for the alternative products that they synthesise. To develop the cultivation of MAP in Slovenia, Directives and Proposal of the National Program for production, processing and quality control for medicinal and aromatic plants had already been prepared a few years ago (8,9). Although the program did not receive financial support as a whole, some parts did–like the National collection/Genebank for MAP which was officially recognized in 1995 and is annually financed by the Ministry for Agriculture, Food and Forestry. The first stage of this program foresees monitoring of natural populations, their *ex situ* characterization and conservation of MAP through cultivation.

In the development of agrosystems for commercial production of MAP, the environmental circumstances of the country have to be considered. Although considered to be a country with an abundant water supply, precipitation is not evenly distributed over the crops' growing period in Slovenia. So, we suffer serious droughts in some parts of the country every year, depending on the water retention characteristics of the soil and the appearance of crop phenophases. Agriculture and livestock, at least in some of localities in Slovenia, are among important sources of environmental pollutants. The problem of high levels of nitrates and nitrites can be found in the most agricultrally-developed areas in Slovenia (Prekmurska, Dravska, and Ptujska regions) (10). These are also the areas of high average water deficit within the summer months (weather station Murska Sobota–June: precipitation 100 mm, evapotranspiration 140 mm; July: precipitation 100 mm, evapotranspiration 140 mm; August: precipitation 100 mm, evapotranspiration 140 mm). This is still more pronounced in dry years like 1983, 1985, 1988, 1992, 1993, 1994 and 2000) (weather station Murska Sobota/1992–June: precipitation 100 mm, evapotranspiration 140 mm; July: precipitation 25 mm, evapotranspiration 155 mm; August: precipitation 10 mm, evapotranspiration 165 mm).

Based on experimental evidence, field trials seem often too rough to estimate the tender relationships among genotype and environmental parameters in agrosystems. This is the reason why evaluation of susceptibility of germplasm descendants of MAP to environmental stress (like drought) or of impact of N-fertilization in pot trials/lysimeters, under the conditions of controlled environment, are a part of routine research work within Genebank for MAP.

This paper focuses on different approaches in evaluation of drought and/or fertilization impacts on the yield and production of secondary metabolites in accessions of fenugreek (*Trigonella foenum-graecum* L.) and deadly nightshade (*Atropa belladonna* L.).

MATERIALS AND METHODS

Trigonella foenum-graecum L. *Pot Experiment*

Plant material and growing conditions. Two different cultivars of fenugreek, acquired from National Seed Development Organization Ltd., Newton

Hall, Newton, Cambridge, UK, and multiplied in Slovenia, were used. Cv. Paul is an intersubspecies crossbreed between subsp. *foenum-graecum* and subsp. *indica*, and cv. Margaret belongs to the subsp. *foenum-graecum*. After germination, the seedlings were grown at a nursery. Well developed, 2 weeks-old seedlings of each of the cultivars were transplanted to experimental pots (V = 12.46 l; 5 plants per pot) in a glasshouse ($23 \pm 2°C$, daylight, $60 \pm 5\%$ relative humidity). The pots were filled with a mixture of 8 kg of alluvial sandy loam soil (Cambisol; the mechanical composition of the soil was as follows: 40% sand, 40% silt, 20% clay) and 3 kg of peat; the apparent specific gravity of the soil mixture was 1.36. The basic soil macronutrient (N-P-K) supply was as follows: total available N, 0.40 g/pot; P_2O_5, 1.40 g/pot; and K_2O, 2.50 g/pot. No fertilizers were added to the soil.

After good acclimation of the plants in a glasshouse, two levels of water treatment were applied (35% and 95% depletion of available soil water) in order to study the impact of drought stress (95% depletion of available soil water) on growth-developmental parameters of the plant and on the diosgenin content. The soil water holding characteristics were determined through the development of a soil desorption curve (field capacity being at 47% on weight basis, and wilting point being at 15% on weight basis) and thereafter the quantity of daily water supply was determined.

Drought stress was applied for 3 weeks. After this period, plants were irrigated following the scheme of optimal water treatment (35% depletion of available soil water). Each of the treatments was replicated five times.

After 21 days of growth under drought stress conditions, one half of treated (stress, control) plants (8 weeks old) were randomly sampled/harvested in order to measure growth-developmental parameters–length and weight of above ground part of the plants, length and weight of the roots, number of flowers and number of pods/plant–in stressed plants with regard to the controls. The remainder of the plants were harvested after recovery from stress, i.e., 8 weeks after normalization of water supply in experimental pots. Parameters such as length and weight of above ground part of plants, length and weight of roots, number of nodes, number and weight of pods, and weight of seeds were measured and the content of diosgenin in fenugreek seeds was quantified in order to observe a response of cultivars to stress. Pods were threshed and the seeds were cleaned and prepared for chemical analysis. The overall pot experiment lasted 117 days.

Glycoside hydrolysis and extraction of diosgenin. Diosgenin is an aglicon (sapogenine) of saponine glycoside dioscin. To determine seeds' diosgenin content in our study, hydrolysis of glycoside dioscin was necessary. Hydrolysis and successive extraction of fenugreek seeds followed the procedure of Sanchez et al. (11). One gram of whole seeds of fenugreek was hydrolyzed with 30 ml of 3 M HCl in a 100 ml flask for 2 hours under reflux. After neutralization with 10 M NaOH, samples were filtered with a vacuum pump. Filter papers with seeds were dried in a vacuum dryer (VS-50 S, Vakuumska tehnika Kambic) under negative pressure at 60°C. Dried seed samples (1.0 g) were extracted with dichloromethane (50 ml) in Erlenmayer flasks on a shaker (Kinetor M, Elektromedicina) for 4 hours. Extracts were filtered and solvent was evaporated under reduced pressure (Büchi Rotavapor®, Switzerland) at 40°C. Before GC analysis, dry extracts were dissolved in 5 ml dichlormethan and samples (1 µl) injected to a gas chromatograph.

Gas chromatography. A 5890 Series II model gas chromatograph (Hewlett Packard) equipped with a flame ionization detector and coupled with an integrator (HP 3396 Series II) was used. Diosgenin determinations were carried out on an Ultra 2 capillary column (25 m × 0.32 mm inner diameter; (5%)-diphenyl-(95%)-dimethylsiloxane phase-0.52 µm), with temperature programming from 230 to 310°C at 1°C/min. The carrier gas (helium) and hydrogen flow rates were 1.50 ml/min and 66.7 ml/min, respectively, and the air-flow rate was 286 ml/min. The injector and detector were thermostated at 280 and 320°C, respectively. The overall chromatographic analysis took 25.4 min. The determination of diosgenin was carried out by the external standard method with diosgenin as the external standard. Calibration curve for diosgenin was measured by analyzing the standard solutions (0.1 to 1.0 mg/ml), six times each standard. The signal was linear in the whole range of measured concentrations (regression coefficient was 0.9998; standard error for diosgenin was 0.377). The relative standard deviation of the peak areas was 3.168%. Quantification limit was 36.2 µg/ml. The concentration (mg/ml) of diosgenin in each of the samples was calculated by comparing the areas under the peaks of the analytes with those of the external standard.

Atropa belladonna L. *Pot Experiment*

Plant material and growing conditions. Seeds of *Atropa belladonna* L. were collected from the autochthonous population at Podzavrh, Slovenia (46° 01' N, 15° 23' W). After germination, the seedlings were grown at a nursery. Well developed, 2-month-old seedlings were transplanted to experimental pots in a glasshouse (23 ± 2°C, daylight). Pots were filled with alluvial sandy loam soil (40% sand, 40% silt, 20% clay; the apparent specific gravity of the soil was 1.24, pH = 7.3). The basic soil macronutrient (N-P-K) supply was as follows: total available N = 0.37 g/pot, P_2O_5 = 1.39 g/pot and K_2O = 2.57 g/pot. Four levels of water treatment (irrigation) were used (35%, 55%, 65% and 95% depletion of available soil water) and also 4 levels of treatment with nitrogen (additional supplies to the basic soil nitrogen level were 0 g/pot, 0.41 g/pot, 0.82 g/pot and 1.23 g per pot) were applied.

Each treatment was replicated four times. After 10 days of growth under drought stress conditions, the plants (32 weeks old) were harvested. The cleaned roots were dried at 60°C for 36 hours, until a constant weight was obtained. The plant material was pulverized (sieve No. 0.75) in a mortar grinder (Waring model 34 BL 65) and preserved in dark glass until the extraction procedure.

Two hundred milligram of dry roots were extracted (Ultrasound chamber Iskra, model U2-2R) with a solvent mixture of chloroform, methanol and 25 v/v% aqueous solution of ammonium hydroxide ($CHCl_3$:MeOH:NH_4OH, 15:5:1) for 10 min. After further maceration at room temperature for 1 h, the extract was filtered and rinsed twice with 1ml of chloroform. Dry plant extracts for chemical analysis of tropane alkaloids were obtained after evaporation of native extract under reduced pressure (Büchi Rotavapor®, Switzerland) at 40°C. Tropane alkaloids (hyoscyamine, scopolamine) were analyzed by means of capillary electrophoresis (CE) (12).

The results of plant treatment responses/measurements of the greenhouse study were evaluated by ANOVA, and the significance of the differences between treatment groups was tested by t-test at $\alpha = 0.05$ (Statgraphics 5.0).

RESULTS

Trigonella foenum-graecum L. *pot experiment.* Results of the study show that fenugreek cultivars responded to the applied water deficit similarly, although a significant interaction between cultivar and stress-treatment was observed in root biomass (p = 0.0040) and in height of above ground parts (p = 0.0096). The results of the measurements made immediately after stress are presented in Tables 1 and 2. Plants exposed to drought stress did not show significant differences in height of above ground parts, in length of the roots and in the number of pod buds/plant, when compared to the control plants (Table 1 and Table 2). However, the dry weight (d.w.) of above ground parts (cv. 'Margaret': p = 0.0596; cv. 'Paul': p = 0.0218) and the roots d.w. (cv. 'Margaret': p = 0.0001; cv. 'Paul': p = 0.0003) were affected by drought. Both cultivars responded to drought with a reduction in flower development, cv. 'Margaret' being more sensitive (cv. 'Margaret': p = 0.0087; cv. 'Paul': p = 0.0618).

When assessing mature plants at the end of the pot experiment, both cultivars grown under drought stress showed a significantly higher number of nodes (average 25) than those grown under the optimal irrigation regime (average 22). After recovery, plants of both cultivars developed more pods/plant than control plants, although this difference was significant only in cv. 'Margaret' (p = 0.0210) (Table 3). The seed yield of normally irrigated plants of cv. 'Paul' was 20.8 g and for cv. 'Margaret' was 19.2 g. Both yields were higher

TABLE 1. Green plants immediately after perception of stress: cv. 'Margaret'.

Parameter	stress		non-stress		p-value
	average (n = 5)	sd	average (n = 5)	sd	
height of above ground part (cm)	38.53	9.66	38.725	13.35	0.9420
d.w. of above ground part (g/plant)	3.79	2.37	5.42	4.60	0.0596
length of the roots (cm)	23.72	7.92	23.82	8.99	0.9582
d.w. of the roots (g/plant)	0.43	0.31	0.89	0.61	0.0001*
number of flowers/plant	2.70	1.90	4.06	1.98	0.0087*
number of pods/plant	2.86	1.93	3.22	1.90	0.4107

* Assigns statistically significant difference.

TABLE 2. Green plants immediately after perception of stress: cv. 'Paul'.

Parameter	stress		non-stress		p-value
	average (n = 5)	sd	average (n = 5)	sd	
height of above ground part (cm)	37.12	10.92	37.40	6.97	0.8895
d.w. of above ground part (g/plant)	4.67	3.32	6.72	4.64	0.0218*
length of the roots (cm)	22.90	9.73	25.55	6.73	0.1466
d.w. of the roots (g/plant)	0.54	0.48	1.04	0.71	0.0003*
number of flowers/plant	3.29	3.19	4.49	2.64	0.0618
number of pods/plant	3.40	2.21	3.91	2.67	0.3482

* Assigns statistically significant difference.

than drought stressed plant yields (Tables 3 and 4). No interaction between cultivar and stress treatment was noticed when assessing the seed yields of both cultivars (p = 0.1838).

In spite of similarity between cultivars, some differences could be observed after recovery. Plants of cv. 'Paul' responded to the normalization of growing conditions with intensive growth in height. Plants of cv. 'Paul' exposed to drought stress were higher than optimally irrigated plants (p = 0.0001) (Table 4). Contrary to this, there was no significant difference in height of plants of cv. 'Margaret' (Table 3). Drought stressed seeds of cv. 'Margaret' contained less diosgenin (p = 0.0021) than normally irrigated plants (0.123%): difference was 0.0774%. A similar content of diosgenin was determined in seeds of cv. 'Paul' (0.1863).

Atropa belladonna L. *pot experiment.* The results of the ANOVA showed

TABLE 3. Mature plants: cv. 'Margaret'.

Parameter	stress		non-stress		p-value
	average (n = 5)	sd	average (n = 5)	sd	
height of above ground part (cm)	69.40	12.48	65.97	9.37	0.1771
d.w. of above ground part (g/plant)	14.86	7.82	13.80	7.25	0.5361
length of the roots (cm)	32.16	17.99	27.54	6.78	0.1384
d.w. of the roots (g/plant)	2.96	1.98	2.22	1.78	0.0873
number of pods/plant	23.61	12.04	17.97	8.69	0.0210*
d.w. of pods (g/plant)	15.92	8.94	14.50	7.06	0.4375
weight of seeds (g/plant)	13.55	5.38	19.19	2.14	0.0156*
seeds diosgenin content (%)	0.12	0.03	0.20	0.02	0.0021*

* Assigns statistically significant difference.

TABLE 4. Mature plants: cv. 'Paul'.

Parameter	stress		non-stress		p-value
	average (n = 5)	sd	average (n = 5)	sd	
height of above ground part (cm)	75.5	17.16	60.88	10.87	0.0001*
d.w. of above ground part (g/plant)	15.84	7.47	16.74	8.21	0.6361
length of the roots (cm)	29.18	7.41	30.97	7.61	0.3284
d.w. of the roots (g/plant)	2.17	1.36	3.17	1.93	0.0167*
number of pods/plant	21.38	9.02	18.59	9.13	0.2088
d.w. of pods (g/plant)	14.46	6.78	16.68	8.33	0.2319
weight of seeds (g/plant)	11.46	3.85	20.79	2.31	0.0001*
seeds diosgenin content (%)	0.19	0.06	0.18	0.04	0.7818

* Assigns statistically significant difference.

that the irrigation regime significantly affected the yield (p < 0.001) of the roots of deadly nightshade as well as their hyoscyamine (p < 0.001) and scopolamine (p < 0.001) content. The root scopolamine content depended on the N-fertilization level ($\alpha = 0.05$). Also, the interaction between the irrigation regime and the nitrogen supply was significant (roots biomass: p = 0.007; hyoscyamine: p < 0.001; scopolamine: p = 0.015).

Maximal average yields of biomass of roots (17.95 g dw/plant) can be expected in plants grown under the highest water supply (35% of water deple-

FIGURE 1. The root biomass (d.w. g/plant) of deadly nightshade under different irrigation regimes and nitrogen supplies.

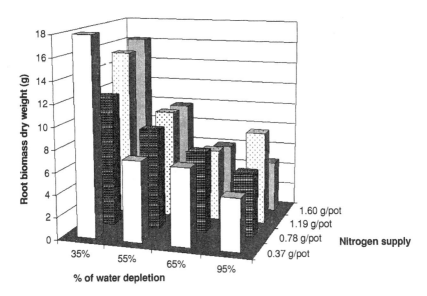

tion) at a very low nitrogen level of 0.37 g/pot (Figure 1). The maximal contents of hyoscyamine (0.41%) and of scopolamine (0.07%) in roots were observed under severe drought stress growth conditions (95% depletion of available soil water) at a nitrogen level of 1.60 g/pot. The same range of hyoscyamine or scopolamine contents in roots can also be expected under less severe drought stress (55% and 65% of soil water depletion) conditions at lower N-fertilization levels (Figure 2).

DISCUSSION

It is well known that nitrogen and phosphorous, as highly mobile macroelements, play a pivotal role in vegetative growth and in basic metabolism of plants, which might be directly or indirectly involved in production of secondary metabolites (13). Moderate N (60 kg/ha) and P fertilization (60 kg/ha) have a beneficial effect on plant height and on the number of pods/plant, which were also associated with high seed yield (14). Although there is no direct relationship between nitrogen and phosphorous fertilization and diosgenin content, concomitant water availability is probably an essential factor in diosgenin biosynthesis in sensitive cultivars. It may be that drought stress in our study influenced the changes in nitrogen as well as in other macronutrients uptake,

FIGURE 2. The interaction between irrigation regime and nitrogen supply on the concentration of hyoscyamine (a) and scopolamine (b) in the roots of *Atropa belladonna.*

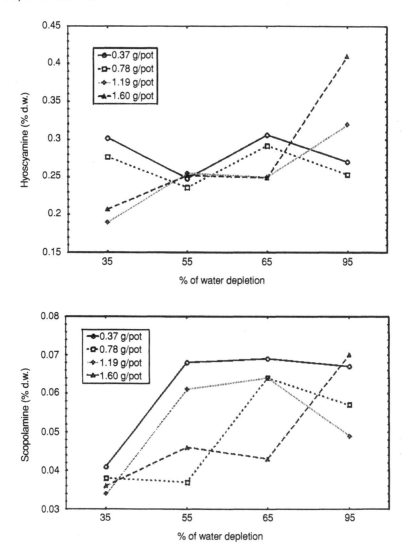

blocked the amino acids biosynthesis, suppressed the vegetative growth and simultaneously caused a shift in diosgenin metabolism in young fenugreek plants, reflected in lower diosgenin content in mature seeds. Further work will be necessary to test this hypothesis.

The observed diminished biomass production counts for the main impact of drought stress on growth and development of deadly nightshade. So, when considering the yield of tropane alkaloids per plant, we can assume that the maximal yield of hyoscyamine (54 mg/plant) and scopolamine (7 mg/plant) is expected to be achieved in plants grown under optimal water supply (35% depletion of available soil water) accompanied with a nitrogen supply of 0.37 g/pot.

The phenomenon of the influence of an excess of nitrogen fertilizers at very low soil water availability on the raised accumulation of hyoscyamine and scopolamine can be explained by the nitrogen turnover and stress response in suffering plants. It is well known that excess nitrate nitrogen in soils, deriving from intensive nitrogen fertilization, leads to an increase in plants of nitrate, a soluble form of nitrogen, that can be further incorporated in N-containing secondary metabolites in intact plants. Moreover, plants exposed to stress growth conditions (all factors which obstruct normal biochemical processes in plant metabolism) accelerate the nitrate accumulation in plant tissue and slow down the protein synthesis (15,16). It is possible that alkaloid plants under these conditions shift metabolism towards accelerated synthesis of nontoxic secondary products–alkaloids–which might represent the form of nitrogen storage in suffering plants (17-19).

REFERENCES

1. Trpin, D. and Vreš, B. (1995). Register flore Slovenije (List of Slovenian flora), Slovenska akademija znanosti in umetnosti (Slovenian Academy of Sciences and Arts), Ljubljana.

2. Lange, D. (1998). Europe's medicinal and aromatic plants, their use, trade and conservation.-TRAFFIC International, Cambridge, UK, 77 pp.

3. Bernath, J. (2000). Possibilities and limitations of medicinal plant production in Hungary. First Conference on Medicinal and Aromatic Plants of Southeast European Countries, Arandzelovac, Yugoslavia, May 29-June 3, 2000. Book of Abstracts, p. 23.

4. Dahler, M. and Pelzmann, H. (1999). Arznei-und Gewürzpflanzen. Anbau, Ernte, Aufbereitung. Österreichischer Agrarverlag, Klosterneuburg, pp. 31-33.

5. Bomme, U. (1998). The Cultivation of Medicinal Plants–New Trends and Investigations. 46th Annual Congress Society for Medicinal Plant Research "Quality of Medicinal Plants and Herbal Medicinal Products," August 31st-Sep 4th, 1998, Vienna, Austria. Abstracts of Plenary Lectures, p. 4.

6. Kišgeci, J. and Sekuloviš, D. (2000). Medicinal plants of Yugoslavia. First Conference on Medicinal and Aromatic Plants of Southeast European Countries, Arandzelovac, Yugoslavia, May 29-June 3, 2000. Book of Abstracts, p. 21.

7. Baričevič, D., Dedek, J., Bartol, T., and Zupančič, A. (1996). Market possibilities of medicinal and aromatic plants in Slovenia. Proceedings of the International symposium–Novi izzivi c poljedelstvu (New Challenges in Agriculture) '96, Biotechnical Faculty, pp. 95-99.

8. Baričevič, D., Spanring, J., Činč, M., Umek, A., Stupica, T., Kus, M., and Šuštar, F. (1994). National program for cultivation, processing and quality control of drug plants and herbal remedies. Directives. University of Ljubljana, Biotechnical Faculty Ljubljana, 11 pp.

9. Baričevič D. and Rode, J. (1996). National program for cultivation, processing and quality control of drug plants and herbal remedies. Proposal. University of Ljubljana, Biotechnical Faculty and IHP Zalec, 12 pp.

10. Baričevič, D. (1998). The pollution of groundwater and surface waters in Slovenia with special reference to nitrates and pesticides. International workshop on protection of natural resources in agriculture of central and eastern Europe, Kralupy nad Vltavou and Prague-Zbraslav, 30 November to 7 December 1998, edited by František Doležal. Prague: Research Institute for Soil and Water Conservation, pp. 58-63.

11. Sanchez, G.L., Acevedo, J.C.M., and Soto, R.R. (1972). Spectrophotometric determination of diosgenin in *Dioscorea composita* following thin-layer chromatography. Analyst, 97, 973-976.

12. Baricevic, D., Umek, A., Kreft, S., Maticic, B., and Zupancic, A. (1999). Effect of water stress and nitrogen fertilization on the content of hyoscyamine and scopolamine in the roots of deadly nightshade (*Atropa belladonna*). Environmental and Experimental Botany, 42, 17-24.

13. Mehra, P. and Kamal, R. (1995). Effect of fertilizes and foliar sprays on yield and diosgenin content of fenugreek. Ad. Plant Sci., 8 (1), 71-77.

14. Parek, S.K. and Gupta, R. (1981). Effect of fertilizer application on seed yield and diosgenin content in fenugreek. Indian J. Agric. Sci., 50(10), 746-749.

15. Larcher, W. (1995). Physiological Plant Ecology. 3rd Edition. Springer-Verlag, Berlin, pp. 189-192.

16. Matičič, B. (1997). Mineral balances and nitrate policies in Slovenia. In: F. Brouwer, W. Kleinhanss (Eds.), The implementation of nitrate policies in Europe. Wissenschaftverlag Vauk., Kiel, pp. 103-112.

17. Nowacki, E., Jurzysta, M., Gorski, P., Nowacka, D., and Waller, G.R. (1976). Effect of nitrogen nutrition on alkaloid metabolism in plants. Biochem. Physiol. Pflanz., 169, 231-240.

18. Bernath, J. (1986). Production Ecology of Secondary Plant Products. In: Craker L.E., Simon J.E. (Eds.), Herbs, Spices and Medicinal Plants. Recent Advances in Botany, Horticulture and Pharmacology. The Oryx Press, Arizona, pp. 185-234.

19. Saker, M.M., Rady, M.R., and Ghanem, S.A. (1997). Elicitation of tropane alkaloids in suspension cultures of Hyoscyamus, Datura and Atropa by osmotic stress. Fitoterapia, 68, 4, 338-342.

Fertility of *Echinacea angustifolia* Moench and *Linum usitatissimum* L. After Herbicide Treatment

Romana Izmaiłow

Maria Pająk

SUMMARY. The purpose of these studies was to evaluate the influence of the herbicide Roundup® on reproductive processes of *Echinacea angustifolia* Moench and *Linum usitatissimum* L. The plants with the flowers and fruits at various developmental stages were treated with Roundup 360 SL at 1% concentration. The effects on microsporogenesis, pollen viability, megasporogenesis, embryo sac development, embryogenesis and endosperm development were studied. All mentioned processes could be disturbed, resulting in decrease of pollen viability and seed set of the treated plants, in comparison with non-treated individuals from the same collection. *[Article copies available for a fee from The Haworth Document Delivery Service: 1-800-HAWORTH. E-mail address: <getinfo@ haworthpressinc.com> Website: <http://www.HaworthPress.com> © 2002 by The Haworth Press, Inc. All rights reserved.]*

KEYWORDS. *Echinacea*, *Linum*, action of herbicide, Roundup, reproductive processes, pollen, tapetum, seed formation

Romana Izmaiłow and Maria Pająk are affiliated with the Department of Plant Cytology and Embryology, Institute of Botany, Jagiellonian University, ul. Grodzka 52, 31-044 Kraków, Poland.

This study was supported by the Polish Committee for Scientific Research (KBN), project No. 6 P04C 028 14.

[Haworth co-indexing entry note]: "Fertility of *Echinacea angustifolia* Moench and *Linum usitatissimum* L. After Herbicide Treatment." Izmaiłow, Romana, and Maria Pająk. Co-published simultaneously in *Journal of Herbs, Spices & Medicinal Plants* (The Haworth Herbal Press, an imprint of The Haworth Press, Inc.) Vol. 9, No. 2/3, 2002, pp. 65-70; and: *Breeding Research on Aromatic and Medicinal Plants* (ed: Christopher B. Johnson, and Chlodwig Franz) The Haworth Herbal Press, an imprint of The Haworth Press, Inc., 2002, pp. 65-70. Single or multiple copies of this article are available for a fee from The Haworth Document Delivery Service [1-800-HAWORTH 9:00 a.m. - 5:00 p.m. (EST). E-mail address: getinfo@haworthpressinc.com].

INTRODUCTION

Plantations of cultivar plants and nurseries, even if not directly treated with pesticides, are often damaged from the uncontrolled influence of these chemicals, used in their neighbourhood or accumulated in soil. The effects of various pesticides applied in agriculture are the subject of numerous cytological researches but rarely of embryological ones.

The aim of the present studies was to determine whether embryological processes of some cultivated plants are sensitive in any degree to the action of the herbicide Roundup®; the side effects of accidental herbicide treatment on their fertility and seed production were analyzed. Among the plants studied were *Echinacea angustifolia* Moench (Asteraceae) and *Linum usitatissimum* L. (Linaceae), commonly used for production of commercial medicines.

MATERIALS AND METHODS

The studied plants of *E. angustifolia* and *L. usitatissimum* were treated with the herbicide Roundup 360 SL (glyphosate isopropylamine salt) at 1% concentration, which is half the dose recommended for application on weeds. The 1% concentration tested before the experiment did not result in destroying or damaging the vegetative organs.

At time of treatment the plants of both studied taxa had flowers and fruits at various developmental stages. They were sprayed once, but the treatment of a part of *Echinacea* inflorescences was repeated after one day. The control material were the plants from non-treated parts of the same collections of *Echinacea* and *Linum*.

The flowers and young fruits to be used for embryological studies were fixed in 96% ethanol/acetic acid (3:1) 24 h and 48 h after treatment; then the customary paraffin method was used. The microtome sections 10 μm thick were stained with Heidenhain's haematoxylin and alcyan blue.

Capability of seed germination was controlled in petri dishes. Viability of *Linum* embryos was tested according to Nielubov's method, after using indygocarmine, which stains necrotic cells.

RESULTS

Echinacea angustifolia Moench

In all flower buds treated with the herbicide, developmental disturbances and destruction processes could be observed. As usual, they concerned degeneration at the stage of microspores, mature pollen grains (Figure 1a) and precocious destruction of tapetum; they were not always associated each other. Also, single microspores could degenerate before separation from the tetrad. A part of the anthers in each of the examined flower buds did not show anomalies either in their structure or in microsporogenesis.

Degenerative processes in the ovules from the treated flower buds could start during megasporogenesis; degeneration of megaspore mother cell and of megaspore tetrad were found in some flowers. In 35% of ovules, hyperplastic growth of the endothelium was observed (Figure 1b). This resulted in starving the endosperm and the embryo of essential nutrients; in later stages the endosperm degenerated and the embryo ceased to grow. Less common phenomena were: extensive vacuolation of proembryo cells (Figure 1c), degenerative changes in the suspensor at the proembryo stage and lack of endosperm in embryo sac containing the globular proembryo. In about 34% of the studied flowers, the ovules were completely degenerated. A significant part of the flowers, especially the older ones, escaped destruction in the absence of twice treatment; ca. 31% of achenes contained embryos which were able to germinate. In the control material, only 3% of the pistils contained the aborted ovules.

Linum usitatissimum L.

Both flower buds and expanded flowers were sensitive to the treatment. In the flower buds some anthers showed disturbances in meiosis (lagging chromosomes during I and II division). At the tetrad stage a significant part of anthers contained degenerated tetrads of microspores, resulting in a distinct reduction of pollen viability (ca. 40%) or, in some anthers, in its total degeneration (Figure 1d). In the control material, viability of pollen reached ca. 85%. Precocious degeneration of the tapetal cells was observed from the beginning of the tetrad stage. Pollen sterility in most anthers might secondarily result from precocious destructive changes in the tapetum.

Degenerative processes in 20% of ovules were observed in the chalazal or micropylar region; the former one separated an ovule from nutrients and the latter made the entrance of a pollen tube into micropyle impossible (Figure 1e). The other types of herbicide effect were degenerative processes of the young embryo sac, the egg apparatus and young proembryo. A less common phenomenon was degeneration of endosperm (Figure 1f). The seeds, when closed in the pericarp, escaped contact with the herbicide and were able to germinate, like in the control material, in ca. 100%. After opening of the fruit, the seeds were directly treated, but the seed coat had probably been non-permeable. Ten percent of these seeds germinated with delay and seedlings were retarded in development. The embryos from the fruits treated before maturity showed small necrotic regions.

FIGURE 1. a-c–*Echinacea angustifolia* Moench; d-f–*Linum usitatissimum* L.

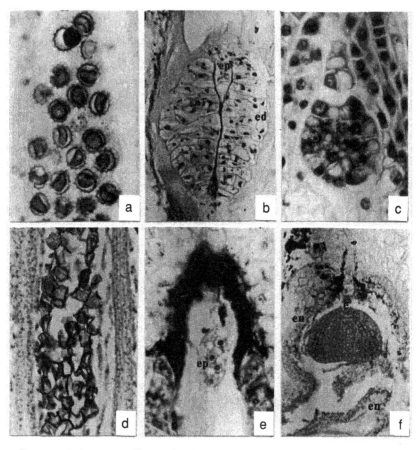

a = Degenerated mature pollen grains (× 440).
b = Hyperplastic growth of the endothelium; the proembryo and endosperm are degenerated. ed = endothelium, ep = proembryo (× 440).
c = Extensive vacuolation of the proembryo cells (× 440).
d = Totally degenerated pollen grains (× 440).
e = Degenerated micropylar part of the nucellus and integuments (black stained); aborted five-celled proembryo. ep = proembryo (× 880).
f = Young embryo surrounded with degenerated endosperm; destroyed somatic cells of the ovule. e = embryo; en = endosperm (× 220).

DISCUSSION

The present embryological research was undertaken with special reference to the action of the herbicide on fertility of cultivated plants which occasionally are treated by chemicals employed in their close vicinity.

Many authors studying the effects of herbicides on plant cells noted the cytogenetic activity of these chemicals. As the result of treatments with different herbicides, various phenomena could be observed in meristematic cells: C-mitosis, decreased mitotic index and chromosome contraction (2-5,8,10).

Abnormalities of the meiotic division in the pollen mother cells after herbicide treatment, resulting in significant pollen sterility, have been well known in different plants for a long time (1,11,12).

We have stated that, in addition to pollen mother cells, the tapetum can also be sensitive. Similarly, in young ovules, various abnormalities or degenerative processes were observed in the funiculus, nucellus, and integuments as well as during megasporogenesis and embryo sac development. In the ovules from older flowers, treated with the herbicide after fertilization, the most sensitive structures were: young proembryo, the endosperm and endothelium. Irregularities in their development or processes of degeneration significantly reduced seed production. The results of our studies on seed development after herbicide treatment correspond to the research on the parasitic plant *Striga* (6,7). In the case of these parasitic plants, treatment with the proper herbicide can be a successful mode of disruption of reproduction process of the weed, which devastates food crops in Tropical Africa. On the other hand, lowered seed set in *Linum* considerably affects crops decreasing of needed medical material; in *Echinacea* it lead to a reduction in the natural regeneration of the plantation. It may be alarming that a significant cytological effect of the chemicals on microsporogenesis was observed also in the 1st or 2nd generation (1). Also, Skorupska (9) suggests, that the herbicides may cause permanent changes in the chromosome structure. The plants obtained by the authors from the seeds treated with the herbicides showed irregularities in the meiotic division in pollen mother cells, and decreased viability of pollen grains. One should therefore take into account that even occasional treatment of seed plantations or nurseries with a herbicide may be a potential cause of cytogenetic changes in future generations.

REFERENCES

1. Amer, S.M. and E.M. Ali. 1974. Cytological effects of pesticides. V. Effects of some herbicides on *Vicia faba. Cytologia* 39: 633-643.

2. Badr, A. 1983. Mitodepressive and chromotoxic activities of two herbicides in *Allium cepa. Cytologia* 48: 451-457.

3. Badr, A. 1986. Effects of the S-Triazine herbicide Turbutryn on mitosis, chromosomes and nucleic acids in root of *Vicia faba*. *Cytologia* 51: 571-577.

4. Badr, A. and A.G. Ibrahim. 1987. Effect of herbicide Glean on mitosis, chromosomes and nucleic acids in *Allium cepa* and *Vicia faba* root meristems. *Cytologia* 52: 293-643.

5. Hess, F.D., J.D. Holmsen, and C. Fedtke. 1990. The influence of the herbicide mefenacet on cell division and cell enlargement in plants. *Weed Research* 30: 21-27.

6. Paré, J. and A. Raynal-Roques. 1993. The genus *Striga* (Scrophulariaceae): taxonomy, geographical distribution. Embryological results in *Striga hermonthica* (Del.) Benth. scourge of food crops in Tropical Africa with special reference to the action of herbicide on seed development. *Acta Biologica Cracoviensia Ser. Botanica* 34-35: 25-35.

7. Paré, J. 1994. Resistance to herbicide (2, 4-D) of *Striga hermonthica* (Scrophulariaceae) in relation to embryonic development. *Polish Botanical Studies* 8: 51-59.

8. Rao, B.V., B.G.S. Rao, and C.B.S.R. Sharma. 1988. Cytological effects of herbicides and insecticides on *Allium cepa*. *Cytologia* 53: 255-261.

9. Skorupska, H. 1976. The effect of some herbicides, chemomutagens and gamma radiation on meiosis in pea (*Pisum sativum*). *Genetica Polonica* 17: 149-157.

10. Topaktas, M. and E. Rencüzogullari. 1991. Cytogenetic effects of herbicides Gesagard and Igran in barley. *Cytologia* 56: 419-424.

11. Unrau, J. and E.N. Larter. 1952. Cytogenetic responses of cereals to 2,4-D. I. A study of meiosis of plants treated at various stages of growth. *Canadian J. Botany* 30: 22-27.

12. Wuu, K.D. and W.F. Grant. 1967. Chromosomal aberrations induced by pesticides in meiotic cells of barley. *Cytologia* 32: 31-41.

Seed Yields, Yield Components and Essential Oil of Selected Coriander (*Coriandrum sativum* L.) Lines

Filiz Ayanoğlu
Ahmet Mert
Neşet Aslan
Bilal Gürbüz

SUMMARY. Forty-three coriander lines which were developed by Ankara University were tested under east Mediterranean conditions for two years to determine the best yielding lines in winter season. The experiment was laid out in a randomized complete block design with three replications.

During this study, the plant height (cm), branch number/plant, umbel number/plant, seed number/umbel, 1000 seed weight (g), seed yield (kg/da), the essential oil content (%) and the essential oil yield were investigated. The seed yields of coriander lines varied between 113.8 (K11) and 229.7 kg/da (K46). The highest seed yields were obtained from the lines K67, K28, K69 and K46. The seed yields of those lines were higher than currently planted cultivars. *[Article copies available for a fee from The Haworth Document Delivery Service: 1-800-HAWORTH. E-mail address: <getinfo@haworthpressinc.com> Website: <http://www.HaworthPress. com> © 2002 by The Haworth Press, Inc. All rights reserved.]*

Filiz Ayanoğlu and Ahmet Mert are affiliated with the Department of Field Crops, Faculty of Agriculture, Mustafa Kemal University, 31034 Antakya, Hatay, Turkey.

Neşet Aslan and Bilal Gürbüz are affiliated with the Department of Field Crops, Faculty of Agriculture, University of Ankara, 06130 Ankara, Turkey.

[Haworth co-indexing entry note]: "Seed Yields, Yield Components and Essential Oil of Selected Coriander (*Coriandrum sativum* L.) Lines." Ayanoğlu, Filiz et al. Co-published simultaneously in *Journal of Herbs, Spices & Medicinal Plants* (The Haworth Herbal Press, an imprint of The Haworth Press, Inc.) Vol. 9, No. 2/3, 2002, pp. 71-76; and: *Breeding Research on Aromatic and Medicinal Plants* (ed: Christopher B. Johnson, and Chlodwig Franz) The Haworth Herbal Press, an imprint of The Haworth Press, Inc., 2002, pp. 71-76. Single or multiple copies of this article are available for a fee from The Haworth Document Delivery Service [1-800-HAWORTH 9:00 a.m. - 5:00 p.m. (EST). E-mail address: getinfo@haworthpressinc.com].

KEYWORDS. *Coriandrum sativum*, coriander, yield, essential oil

INTRODUCTION

Coriander (*Coriandrum sativum* L.) is an aromatic umbellifer native to the East Mediterranean Region of Turkey. Its seeds and oil have economical importance in the world exchanges as both medicinal and spice. It is cultivated throughout the world including the temperate countries of central and western Europe, the Mediterranean region, North and South America and India (1). India has been one of the major producers of this spice.

Dry fruits as well as fresh green leaves of coriander are used as spices and condiments. Therefore, a variety with a high grain and leaf yield would be highly desirable (2). Oil yield based on dry weight of seeds was 0.66% (3).

This study was conducted to determine seed yield, yield components and essential oil of some coriander lines and to select them for cultivar candidates.

MATERIALS AND METHODS

Forty-three coriander lines were tested under east Mediterranean conditions for two years to determine the best yielding lines in winter season. These lines had been selected from twelve population from Turkey, Germany and Bulgaria by using the "single plant selection method" at Ankara University. The experimental design was a completely randomized block with three replications. During this study, the plant height (cm), branch number/plant, umbels/plant, seed/umbel, one thousand seed weight (g), seed yield (kg/da), essential oil content (%) and essential oil yield (l/da) were investigated.

RESULTS

The plant height of coriander lines differed significantly (p < 0.01) in both years of the study and ranged from 68.3 cm (line labeled K4) to 127.4 cm (line labeled K57). The highest plant height was obtained in line K57 (127.4 cm) in the first growing season. The lines K8, K46 and K57 gave the highest plant height with 126.0, 124.0 and 122.0 cm, respectively, in the second growing season (Table 1).

Branch number per plant differed significantly (p < 0.01) in both growing seasons and ranged from 3.27 (K4) to 8.75 (K47). Highest branch number was

TABLE 1. Plant height, number of branches per plant, number of umbels per plant, and number of seeds per umbel of coriander lines in two growing seasons (1996-1998).

Lines	Plant height (cm)		Branch num./plant		Umbel num./plant		Seed num./umbel	
	1996/97	1997/98	1996/97	1997/98	1996/97	1997/98	1996/97	1997/98
K3	111.6 a-e	114.3 c-f	5.50 j-n	5.00 e-j	7.37 e-g	11.23 k-n	34.73 a-d	27.00 e-k
K4	68.3 n	77.4 o	5.13 l-o	3.27 k	11.55 a-f	6.00 n	36.71 ab	27.73 c-j
K6	101.1 b-j	102.6 h-k	4.82 m-p	4.80 f-j	13.50 a-g	19.93 b-g	33.82 a-e	23.40 j-m
K8	100.6 b-j	126.0 a	5.53 i-m	5.80 c-h	11.53 b-g	17.47 b-k	25.53 d-g	29.80 a-g
K9	91.2 e-m	117.6 a-c	5.32 k-o	6.20 c-e	11.02 c-g	23.70 ab	32.36 a-g	27.70 c-j
K10	72.4 mn	95.2 i-m	5.22 l-o	5.00 e-j	11.38 b-g	15.90 d-l	27.78 b-g	29.60 a-h
K11	90.8 e-m	105.7 f-h	5.75 h-l	5.87 c-h	9.63 d-g	18.90 b-h	28.05 b-g	31.70 a-e
K12	76.8 k-n	92.5 l-m	5.20 l-o	4.77 f-j	9.80 d-g	12.32 h-n	25.73 d-g	29.48 a-h
K14	107.6 a-gf	114.5 c-f	5.13 l-o	5.75 c-h	15.87 a-e	22.25 a-e	36.67 ab	27.37 e-k
K15	106.6 b-g	107.1 d-h	6.72 d-f	6.69 a-c	15.84 a-e	18.00 b-j	28.93 b-g	32.64 a-c
K16	96.2 c-k	101.0 h-l	6.29 f-h	6.00 c-f	15.56 a-f	21.37 a-f	40.53 a	34.00 a
K17	82.9 h-n	86.7 mn	5.47 j-n	4.60 h-j	8.20 d-g	10.80 k-n	26.13 d-g	23.40 j-m
K18	112.9 a-d	101.6 h-l	6.20 f-i	5.25 d-j	14.80 a-f	11.93 i-n	32.07 a-g	20.43 m
K20	101.6 b-j	103.3 h-k	5.13 l-o	5.82 c-h	19.70 ab	18.50 b-i	33.50 a-f	24.00 i-m
K26	83.4 h-n	100.4 h-l	5.00 m-o	4.82 f-j	9.07 d-g	10.13 l-n	29.60 b-g	29.24 a-h
K28	75.2 l-n	93.9 k-m	5.13 l-o	4.66 g-j	7.00 fg	11.60 j-n	23.57 g	28.77 b-i
K31	81.6 j-n	100.2 h-l	4.20 p	4.40 i-j	5.76 g	19.57 b-g	28 .04 b-g	31.67 a-e
K33	111.1 a-f	116.3 b-d	5.22 l-o	7.57 ab	9.87 d-g	27.37 a	29.98 b-g	30.57 a-g
K34	103.2 b-i	106.5 e-h	4.80 n-p	4.30 j	7.53 e-g	14.57 g-m	24.80 e-g	24.57 h-m
K35	111.0 a-f	113.3 c-g	5.40 k-o	5.20 e-j	7.33 e-g	18.00 b-j	26.57 c-g	32.57 a-d
K38	84.3 h-n	107.4 d-h	5.27 l-o	4.88 f-j	7.80 e-g	15.15 f-m	24.07 e-g	28.78 b-i
K40	82.1 i-n	95.3 i-m	6.00 g-k	5.73 c-h	12.87 b-g	13.20 g-m	25.00 d-g	25.67 g-l
K42	99.6 b-j	104.7 g-i	6.20 f-i	5.13 e-j	18.60 a-c	16.33 c-l	33.53 a-f	27.45 d-k
K43	119.8 ab	114.2 c-f	7.50 b-c	6.50 b-d	13.47 a-g	18.83 b-h	36.13 a-c	25.83 f-l
K44	101.6 b-j	115.4 b-e	6.40 f-h	5.23 d-j	14.40 a-g	15.50 f-m	32.00 a-g	31.25 a-e
K46	97.4 c-k	124.0 ab	7.20 b-d	5.33 d-j	14.80 a-f	18.00 b-j	33.07 a-g	29.20 a-h
K47	103.9 b-h	97.9 h-l	8.75 a	6.00 c-f	9.87 d-g	16.33 c-l	25.30 d-g	27.93 c-j
K48	96.4 c-k	103.8 h-j	6.00 g-k	5.59 c-i	12.67 b-g	15.30 f-m	32.27 a-g	33.54 ab
K49	92.0 d-m	94.3 j-m	6.00 g-k	5.93 c-g	9.60 d-g	15.43 f-m	32.80 a-g	27.80 c-j
K52	113.4 a-c	114.4 c-f	6.50 e-g	5.60 c-i	15.87 a-e	15.57 e-m	27.67 b-g	27.30 e-k
K55	88.1 g-n	99.1 h-l	6.27 f-h	5.80 c-h	12.60 b-g	11.38 j-n	30.20 b-g	30.00 a-g
K56	94.1 c-l	98.9 h-l	4.87 m-o	5.20 e-j	10.20 c-g	16.13 c-l	25.90 d-g	27.80 c-j
K57	127.4 a	122.0 a-c	6.60 d-g	5.37 d-j	21.40 a	18.80 b-h	31.53 a-g	25.87 f-l
K59	99.0 b-j	118.1 a-c	7.60 b-c-	7.40 ab	11.27 b-g	22.73 a-e	32.00 a-g	28.20 c-j
K60	97.4 c-k	82.6 no	5.30 l-o	5.00 e-j	9.93 d-g	9.00 mn	28.02 b-g	30.33 a-g
K61	106.6 b-g	102.9 h-k	6.09 f-j	5.40 d-j	18.71 a-c	15.80 d-l	29.33 b-g	31.53 a-e
K62	97.2 c-k	101.6 h-l	6.40 f-h	5.67 c-i	11.87 b-g	10.73 k-n	29.95 b-g	22.53 k-m
K63	89.9 f-m	103.3 h-k	4.73 op	5.20 e-j	9.60 d-g	15.33 f-m	25.40 d-g	25.77 g-l
K65	80.5 j-n	114.0 c-f	5.00 m-o	6.00 c-f	7.80 e-g	21.63 a-f	24.00 fg	21.20 lm
K67	106.1 b-g	101.3 h-l	7.73 b	7.47 ab	16.93 a-d	18.37 b-i	29.70 b-g	27.24 e-k
K68	113.0 a-d	101.9 h-l	5.20 l-o	5.27 d-j	9.93 d-g	12.40 h-n	32.40 a-g	29.60 a-h
K69	88.0 g-n	115.0 b-f	5.00 m-o	5.32 d-j	8.00 e-g	16.75 c-l	25.20 d-g	32.00 a-e
K71	94.4 c-l	113.0 c-g	7.07 c-e	7.78 a	13.67 a-g	22.40 a-d	36.20 a-c	30.97 a-f
LSD (0.05)	17.10	7.889	0.5854	1.031	6.974	5.447	7.767	4.145

obtained from K47 (8.75) in the first growing season and K71 (7.78) in the second growing season.

Umbel number per plant differed significantly ($p < 0.01$) in both growing seasons and ranged between 5.8 (K31) and 27.4 (K33). The highest umbel number per plant was obtained from K57 in the first year and K33 in the second year.

Differences in seed number per umbel among the lines were significant ($p < 0.01$) in both growing seasons and ranged between 20.4 (K18) and 40.5 (K16). In both growing seasons line K16 gave the highest seed number per umbel.

Differences in 1000 seed weight of lines were significant ($p < 0.01$) in both growing seasons and varied between 7.05 g (K14) and 17.55 g (K69). In both growing seasons lines K44, K69 and K46 gave significantly higher 1000 seed weight than all the remaining lines (Table 2).

Seed yield of coriander lines showed significant ($p < 0.01$) differences in both years and varied from 113.8 kg/da (K11) to 229.7 kg/da (K46). In both crop seasons, line K67 gave higher seed yield than all the remaining lines. K28 and K69 also gave significant amount of seed yield in both growing seasons. K46 and K14 can be considered high yielding lines.

Essential oil content of lines showed significant ($p < 0.01$) differences in both years and ranged from 0.18% (K26) to 0.60% (K59). In both years of the study the highest essential oil content was obtained from the line K59. K49, K52 and K15 also gave higher essential content than all the remaining lines.

Essential oil yield of lines showed significant ($p < 0.01$) differences in both years and ranged from 0.27 l/da (K26) to 1.09 l/da (K52). The lowest essential oil yield was obtained from K26 in both years. K67, K49, K52, K63 and K6 yielded higher essential oil yields than all the other lines when two years of results were taken into account.

DISCUSSION

Although biometric analysis of coriander lines demonstrated high variability for all measured characters, particularly for characteristics useful in selecting cultivars for commercial production, certain principles could be established. In the study, lines labeled K67, K28, K69 and K46 were selected as cultivar candidates because of their high seed yield performance. The seed yield of these lines was very good for Hatay province and higher than currently growing cultivars (4). The 1000 seed weight obtained from some lines of our study was considerably higher than some coriander cultivars regarded as having a large seed (5). Lines labeled K59, K49 and K3 should be considered because of their high essential oil content.

TABLE 2. 1000 seed weight, seed yield, essential oil content and essential oil yield of coriander lines in two growing seasons (1996-1998).

Lines	1000 seed weight (g)		Seed yield (kg/da)		Essential oil content (%)		Essential oil yield (l/da)	
	1996/97	1997/98	1996/97	1997/98	1996/97	1997/98	1996/97	1997/98
K3	9.94 f-i	9.32 i-n	131.0 g-j	150.1 g-l	0.56 a-b	0.50 a-c	0.73 e-k	0.75 f-l
K4	8.37 i-o	10.90 f-h	155.3 d-jn	163.9 d-l	0.33 j-k	0.38 h-k	0.51 i-p	0.62 k-r
K6	9.34 f-l	10.05 h-k	183.8 a-i	185.I b-j	0.55 a-c	0.48 b-d	1.01 a-d	0.88 b-h
K8	8.54 h-o	8.92 k-n	152.2 d-j	132.1 I	0.36 j	0.38 h-j	0.55 i-o	0.50 n-u
K9	7.15 no	8.96 m-o	131.6 g-j	181.3 b-k	0.45 e-h	0.45 d-f	0.59 i-n	0.81 d-k
K10	11.54 e	12.19 d-f	115.0 j	179.1 b-l	0.33 j-k	0.34 k-n	0.37 m-p	0.58 l-r
K11	8.89 g-m	10.47 g-j	113.8 j	163.7 d-l	0.28 kl	0.30 no	0.31 op	0.49 o-u
K12	14.20 cd	10.95 f-h	157.1 c-j	144.2 i-l	0.50 c-f	0.49 a-c	0.78 b-i	0.72 g-m
K14	7.05 o	10.07 h-k	162.7 b-j	225.7 ab	0.43 g-h	0.45 d-f	0.69 f-k	1.02 a-d
K15	9.46 f-k	8.93 k-n	135.0 g-j	182.8 b-j	0.48 d-g	0.53 a	0.66 g-l	0.97 a-e
K16	8.34 i-o	9.75 h-m	165.5 g-j	192.8 a-h	0.36 j	0.35 j-m	0.60 i-n	0.67 i-p
K17	10.04 f-h	9.80 h-l	127.0 h-j	139.9 j-l	0.33 j-k	0.37 i-l	0.41 l-p	0.51 m-t
K18	8.24 j-o	9.21 j-n	178.9 a-j	174.0 c-l	0.50 c-f	0.48 b-d	0.87 a-h	0.84 c-j
K20	8.82 g-n	9.21 j-n	185.2 a-i	166.8 d-l	0.51 b-e	0.49 a-d	0.94 a-f	0.81 e-k
K26	10.42 e-g	11.85 e-g	149.8 d-j	138.9 j-l	0.18 n	0.22 qr	0.27 p	0.31 u
K28	14.67 bc	13.60 b-d	223.8 a-c	220.5 a-c	0.23 l-n	0.27 op	0.54 l-p	0.59 l-r
K31	8.11 k-o	8.84 k-n	146.2 f-j	162.8 e-l	0.35 j	0.37 i-l	0.51 i-p	0.60 l-r
K33	7.70 l-o	8.06 no	214.6 a-e	206.2 a-f	0.45 f-h	0.43 e-g	0.97 a-e	0.88 b-h
K34	14.49 cd	14.68 b	186.4 a-i	140.1 j-l	0.32 jk	0.35 j-m	0.59 i-n	0.49 o-u
K35	13.07 d	14.37 bc	192.1 a-h	175.7 c-l	0.25 lm	0.25 pq	0.48 k-p	0.44 r-u
K38	14.86 a-c	14.62 b	136.2 g-j	133.5 kl	0.24 lm	0.27 op	0.33 n-p	0.35 t-u
K40	9.84 f-j	9.38 j-n	166.4 a-j	162.7 e-l	0.37 ij	0.38 h-j	0.60 l-n	0.56 l-s
K42	8.49 h-o	8.92 k-n	156.5 c-j	200.2 a-f	0.43 gh	0.45 d-f	0.68 f-l	0.90 a-g
K43	10.70 ef	11.79 e-g	196.0 a-g	194.7 a-g	0.32 jk	0.35 j-m	0.62 h-m	0.68 h-p
K44	16.36 a	16.23 a	183.8 a-i	172.8 c-l	0.22 mn	0.21 r	0.41 l-p	0.37 s-u
K46	14.98 a-c	17.04 a	229.7 ab	202.2 a-f	0.33 jk	0.33 l-n	0.75 d-k	0.67 j-q
K47	8.27 j-o	9.08 j-n	157.4 c-j	148.2 g-l	0.32 jk	0.32 mn	0.50 j-p	0.47 q-u
K48	15.55 a-c	13.06 c-e	190.3 a-i	159.2 f-l	0.22 mn	0.30 no	0.40 l-p	0.48 p-u
K49	7.47 m-o	11.03 f-h	180.3 a-j	250.4 a-f	0.57 ab	0.52 ab	1.02 a-c	1.07 ab
K52	7.89 k-o	8.79 k-o	175.0 a-j	210.9 a-e	0.53 b-d	0.52 ab	0.93 a-f	1.09 a
K55	7.87 k-o	9.31 i-n	136.0 g-j	149.3 g-l	0.48 d-h	0.50 a-c	0.65 g-l	0.75 f-l
K56	8.00 k-o	8.08 no	123.8 ij	169.5 d-l	0.42 hi	0.40 g-i	0.52 i-p	0.68 h-p
K57	8.20 j-o	9.04 j-n	172.2 a-j	188.3 b-i	0.45 e-h	0.45 d-f	0.77 c-j	0.85 c-j
K59	7.81 k-o	8.39 l-o	140.6 d-j	166.6 d-l	0.60 a	0.53 a	0.91 a-g	0.88 b-i
K60	8.28 j-o	7.30 o	147.7 e-j	145.5 h-l	0.50 c-f	0.50 a-c	0.74 e-k	0.73 g-l
K61	8.12 k-o	9.12 j-n	164.1 b-j	169.5 d-l	0.33 jk	0.35 j-m	0.55 i-o	0.59 l-r
K62	7.90 k-o	9.33 i-n	171.7 a-j	189.6 b-i	0.47 e-h	0.47 c-e	0.92 a-g	0.88 b-i
K63	8.80 h-n	11.09 f-h	189.5 a-i	193.7 a-g	0.55 a-c	0.48 b-d	1.04 ab	0.93 a-f
K65	9.31 f-l	10.83 f-i	137.1 g-j	211.7 a-d	0.43 gh	0.42 f-h	0.59 i-n	0.88 b-h
K67	8.46 h-o	8.97 j-n	233.3 a	238.4 a	0.45 e-h	0.43 e-g	1.05 a	1.03 a-c
K68	8.45 h-o	9.34 i-n	211.2 a-f	204.1 a-f	0.32 jk	0.33 l-n	0.66 g-l	0.39 g-o
K69	16.14 ab	17.55 a	216.0 a-d	218.2 a-c	0.23 l-n	0.27 op	0.51 i-p	0.58 l-r
K71	8.83 g-n	8.23 m-o	173.3 a-j	209.7 a-e	0.32 lm	0.33 l-n	0.55 i-o	0.70 g-n
LSD (0.05)	1.358	1.522	54.23	38.77	0.05135	0.03631	0.2238	0.1703

REFERENCES

1. Kothari, S.K., J.P. Singh, and K. Singh. 1989. Chemical weed control in Bulgarian coriander (*Coriandrum sativum* L.). *Tropical Pest Management*, 35(1):2-5.

2. Bhati, D.S. 1988. Effect of leaf plucking on growth, yield and economics of coriander varieties under semi-arid conditions. *Indian J. Agron.* 33(3):242-244.

3. Pino, J.A., A. Rosado, and V. Fuentes. 1996. Chemical composition of the seed oil of *Coriandrum sativum* L. from Cuba. *J. Essent. Oil Res.*, 8:97-98.

4. Kirici, S., A. Mert, and F. Ayanoğlu. 1997. Hatay ekolojisinde azot ve fosforun kişniş (*Coriandrum sativum* L). de verim değerleri ile uçucu yağ oranlarina etkisi. (The effects of N and P on the yield and essential oil of coriander in Hatay ecological conditions.) Türkiye II. Tarla Bitkileri Kongresi, Samsun. 347-351.

5. Arganosa, G.C., Sosulski, F.W., and Slikard A.E. 1998. Seed yields and essential oil of northern-grown coriander (*Coriandrum sativum* L.). *Journal of Herbs Spices and Medicinal Plants* 6(2):23-32.

Changes in Germination Capacity of *Foeniculum vulgare* Mill. Taxa

Eszter Tóth
Katalin Göncz
Zsuzsanna Pluhár

SUMMARY. Protection of propagation material of several medicinal plants in the frame of a genebank project is being carried out in our department. In the present work we have examined the changes in germination capacity of *Foeniculum vulgare* Mill. (Apiaceae) of different origins. The four years' results revealed different reactions of the tested seeds in respect of germination capacity. The effects of storage temperature, storage duration, ripening phase, intraspecific taxa and vegetation year were examined in detail. *[Article copies available for a fee from The Haworth Document Delivery Service: 1-800-HAWORTH. E-mail address: <getinfo@ haworthpressinc.com> Website: <http://www.HaworthPress.com> © 2002 by The Haworth Press, Inc. All rights reserved.]*

KEYWORDS. *Foeniculum vulgare* Mill., germination capacity, germination dynamics

Eszter Tóth, Katalin Göncz, and Zsuzsanna Pluhár are affiliated with the Department of Medicinal and Aromatic Plants, Faculty of Horticultural Sciences, Szent István University, Budapest, 29-35 Villányi str., P.O. Box 53, Budapest H-1118, Hungary.

[Haworth co-indexing entry note]: "Changes in Germination Capacity of *Foeniculum vulgare* Mill. Taxa." Tóth, Eszter, Katalin Göncz, and Zsuzsanna Pluhár. Co-published simultaneously in *Journal of Herbs, Spices & Medicinal Plants* (The Haworth Herbal Press, an imprint of The Haworth Press, Inc.) Vol. 9, No. 2/3, 2002, pp. 77-81; and: *Breeding Research on Aromatic and Medicinal Plants* (ed: Christopher B. Johnson, and Chlodwig Franz) The Haworth Herbal Press, an imprint of The Haworth Press, Inc., 2002, pp. 77-81. Single or multiple copies of this article are available for a fee from The Haworth Document Delivery Service [1-800-HAWORTH 9:00 a.m. - 5:00 p.m. (EST). E-mail address: getinfo@haworthpressinc.com].

INTRODUCTION

Preservation of valuable plant chemotaxa, which may serve as the bases for production of biologically active agencies, is a world-wide concern. However, information on genebanks, especially seed storage technologies of medicinal plant species, are insufficient. The scientific examinations reported here were carried out in the frame of our genebank activity at the Department of Medicinal and Aromatic Plants of SZIU. In the present work we have examined the changes in germination capacity of *Foeniculum vulgare* Mill. (Apiaceae).

MATERIALS AND METHODS

The seeds used in the experiments originated from cultivated populations as well as genebank collections of the years 1994, 1995 and 1998. In the case of samples harvested in 1994 and 1995, we examined the effect of the following treatments:

- ripening stage (unripe, half-ripe and ripe),
- storage temperature (room temperature and cooled condition at 4°C),
- age of populations.

We have carried out germination trials from harvesting time to sowing time weekly in the case of seeds harvested in 1998. The aim of this investigation was to observe the after-ripening phase and variability of germination capacity.

Both the starting and further germination trials were carried out according to the Hungarian Standard and international regulations (ISTA, AOSA), with 100 seeds/petri dishes in 4 repetitions.

RESULTS AND DISCUSSION

As a result of our experiments it was found that, among the examined factors, the ripeness of seeds exhibited the strongest effect on germination. The germination capacity of unripe seeds after harvesting was 21%. The half-ripe seeds germination rate was 37% and the value for the ripe phase was 56%. After a half year storage period, the germination rate increased up to 58, 65 and 78%, respectively. This fact supports the investigations of Jámbor (1), and it varied approximately 20-25% during the further storage (Figure 1). During the four-year storage period, the storage temperature had no significant effect on germination capacity, but storage at 4°C had positive effect on germination dynamics: the cold-stored seeds germinated 3-5 days earlier (Figure 2).

FIGURE 1. Germination capacity of fennel seeds in different ripening stages.

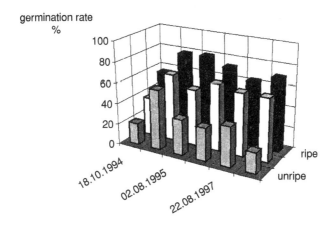

FIGURE 2. Germination dynamics of fennel seeds after 46 months storage period. Symbols: diamonds, beginning of experiment; squares, stored cooled; triangles, stored at room temperature.

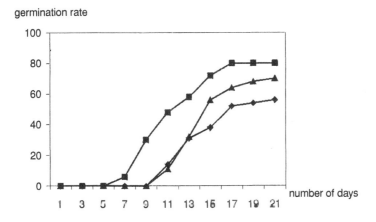

The genetic differences of the intraspecific taxa appeared in the level of germination capacity. It can be explained by complex factors: fertility relations, the structure of the umbels, flowering dynamics, etc. We could not detect significant differences between the germination rate of the one-year-old and two-year-old seeds (Table 1), but the farmers obtain the sowing-seed from two-year-old or older populations (2). It is interesting to compare the germination capacity of populations C and D, or C' and D'. The populations D and D'

showed lower values than the populations C and C'. Probably this can be explained by the effect of genotype. The average difference was notable, 13% in case of the one-year-old population, and 26% in the two-year-old population.

We have observed during our examinations that considerable changes exist in the germination capacity of fennel from October to April (Figure 3). In this

TABLE 1. Germination capacity of seeds of different fennel populations and different ages.

Date	A Two-year old (1994)	B Two-year old (1994)	C One-year old (1994)	D One-year old (1994)	S Three-year old (1995)	C' Two-year old (1995)	D' Two-year old (1995)
1	64,66	56,00	84,66	91,33	53,33	97,33	62,66
2	78,00	78,66	93,33	73,99	71,33	88,00	66,66
3	83,83	78,00	90,00	77,00	75,33	94,66	68,00
4	87,99	78,33	88,16	75,33	71,33	85,00	61,66
5	86,33	66,66	85,00	64,66	67,33	91,66	67,33
6	88,66	75,33	83,33	61,99	--	--	--
average	81,57	72,16	87,41	74,04	67,72	91,33	65,26

FIGURE 3. Changes in germination capacity of one- and two-year-old samples. Symbols: squares, seeds from one-year-old population; diamonds, seeds from two-year-old population.

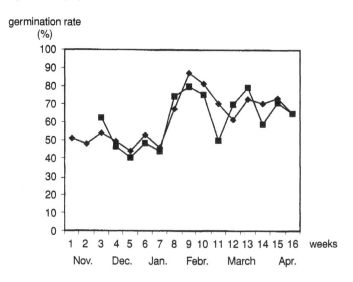

period we have detected the lowest germination rate (50-60%) at the beginning of the storage period. In the middle of February, the rate of germination increases up to 80-90% in the case of both populations. Afterwards, it decreases, even though it remains at a higher level than in the beginning of storage.

After-ripening was detected by (3), and our investigations supported his observations too. We have also detected about 20-35% increasing of germination capacity at the beginning of the storage. It is interesting that the germination dynamics of the two different aged populations are quite similar. It can be seen that the germination rate of the two-year-old sample is better than in case of the one-year-old sample.

REFERENCES

1. Jámbor, R. 1960. Über die Keimfhigkeit von gelagerten Arzneipflanzensamen. Planta Med. 2, 157-159.

2. Földesi, D. 1993. *Foeniculum vulgare*–közönséges édeskömény. In: Bernáth, J. (szerk.):Vadon termő és termesztett gyógynövények. Budapest. Mezőgazda Kiadó. 256-261.

3. Heeger, E.F. 1956. Handbuch des Arznei- und Gewürzpflanzenbaues. Drogengewinnung. Deutscher Bauernverlag. Berlin. 692-699.

Investigations on the Infraspecific Variability of *Hypericum perforatum* L.

Zsuzsanna Pluhár
Jenő Bernáth
Éva Németh

SUMMARY. The natural variability of *Hypericum perforatum* L. (Hypericaceae) serves as a sound basis for breeding programs. The aims of this work were to study its infraspecific diversity from a biological and chemical points of view, and to promote the breeding of varieties which would be suitable for growing under Hungarian climatic conditions.

At the beginning of our investigations, 18 accessions of different origin had been tested for two years (1996-1997), evaluating the overall population properties and the individual divergences. Outstanding individuals could be selected from the basic populations according to their promising production-biological particulars. Their lines have been studied through two subsequent vegetation cycles (1998-1999). During the investigation of morphological, phenological and production properties of taxa involved, degrees of seasonal and ontogenetical variability have been estimated within and among populations. Data of genetical stability and heritability of taxon characteristics have also been obtained.

Great individual divergences characterized the accumulation levels of hypericin derivatives, while flavonoid content was more stable within

Zsuzsanna Pluhár, Jenő Bernáth, and Éva Németh are affiliated with the Department of Medicinal and Aromatic Plants, Faculty of Horticultural Sciences, Szent István University, Budapest, 29-35 Villányi str., P.O. Box 53, Budapest II-1118, Hungary.

[Haworth co-indexing entry note]: "Investigations on the Infraspecific Variability of *Hypericum perforatum* L." Pluhár, Zsuzsanna, Jenő Bernáth, and Éva Németh. Co-published simultaneously in *Journal of Herbs, Spices & Medicinal Plants* (The Haworth Herbal Press, an imprint of The Haworth Press, Inc.) Vol. 9, No. 2/3, 2002, pp. 83-88; and: *Breeding Research on Aromatic and Medicinal Plants* (ed: Christopher B. Johnson, and Chlodwig Franz) The Haworth Herbal Press, an imprint of The Haworth Press, Inc., 2002, pp. 83-88. Single or multiple copies of this article are available for a fee from The Haworth Document Delivery Service [1-800-HAWORTH 9:00 a.m. - 5:00 p.m. (EST). E-mail address: getinfo@haworthpressinc. com].

populations, influenced principally by genotype. The highest total content of hypericin derivatives (4.0-15.8 mg/g) have been found in the second year of cultivation, while that of the flavonoid content (23.6-30.1 mg/g) could have been measured in the third vegetation cycle. The taxa studied only in very few cases possessed excellent hypericin and flavonoid contents at the same time. *[Article copies available for a fee from The Haworth Document Delivery Service: 1-800-HAWORTH. E-mail address: <getinfo@haworthpressinc.com> Website: <http://www.HaworthPress.com> © 2002 by The Haworth Press, Inc. All rights reserved.]*

KEYWORDS. *Hypericum perforatum* L., hypericin, flavonoids, breeding

INTRODUCTION

Hyperici herba, its standardized extract as well as the St. John's wort preparations, have become important commercial products worldwide in the last few years (1). These preparations are widely utilized in the treatment of mild and moderate forms of depression. Although the isolated active substances of *Hyperici herba* have been tested systematically, the most effective antidepressive agents of the complex have not yet been revealed. Generally, a water-alcoholic extract is prepared and standardized for 0.24-0.32% hypericin, containing all the compounds–naphthodianthrones, flavonoid glycosides, biflavones, xanthones, floroglucin derivatives–showing antidepressant activity (2).

Beside collection from wild habitats, *Hypericum perforatum* L. was introduced into intensive cultivation in Hungary in 1997 and 1998. Under growth conditions, controlled drug production can be achieved; however, the applied varieties ('Topas', 'Anthos') or populations could not fulfil completely the quantitative and/or qualitative requirements.

The natural variability of the species is conspicuous when evaluating the drug quality harvested in the wild, though the degree of diversity and its biological-chemical background is only partly known (3-5). This natural source of heterogeneity serves as a proper basis for breeding programs and the apomictic way of reproduction of the species also promotes the selection.

The theoretical purpose of our work was to clarify certain details of infraspecific diversity of *H. perforatum* from biological and chemical points of view. A practical aspect of our experiments was to advance the breeding of varieties which would be suitable for growing under Hungarian climatic conditions.

In 1996, a complex research program was started in order to investigate ontogenetical, morphological and production biological variability in St. John's wort taxa of different origin. Degrees of seasonal and ontogenetical variability have been estimated within and among populations. Data of genetical stability and heritability of taxon characteristics has also been obtained.

MATERIALS AND METHODS

The open field experiments were carried out at the Research Station of the Department of Medicinal and Aromatic Plants in Budapest, 1996-99. The origins of the basic populations examined during three successive years (1996-98) were the following:

a. population of the original Polish cultivar 'Topas' as control,
b. seven accessions of 'Topas' origin (marked 3, 10, 14, 16, 17, 20 and 28),
c. two Hungarian wild origins (marked 23 and 24),
d. eight selected German materials (marked 30, 31, 32, 34, 35, 36, 37 and 38).

Ten lines of individually selected mother plants–24/5, 24/7, 28/5, 28/6, 31/2, 31/8, 34/3, 34/7, 35/3 and 35/8–were investigated during further two vegetation periods (1998-99).

The establishment of the populations of 25-50 individuals was carried out under greenhouse conditions. Plant density of 5 plant/m^2 was applied.

Morphological properties considered to be important–plant height, plant width, number of flowering shoots, length of inflorescences, leaf type and plant habit–were determined in the phase of flowering. Within populations, 7-10 individuals exhibiting advantageous morphological features were isolated (1/3 plant) for self-pollination and individual seed production. Production properties, e.g., herba and flower yields, active substance levels, were investigated at isolated plants only. Content and composition of different chemical compounds–naphthodianthrones, flavonoids, floroglucin derivatives–were established by an acetone-methanol extraction followed by HPLC-analysis, according to the modified method of Hölzl and Ostrowski (6).

Morphological and production properties were evaluated by one-way analysis of variance. Cluster analysis was applied to compare and group individuals according to their active agent content. Homogeneity within populations concerning different characteristics was determined by coefficients of variance (CV%). Statistica 4.5, Statgraphics 5.1 and Microsoft Excel 5.0 programs were used for data analysis.

RESULTS AND DISCUSSION

Distances among populations regarding the examined properties were determined as quite constant and heritable phenomena, generally. Considering ontogenetical aspects, four groups of flowering behaviour (early, medium-early, medium and late types) were exactly distinguished among populations, and the order of blooming remained constant even at progenies at first and second cutting. Individual flowering divergences found in some basic populations disappeared in progenies, showing the effectiveness of individual selection in unifying flowering time.

H. perforatum is a perennial plant with a lifecycle of some 3-5 years. According to growing practice, its intensively cultivated fields can be maintained through two or three years, where the second vegetation cycle is the most productive. In our experiments, some narrow leaved taxa (e.g., 23 and 24) had high generativity, early flowering ability and proved to be utilizable in the first two vegetation periods, while the lifelength of other narrow leaved and that of the broader leaved taxa was three years.

Within original populations, the values of morphological properties influencing production increased by age. Populations and individuals could be found possessing a large number of flowering shoots, plenty of flowering branches (e.g., 16, 34, 35, 37). The greatest morphological heterogeneity was experienced in the accessions of wild origin (No. 23 and 24) as was expected. When comparing mother populations to progenies of the same age, an increase can be experienced in each trait measured, affected mainly by seasonal differences. However, the selected lines retained the original distances that existed among basic populations. Plant height and inflorescence length proved to be rather homogenous morphological properties within a population, with the lowest CV% values, which could have been improved significantly by individual selection.

The heterogeneity of yields experienced was due to individual differences in the number of flowering shoots; however, high productivity taxa have been found based on population averages. Limits of herba masses in population averages, experienced by one cutting per year, were as follows: 30-160 g/m^2 in the first, 170-400 g/m^2 in the second and 300-500 g/m^2 in the third vegetation cycle. The flower yields obtained did not show considerable annual changes; the most prominent populations showed values of approximately 100 g/m^2 (Table 1).

Morphological changes between basic populations and progenies induced a 1.5-2 fold incease in herba and flower masses. In our opinion, this alteration is not only caused by selection but also affected by environmental factors (more precipitation in 1999). Improved homogeneity levels (CV%) could be experienced in the case of yields of progenies.

Great individual divergences characterized the accumulation levels of hypericin derivatives, while flavonoid content was more stable within populations, influenced principally by genotype. The highest total content of hypericin derivatives (4.0-15.8 mg/g) was found in the second year of cultivation, while

TABLE 1. Production-biological parameters obtained during their three year lifecycle at the investigated *H. perforatum* L. basic populations of different origin.

Properties	Limits (min-max) of the detected values expressed in population average		
	1st vegetation cycle: 1996	*2nd vegetation cycle:* 1997	*3rd vegetation cycle:* 1998
Yields			
*Fresh herba yield, g/m^2	90-390	500-1500	990-1800
*Dry herba yield, g/m^2	30-160	170-400	300-500
*Dry flower yield, g/m^2	-	43-122	60-120
Active principles			
Naphthodianthron derivatives			
Hypericin content, mg/g	0.18-2.16	0.64-3.88	0.42-2.08
Pseudohypericin content, mg/g	1.40-7.30	3.18-11.88	2.33-6.3
Hypericin/pseudohypericin ratio	1:1.83-9.50	1: 2.03-8.2	1:2.70-8.07
Total hypericin content, mg/g	1.88-9.90	4.00-15.76	2.83-8.38
Total hypericin yield, g/m^2	0.08-0.65	1.00-6.00	0.69-4.42
Flavonoid components			
Flavone-glycosides			
Rutin content, mg/g	0.08-24.00	1.69-6.22	1.23-16.8
Hyperoside content, mg/g	3.80-9.60	4.34-9.62	6.6-14.48
Isoquercitrin content, mg/g	1.40-7.80	1.58-6.74	2.3-11.23
Quercitrin content, mg/g	0.40-6.40	1.3-2.78	0.86-2.62
Total flavon-glycoside content, mg/g	16.00-36.00	15.06-20.84	23.55-29.55
Total flavon-glycoside yield, g/m^2	0.47-2.75	3.78-9.02	5.76-15.62
Flavonoid-aglicon			
Quercetin content, mg/g	0.11-0.39	0.26-1.2	0.14-0.49
Biflavonoids			
Biapigenin content, mg/g	0.33-0.67	0.98-1.82	0.53-1.07
Amenthoflavone content, mg/g	0.01-0.03	0.03-0.07	0.01-0.04
Total biflavone content, mg/g	0.32-0.84	1.02-1.89	0.59-1.11
Total biflavone yield, g/m^2	0.01-0.07	0.25-0.77	0.15-0.52
Flavonoids (Summarized)			
Total flavonoid content, mg/g	16.01-26.04	16.63-23.91	23.6-30.1
Total flavonoid yield, g/m^2	0.65-2.83	4.11-10.19	5.78-15.3
Floroglucin derivative			
Hyperforin content**	2414-11245	3636-11004	1548-4683

* Yields obtained by one harvest per year.
** Hyperforin content expressed as peak area on the HPLC chromatogram.

that of the flavonoid content (23.6-30.1 mg/g) was measured in the third vegetation cycle. From the two most important elements of flavonoid spectrum, rutin content varied on the widest range (0.08-24.0 mg/g), while hyperoside showed the limits of 3.8-14.5 mg/g in mean. There could be found taxa (24, 35) accumulating constantly low rutin amounts, while their hyperoside levels were increased as a compensation. Comparative data about the unstable hyperforin can be expressed in peak area. Their values varied between 2000-11000 in the first two years and decreased significantly in the third growing season (Table 1).

Chemotypes accumulating basically hyperoside + isoqercitrin (24, 34, 35, 37) and others with similarly high levels of rutin and hyperoside (3, 10, 17, 20, 28, 31, 32, 38) could have been described. Amongst the taxa studied, only a very few cases (e.g., 35) possessed excellent hypericin and flavonoid contents at the same time.

REFERENCES

1. Brevoort, P. 1998. The Booming U.S. Botanical Market. A New Overview. *Herbalgram* 44: 18-20, 33-45.

2. Upton, R. (ed.) 1997. St. John's Wort Monograph In: American Herbal Pharmacopoea. *Herbalgram,* 40: 2-31.

3. Büter, B., Orlacchio, C., Soldati, A., and Berger, K. 1998. Significance of Genetic and Environmental Aspects in the Field Cultivation of *Hypericum perforatum. Planta Medica*, 64: 431-437.

4. Franke, R., Schenk, R., and Bauermann, U. 1999. Variability in *Hypericum perforatum* L. Breeding Lines. Proceedings of WOCMAP In: *Acta Horticulturae*, 502: 167-173.

5. László-Bencsik, Á. 1996. Hypericum populációk összehasonlító vizsgálata (Comparative Investigations of *Hypericum* Populations). PhD Thesis, UHFI

6. Hölzl, J. and Ostrowski, E. 1987. Johanniskraut (*Hypericum perforatum* L.), HPLC Analyse der wichtigen Inhaltstoffe und deren Variabilität in einer Population, *Deutsche Apotheker Zeitung*, 127 (23): 1227-1230.

First Field Trials
of Borage (*Borago officinalis* L.)
in Andalusia (Southern Spain) as a Source
of "Biological" Gamma Linolenic Acid

C. Gálvez

A. De Haro

SUMMARY. Borage (*Borago officinalis* L.) is cultivated in the north of Spain for fresh edible production and is gathered from natural populations for its fresh flowers with saline, cucumber-like flavor. It is considered as a weed in the rest of Spain. Nevertheless, the agroecological mediterranean conditions in the south of Spain are adequate for growing borage as an oilseed crop for γ-linolenic acid (GLA) production.

One ha of borage was sown on a farm in the Guadalquivir Valley in the Autumn of 1999. The farm was located in Carmona (province Seville) and cultivated under biological production conditions, controlled by the "Comité Andaluz de Agricultura Ecológica" (CAAE), the Andalusian biological control organization (an IFOAM member).

The crop was developed by dry farming, fertilized with common vetch, and false technical sowing was employed to avoid weeds. The harvest was collected mechanically and the seed was produced using

---

C. Gálvez is affiliated with Semillas Silvestres, S.L., C/Aulaga n° 24, 14012 Cordoba, Spain.

A. De Haro is affiliated with the Instituto Agricultura Sostenible, CSIC, Apartado 4084, 14080 Cordoba, Spain.

[Haworth co-indexing entry note]: "First Field Trials of Borage (*Borago officinalis* L.) in Andalusia (Southern Spain) as a Source of "Biological" Gamma Linolenic Acid." Gálvez, C., and A. De Haro. Co-published simultaneously in *Journal of Herbs, Spices & Medicinal Plants* (The Haworth Herbal Press, an imprint of The Haworth Press, Inc.) Vol. 9, No. 2/3, 2002, pp. 89-93; and: *Breeding Research on Aromatic and Medicinal Plants* (ed: Christopher B. Johnson, and Chlodwig Franz) The Haworth Herbal Press, an imprint of The Haworth Press, Inc., 2002, pp. 89-93. Single or multiple copies of this article are available for a fee from The Haworth Document Delivery Service [1-800-HAWORTH 9:00 a.m. - 5:00 p.m. (EST). E-mail address: getinfo@haworthpressinc.com].

densimetrical method. A yield of 544 kg of clean seed/ha with 27.3-33.3% oil content was obtained. This yield is similar to that obtained under conventional agronomic conditions.

That it is possible to cultivate borage under biological conditions in the south of Spain for GLA commercial production with biological label quality has therefore been demonstrated. *[Article copies available for a fee from The Haworth Document Delivery Service: 1-800-HAWORTH. E-mail address: <getinfo@haworthpressinc.com> Website: <http://www.HaworthPress. com> © 2002 by The Haworth Press, Inc. All rights reserved.]*

KEYWORDS. Borage, gamma-linolenic acid, GLA, fatty acids, oilseeds

INTRODUCTION

Borage (*Borago officinalis* L.) is cultivated in the north of Spain for fresh edible production (4) and is gathered from natural populations for its fresh flowers with saline, cucumber-like flavor (free of toxic pyrolizidine alkaloids) (3). It is considered as a weed in the rest of Spain and up to this day it has not been grown for the production of oilseeds rich in γ-linolenic acid.

On the other hand, the added value obtained by the active medicinal principles produced by ecological techniques creates the possibility of producing borage oil for the supply of national and international markets at highly profitable prices for the farmer and producer.

The purpose of this paper is to demonstrate the possibility of producing borage seed ecologically and by dry farming, with a high output in oil and GLA, under the agroecological mediterranean conditions of Andalusia (southern Spain).

MATERIALS AND METHODS

Setting. The farm used in this trial was located in the Guadalquivir Valley, near Carmona, in the province of Seville (Andalusia). This farm had been under the ecological management and supervision of the Andalusian Committee of Biological Agriculture (CAAE) for eight years. CAAE is the organization responsible for biological production control within Andalusia (and a member of the IFOAM).

Soil preparation. Samples of the soil were taken before seeding to establish its properties. The results obtained from these tests are shown in Table 1. The land was seeded the year before with "veza común" (covered in greenery) in

TABLE 1. Breakdown of soil properties in the field trial of borage in Carmona (southern Spain).

CHARACTERISTIC	UNITS	METHOD
CATION EXCHANGE CAPACITY	26.26 (meq/100g)	Photometric
EXCHANGEABLE FRACTION OF CALCIUM	22.19 (meq/100g)	Volumetric
EXCHANGEABLE FRACTION OF MAGNESIUM	2.47 (meq/100g)	Volumetric
EXCHANGEABLE FRACTION OF SODIUM	0.36 (meq/100g)	Spectrophotometric
EXCHANGEABLE FRACTION OF POTASSIUM	1.24 (meq/100g)	Spectrophotometric
AVAILABLE PHOSPHORUS (OLSEN)	17.6 ppm	Colorimetric
ORGANIC NITROGEN	0.07 (%)	Volumetric
ORGANIC MATTER	1.17 (%)	Volumetric
pH (1/2.5)	8.18	pHmetric
pH (1/2.5 en C1K)	7.31	pHmetric
ASSIMILABLE POTASSIUM	490 ppm	Photometric
CLAY	38.7 (%)	Densimetric
SAND	28.3 (%)	Densimetric
LOAM	33.0 (%)	Densimetric
TEXTURE	Loam-clay	

the month of February. Before seeding, a scerifier was used on two occasions to eliminate weeds.

Conditions of seeding and cultivation. Five kilogram of seed (German origin) were used in the seeding of 1 ha. This was done with a cereal seeder (with a steady stream) leaving a separation of 75 cm between lines. Seeding took place on the 25th of November after the first rains of autumn. Absolutely no work was done on the plant while it was growing under dry conditions.

Harvesting. The crop was cut on the 20th of May and left on the ground for four days to facilitate the maturing of the seed. Afterwards the plants were lifted off the ground and the seeds were separated by a combine cereal harvester. The resulting seed was spread on the ground to dry for three more days.

Cleaning of the seed. The seed obtained from the combine-harvester was processed by a siever-winnowing machine to eliminate soil remains and other impurities. Finally a densimetric table was employed to eliminate the empty seeds and to separate three qualities according to different densities.

Oil content and fatty acid composition. Oil content of the seed, previously dried at 65°C for 60 hours, was determined by nuclear magnetic resonance (NMR). The NMR analyzer was standardized by using a clean high quality sample of borage oil extracted by the Soxhlet method. The fatty acid composition of the oil was determined on a bulk of 10 seeds/entry by simultaneous extraction and methylation, followed by gas-liquid chromatography (GLC) on a Perkin Elmer Autosystem (Perkin Elmer, Norwalk, CT) equipped with a flame ionization detector (FID) and a 2 m column packed with 3% SP-2310/2% SP-2300 on Chromosorb WAW. Fatty acids were identified by comparing the retention times of the borage methyl esters with those of known mixtures of methyl esters run on the same column under the same conditions. The GLA standard was purchased from the Sigma Chemical Co. (L2378).

RESULTS AND DISCUSSION

The results obtained from the trial are shown in Table 2. Out of the 544 kg of clean seed harvested, only 350 kg were fully-developed seeds; that is to say, the rest were empty seeds which had not achieved maturity correctly after the plants had been cut, because of the absence of pollinating insects, or alternatively, because of mutual incompatibility. The fully-developed seed had an average oil content of 30.33%, ranging between 33.3 and 27.3% in the 2nd and 3rd categories, whereas the GLA content oscillated between 20.64% and 17% (with an average value of 18.81%) from the first to the third seed categories.

Although the percentage of green seed (supposedly immature) was notably bigger in the third quality (90%) as opposed to the first quality (35%), the content in oil and GLA had not gone down in the same proportion. This indicates that it is possible to adjust the timing of the cutting by counting the fully-developed green seeds as acceptable for the production of oil and GLA.

The destructive harvesting system employed (cutting, maturing on the ground and harvesting by combine-harvester) proved itself effective. The production of seeds was similar to that provided by others who worked with expensive harvesting prototypes by aspiration and different non-destructive methods (2) or by using diquat to facilitate the harvesting (1). The use of ma-

TABLE 2. Seed, oil, γ-linolenic acid yield and seed characteristics from 1 ha of borage cultivated in Carmona (southern Spain) in 1999-2000 crop.

QUALITY	SEED YIELD (KG)	OIL CONTENT (%)	γ-LINOLENIC CONTENT (%)	OIL YIELD (KG)	SEED Nº/KG	BLACK SEEDS (%)
1	22	32.9	20.64	7.23	52,631	65
2	168	33.3	18.46	55.94	62,500	35
3	162	27.3	17.35	44.22	71,428	10
EMPTY SEEDS	194	-	-	-	139,200	0
TOTAL	544	-	-	107.39	-	-
MEAN	-	30.33	18.81	-	81,439	-

chinery not specifically designed for the purpose and the little manual labour involved guarantees low-cost production and also allows for the obtaining of a high quality and highly profitable product. The increase in the density of the plants, the more accurate timing of the cutting, and the support for pollination by placing beehives, could notably increase the production. Nevertheless, after this trial, the biological growing of borage for production of oil rich in gamma-linoleic acid is considered possible in dry-farming in Andalusia (southern Spain).

To sum up, this oil-seed crop can be considered an alternative to cereal and sunflower, and a new highly profitable commercial opportunity, especially if it obtains the added value that ecological products have in today's market.

REFERENCES

1. Helme, J.P. 1992. Onagre, borrache, pépins de cassis. (Karleskind, A. ed.) *Manuel des Corps Grass.* Tecnique et Documentation. Lavoisier. Paris. Vol 1.

2. Janick, J., Simon, J.E., Quinn J., and Beaubaire J. 1989. Borage: A Source of Gamma Linolenic Acid, in Cracker, L.E. & J.E. Simon (eds.) *Herbs, Spices and Medicinal Plants: Recent Advances in Botany, Horticulture and Pharmacology.* Oryx Press, Arizona, USA.

3. Leung, A.Y. and Foster, S. 1996. *Encyclopedia of Common Natural Ingredients Used in Food, Drugs, and Cosmetics.* (2nd ed.), John Wiley & Sons, Inc., New York.

4. Villa Gil, F. and Alvarez, J.M. 1994. El cultivo de la borraja en Aragón. Centro de Transferencia Tecnológica Vegetal. Servicio de Investigación Agraria. Gobierno de Aragón. *Informaciones Técnicas* 6/94.

Ontogenetic Variation Regarding Hypericin and Hyperforin Levels in Four Accessions of *Hypericum perforatum* L.

Karin Berger Büter
Bernd Büter

SUMMARY. Hypericin and hyperforin contents of isolated flowers and capsules, collected from 4 different *Hypericum perforatum* accessions at four developmental stages, i.e., (A) the stage of closed buds with yellow petals already visible, (B) the stage of fully opened flowers, (C) the stage of green capsules and (D) the stage of brown capsules, were determined via HPLC. Principally, hypericin contents increased with advancing of the stages, whereas hyperforin decreased. However, genotype specific deviations were observed, e.g., one genotype showing a low and relatively stable hyperforin content throughout all stages of development or another genotype revealing similar hypericin contents in stage (A) and (B). *[Article copies available for a fee from The Haworth Document Delivery Service: 1-800-HAWORTH. E-mail address: <getinfo@haworthpressinc.com> Website: <http://www.HaworthPress.com> © 2002 by The Haworth Press, Inc. All rights reserved.]*

Karin Berger Büter is affiliated with the Institute of Pharmaceutical Biology, University of Basel, Benkenstr. 254, CH 4108 Witterswil, Switzerland (E-mail: karin. berger@uni-bas.ch).

Bernd Büter is affiliated with VitaPlant AG, Benkenstr. 254, CH 4108 Witterswil, Switzerland (E-mail: bbueter@vitaplant.ch).

[Haworth co-indexing entry note]: "Ontogenetic Variation Regarding Hypericin and Hyperforin Levels in Four Accessions of *Hypericum perforatum* L." Büter, Karin Berger, and Bernd Büter. Co-published simultaneously in *Journal of Herbs, Spices & Medicinal Plants* (The Haworth Herbal Press, an imprint of The Haworth Press, Inc.) Vol. 9, No. 2/3, 2002, pp. 95-100; and: *Breeding Research on Aromatic and Medicinal Plants* (ed: Christopher B. Johnson, and Chlodwig Franz) The Haworth Herbal Press, an imprint of The Haworth Press, Inc., 2002, pp. 95-100. Single or multiple copies of this article are available for a fee from The Haworth Document Delivery Service [1-800-HAWORTH 9:00 a.m. - 5:00 p.m. (EST). E-mail address: getinfo@haworthpressinc.com].

KEYWORDS. *Hypericum perforatum*, hypericin, hyperforin, genetic variation

INTRODUCTION

Various pharmacological and clinical studies regarding the effects of *Hypericum perforatum* (= HP) extracts have been published confirming the antidepressive action of these extracts (summarized in 1). However, much less is known about the plant itself, its physiological development, its growth requirements, the genotypic variation, reproductive biology and hence the applicable breeding method. In addition it is still not clearly known which compounds contribute to the clinical efficacy. Among the known constituents, hypericin and hyperforin are considered to contribute to the antidepressant activity (2,3). Furthermore, a recent study indicates that hyperforin might be involved in the interaction of *Hypericum* extracts with other drugs (4) such as cyclosporin, oral contraceptives or Indinavir as reported during the last months (summarized in 5).

Both hypericin and hyperforin are present in the flowering segment of the HP plants; for hypericin it has been reported that the contents might vary depending on the genotype (6), growing conditions (7), and developmental stage of the plants at the time of harvest (8). For hyperforin, similar interactions might be assumed; however, the effect of the factors listed above has not been thoroughly investigated so far.

In the present study we have investigated the influence of the developmental stage, i.e., the harvesting time, on the hyperforin and hypericin content. For this purpose we have followed the changes with respect to the hyperforin and hypericin contents in different HP accessions during the course of flowering and capsule formation.

MATERIALS AND METHODS

Plant material. During 1998 and 1999, four *H. perforatum* accessions (HP 49, wild type, collected in Canada; HP54, cultivar 'Topas', Inst. of Medicinal Plants, Posnan, Poland; HP 52-S, cultivar 'Vitan', VitaPlant, Switzerland; HP 28-S, selected line, VitaPlant, Switzerland) were cultivated on the experimental field of VitaPlant in Witterswil, Switzerland.

Plant materials were collected in the second year of cultivation (1999), at the following developmental stages:

A. closed buds with yellow petals already visible
B. fully opened flowers

C. green capsules
D. brown capsules

All harvested plant materials were dried at 35°C and pulverized in a centrif-ugal mill.

Phytochemical analysis: For analysis, only separate flowers and capsules were used. The plant material was extracted twice with methanol and acetone (sonication). The combined extracts were filtered and, after evaporating, the solvents were redissolved in methanol. An aliquot was filtered through a RC membrane (0.45 Φm) and analyzed for total hypericin and hyperforin content by HPLC according to Hölzl and Ostrowski (9) using a Hypersil 120-5 ODS column (250 × 4.6 mm, Macherey-Nagel). External standards of Hypericin (Extrasynthese, France) and Hyperforin (Schwabe, Germany) were used to determine the contents of pseudohypericn, hypericin, adhyperforin and hyper-forin. Total hypericin concentration was calculated as the sum of hypericin and pseudohypericin, total hyperforin content correspond to the sum of adhyper-forin and hyperforin.

Statistical analysis: Ten individual plants of each accession were investi-gated. From each plant, 10 flowers or capsules respectively were removed at each of the 4 stages and combined in order to obtain one extract per accession and stage. Each extract was analyzed twice. Means and SE were determined for each accession at each developmental stage.

RESULTS

Generally, the total hypericin contents decreased with advanced develop-mental stages, whereas the hyperforin levels increased (Figure 1). However, genotype specific features were observed, e.g., in HP 49 the hypericin level maintained unchanged during the first and the second developmental stage whereas in all others the hypericin level dropped continuously (Figures 2 and 3).

Genotypic effects also were observed with regard to the hyperforin con-tents: two accessions (HP49, HP52-S) reached the maximum level at the green capsule and two (HP 28-S, HP 54) at the brown capsule stage. One of the ac-cessions, i.e., HP28-S, showed relatively low hyperforin contents; in this ac-cession, even in the green capsule stage the hyperforin content was found to be considerably lower than the content at the closed flower buds stage in acces-sion HP 54 and HP 49.

FIGURE 1. Contents of total hypericin and hyperforin in isolated flowers/capsules of *Hypericum perforatum* harvested at different times during the generative stage (means across 4 *Hypericum* accessions).

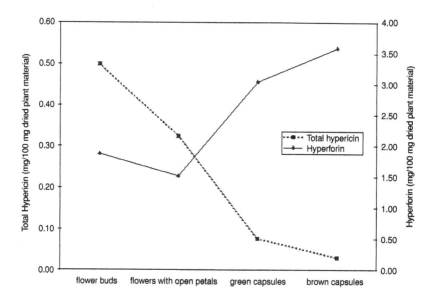

DISCUSSION

Our results indicate that employing the correct developmental stage, i.e., keeping the appropriate harvesting time, is imperative for the production of a *Hyperici herba* drug with a satisfying hypericin or hyperforin content. In agricultural practice it will be difficult though to determine the optimum time of harvest if high contents of both constituents are required, since the two constituents show their maximum levels at quite different stages of the generative phase of HP. A split harvest, i.e., harvesting part of the plants at an early harvest just before or during opening of the flowers and harvesting the remaining part of plants during the green or brown capsule stage, or the use of a genotype with still high hypericin contents in stage (B) and already high hyperforin contents in stage (C) could present ways to solve the problem.

A comparison of the 4 accessions further reveals that there exist significant genotypic differences regarding the total amounts of hypericin and hyperforin as well as regarding the ontogenetic variability of the hypericin and hyperforin production. Thus, the careful combination of accession and harvesting time opens the possibility to adjust the hypericin and hyperforin content in the hyperici herba drug according to the quality and/or market demands.

FIGURE 2. Contents of total hypericin in flowers or capsules of *Hypericum perforatum* harvested at different times during the generative stage (means ± SE).

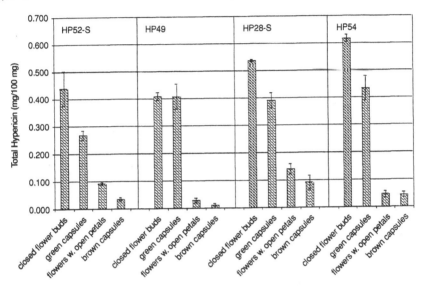

FIGURE 3. Contents of total hyperforin in flowers or capsules of different accessions of *Hypericum perforatum* harvested at different times during the generative stage (means ± SE).

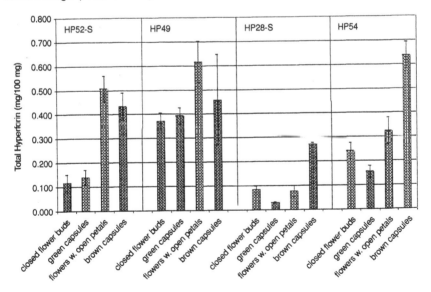

REFERENCES

1. Linde, K., G. Ramirez., C.D. Mulrow, A. Pauls, W. Weidenhammer, and D. Melchart. 1996. St. John's wort for depression–an overview and metanalysis of randomized clinical trials. *Brit. Med. J.* 313: 253-258.

2. Simmen, U., W. Burkard, K. Berger, W. Schaffner, and K. Lundstrom. 1999. Extracts and constituents of *Hypericum perforatum* inhibit the binding of various ligands to recombinant receptors expressed with the semliki forest virus system. *J. of Receptor & Signal Transduction Research* 19 (1-4): 59-74.

3. Chatterjee, S.S., M. Nöldner, E. Koch, and C. Erdelmeier. 1998. Antidepressant activity of *Hypericum perforatum* and hyperforin: the neglected possibility. *Pharmacopsychiat.* 31 (suppl. 1): 30-35.

4. Moore, L.B., B. Goodwin, S.A. Jones, G.B. Wisely, C.J. Serabjit, T.M. Willson, J.L. Collins, and S.A. Kliewer. 2000. St. Johns wort induces hepatic drug metabolism through activation of the pregnane $\times$ receptor. *PNAS* 97 (13): 7500-7502.

5. Jungmayr, P. 2000. Johanniskraut-nicht ohne Wechselwirkungen. *Deutsche Apotheker Zeitung* 140 (18): 2048-2050.

6. Büter, B., C. Orlacchio, A. Soldati, and K. Berger. 1997. Significance of genetic and environmental aspects in the field cultivation of *Hypericum perforatum*. *Planta Medica* 64: 431-437.

7. Lieres, A.L. 1994. Düngungsversuche mit *Hypericum perforatum* (Fertilizer trials with *Hypericum perforatum*). *Hessische Landwirtschaftliche Versuchsanstalt*, Kassel-Harleshausen.

8. Brantner, A., T. Kartnig, and F. Quehenberger. 1994. Comparative phytochemical investigations of *Hypericum perforatum* L. and *Hypericum maculatum* Crantz. *Scientia Pharmaceutica* 62: 261-276.

9. Hölzl, J. and E. Ostrowski. 1987. Johanniskraut (*Hypericum perforatum* L.)–HPLC-Analyse der wichtigen Inhaltsstoffe und deren Variabilität in einer Population. *Deutsche Apotheker Zeitung* 127 (23): 1227-1230.

Breeding Improvement of *Laurus nobilis* L. by Conventional and *In Vitro* Propagation Techniques

Naoufel Souayah
Mohamed Larbi Khouja
Abdelhamid Khaldi
Mohamed Nejib Rejeb
Sadok Bouzid

SUMMARY. Originating from Asia Minor and later transmitted through Greece, *Laurus nobilis* L. (Lauraceae) has overrun Europe and the Mediterranean area. The problems associated with the great variability of this dioïceous species, the difficulties in sexual reproduction and seed germination, could be overcome, using micropropagation and *in vitro* culture techniques, by mass clonal production, as was demonstrated in our laboratory with other species. Here we present preliminary results of the breeding improvement and micropropagation by axillary buds of selected mature individuals.

An analysis of variance of the results showed significant differences in rooting due to the type of cutting. The buds also have an effect in this morphogenetic process, especially in the induction of the first root. In

---

Naoufel Souayah, Mohamed Larbi Khouja, Abdelhamid Khaldi, and Mohamed Nejib Rejeb are affiliated with the Laboratory of Ecology, Institute of Research in Rural Engineering, Water and Forestry (INRGREF), B.P. 2 2080, Ariana, Tunisia.

Sadok Bouzid is affiliated with the Laboratory of Experimental Morphogenesis, Faculty of Sciences of Tunis, Tunisia.

[Haworth co-indexing entry note]: "Breeding Improvement of *Laurus nobilis* L. by Conventional and *In Vitro* Propagation Techniques." Souayah, Naoufel et al. Co-published simultaneously in *Journal of Herbs, Spices & Medicinal Plants* (The Haworth Herbal Press, an imprint of The Haworth Press, Inc.) Vol. 9, No. 2/3, 2002, pp. 101-105; and: *Breeding Research on Aromatic and Medicinal Plants* (ed: Christopher B. Johnson, and Chlodwig Franz) The Haworth Herbal Press, an imprint of The Haworth Press, Inc., 2002, pp. 101-105. Single or multiple copies of this article are available for a fee from The Haworth Document Delivery Service [1-800-HAWORTH 9:00 a.m. - 5:00 p.m. (EST). E-mail address: getinfo@haworthpressinc.com].

101

micropropagation trials, we could control the delicate stage of sterilization. Aseptic and vigorous cultures were obtained after treatment by calcium hypochlorite. Shoot multiplication and elongation were obtained
by addition of benzyl-aminopurine combined with gibberellic acid.
Root induction was obtained in MS medium salts at 1/3 strength with
naphtalene-acetic acid. *[Article copies available for a fee from The Haworth
Document Delivery Service: 1-800-HAWORTH. E-mail address: <getinfo@
haworthpressinc.com> Website: <http://www.HaworthPress.com> © 2002 by
The Haworth Press, Inc. All rights reserved.]*

KEYWORDS. Breeding, cuttings, rhizogenesis, axillary buds, *in vitro*
culture, *Laurus nobilis* L.

INTRODUCTION

In Tunisia, medicinal and aromatic plants represent a national wealth that
should be valorized and preserved. With their active substances, they are the
bases of many pharmaceutics, homeopathics and food products. The lack of a
real horticulture and the rising infatuation for natural therapeutics, without ra-
tional management, has sped up the degradation and the genetic erosion of this
flora. Conservation and amelioration programmes are urgently needed to safeguard medicinal and aromatic plants like bay-laurel. Originating from Asia
Minor and later transmitted through Greece, *Laurus nobilis* L. (Lauraceae)
spread over Europe and the Mediterranean area (1). This shrub or small tree is
sought especially for its culinary role. This aromatic plant, with essential oils,
also has many phytotherapy virtues. In fact, it is used as an antiseptic, sedative
and analgesic. The massive exploitation by the population, and the increased
interest of the pharmacology and food industries, makes intensive multiplication of this plant imperative. Due to the great variability of this dioïc species
and the difficulties in sexual reproduction and seed germination, conventional
propagation and *in vitro* culture techniques and improvement by mass clonal
production could overcome these problems, as was proved in our laboratory
with other species (2).

Here we present preliminary results of the breeding promoted by cuttings
and micropropagation with axillary buds of selected mature individuals.

MATERIALS AND METHODS

In breeding trials, cuttings were collected in February-March and April-
May from mature shrubs. They were approximately 10-12 cm long, herba-

ceous, semiligneous and ligneous, with mature leaves or without. The bases of cuttings were soaked in AIB solution (0, 1, 2, and 4 $g \cdot l^{-1}$). Thirty explants were randomly used in each treatment. They were planted in a box (temperature and humidity controlled) filled with perlite.

In micropropagation trials, nodal segments (1-2 cm) from mature plants bearing one bud were used as primary explants. Surface sterilization was accomplished by dipping the explants in sodium hypochlorite (80 $g \cdot l^{-1}$) for 20 min, and then rinsing several times with sterile distilled water. The cultures were initiated in Murashige and Skoog salts basal medium, plus 3% sucrose and 0.7% agar. Growth regulators (BAP, BAP + ANA, BAP + GA3) were used with charcoal (2 $g \cdot l^{-1}$). In the rhizogenesis stage, vitroshoots were cultured in MS medium salts with ANA (10, 1, 0.5 and 0.1 $g \cdot l^{-1}$). All media were adjusted to pH 5.5 ± 0.2 and then autoclaved for 15 min at 120°C. The explants, at least 24 per experiment, were cultured in a chamber at a temperature of 26 ± 2°C under a photoperiod of 18 h with a light intensity of 2.0 klx.

RESULTS

An analysis of variance of the results showed significant differences in rooting due to the type of cuttings. Maximum rooting was obtained with semi-ligneous cuttings treated with 2 AIB (Figures 1 and 2). With this auxin, radial rhizogenesis is observed. The period February-March is the best for rhizogenesis. The buds also have an effect in this morphogenetic process, especially in the induction of the first root. Meanwhile, the cuttings with removed leaves had the best rate, opposite to other species which need the presence of mature leaf to induce roots (2).

In micropropagation trials, aseptic cultures were obtained. The best shoot development was observed in MS medium added to BAP and GA3 ($1 g \cdot l^{-1}$). In the presence of BAP with ANA, significant callogenesis was obtained. The best rate of rhizogenesis was observed with ANA (0.1 $g \cdot l^{-1}$) (Figure 3). The addition of charcoal ameliorates the development of shoots, roots and callus.

CONCLUSIONS

These results indicate that the success of conventional breeding and the *in vitro* micropropagation of *Laurus nobilis* L. depend on the type of cuttings, the date of sampling and culture conditions. The most effective explants are the semiligneous. The presence of auxins in the media significantly increases the rate of rhizogenesis and permits a radial development of the roots. The pres-

FIGURE 1. Effect of physiological age of cutting on rhizogenesis.

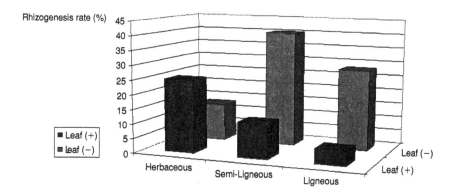

FIGURE 2. Effect of growth hormone (AIB) on the rate of cutting rhizogenesis.

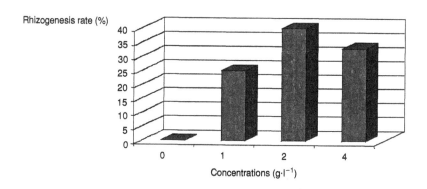

ence of mature leaves does not have any positive effect on this morphogenetic process. *In vitro*, the presence of charcoal is indispensable to induce a good development of shoots and roots, clamping the reaction of the tannin. In fact, the tannin constitutes a physiological inhibitor. Thus, this aromatic and medicinal plant has a considerable potential morphogenic capacity. This potential can be optimized by searching the performing factors in each stage of the breeding technique: by cuttings (the date of sampling, the age of the mother tree, the type of the auxin and the concentrations), by micropropagation (optimizing the different stages) and by using other *in vitro* techniques, like somatic embryogenesis (3).

FIGURE 3. Growth regulator (ANA) and charcoal effects on *in vitro* stem rhizogenesis.

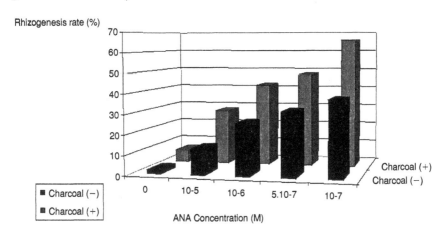

REFERENCES

1. Rivolier, C. (1985). Secrets et vertus des plantes médicinales. Sélection du Reader's Digest. p. 179.

2. Souayah, N., Ammar, S., and Bouzid, S. (1997). Clonage de génotypes adultes d'*Atriplex* (*Atriplex halimus* L.), par bouturage et microbouturage *in vitro*. Actes du 3ème séminaire du Réseau ATRIPLEX-STD3, Marrackech, Maroc. pp. 78-82.

3. Rodriguez, R., Rey, R., Cuozzo M., and Ancora, G. (1990). *In vitro* propagation of caper (*Capparis spinosa* L.). *In Vitro Cell. Dev. Biol.* 26: 531-536.

Breeding
for a *Hypericum perforatum* L. Variety
Both Productive
and *Colletotrichum gloeosporioides* (Penz.)
Tolerant

Myriam Gaudin
X. Simonnet
Nicole Debrunner
A. Ryser

SUMMARY. The acreage of St. John's wort *(Hypericum perforatum* L.), a drug-yielding plant used for its antidepressive properties, considerably increased in Europe over the last few years. In Switzerland, this acreage regularly suffers anthracnose, a disease caused by the *Colletotrichum gloeosporioides* (Penz.) fungus. Our tests were designed to compare 21 wild and 3 commercial varieties on 3 sites with distinct soil climates. This article emphasizes a high genotype variation for this species. We were able to select a genotype that is in agronomical terms more

Myriam Gaudin, X. Simonnet, and Nicole Debrunner are affiliated with Station Fédérale en Production Végétale de Changins, Centre d'Arboriculture et d'Horticulture des Fougères, CH-1964 Conthey, Switzerland.

Nicole Debrunner is currently affiliated with Valplantes, Case Postale 18, CH-1933 Sembrancher, Switzerland.

A. Ryser is affiliated with the Abt. Heilpflanzenanbau, Bioforce AG, Postfach 76, CH-9325 Roggwil, Switzerland.

The authors thank C. Vergères and Huguette Hausammann for technical collaboration.

[Haworth co-indexing entry note]: "Breeding for a *Hypericum perforatum* L. Variety Both Productive and *Colletotrichum gloeosporioides* (Penz.) Tolerant." Gaudin, Myriam et al. Co-published simultaneously in *Journal of Herbs, Spices & Medicinal Plants* (The Haworth Herbal Press, an imprint of The Haworth Press, Inc.) Vol. 9, No. 2/3, 2002, pp. 107-120; and: *Breeding Research on Aromatic and Medicinal Plants* (ed: Christopher B. Johnson, and Chlodwig Franz) The Haworth Herbal Press, an imprint of The Haworth Press, Inc., 2002, pp. 107-120. Single or multiple copies of this article are available for a fee from The Haworth Document Delivery Service [1-800-HAWORTH 9:00 a.m. - 5:00 p.m. (EST). E-mail address: getinfo@haworthpressinc.com].

107

satisfactory than the reference variety (Topas). It is dieback tolerant, high-yielding, easy to harvest and should subsequently prove more cost-effective. It blooms early and is thus particularly suitable for growth at high altitude. Finally, its flavonoid and hypericin contents are pharmaceutically promising. It has also been noted that anthracnose is not so virulent at high altitudes and the soil type has an influence on flower production but does not reduce their secondary metabolite contents. *[Article copies available for a fee from The Haworth Document Delivery Service: 1-800-HAWORTH. E-mail address: <getinfo@haworthpressinc.com> Website: <http://www.HaworthPress.com>* © *2002 by The Haworth Press, Inc. All rights reserved.]*

KEYWORDS. St. John's wort, anthracnose, breeding, tolerance

INTRODUCTION

St. John's wort (*Hypericum perforatum*), a member of the Guttiferae family (syn. Hypericaceae, Clusiaceae), has been used for its medical properties throughout the ages (5). It is currently recommended in plant therapy for its antiviral, vulnerary and antidepressive properties (10-13,19,20), and this ability to fight depression is naturally attracting the pharmaceutical industry's interest (11). *H. perforatum* based formulations are used for light or mild depression.

H. perforatum has been the subject of many plant chemistry studies (6,18). However, the flower extract molecules that help fight depression are still unknown. Flavonoids, naphtodianthrons (hypericin and pseudo-hypericin), phloroglucinols (hyperforin) and xanthones are concentrated in flowers and there could be many secondary metabolites to which this medical property might be attributed. This property was for a long time attributed to hypericin (2); that very often remains the analytical reference in the standardization of *H. perforatum* extracts. Recent research work also emphasizes the probable significance of hyperforin (4,8,15).

Antidepressants represent a huge market that provided the impetus for *H. perforatum* development. Although it was still limited a few years ago, the acreage now covers several dozen hectares in France, some hundred hectares in Italy and more than 400 hectares in Germany. A few selected varieties, e.g., Hyperimed, Elixir and Topas, already are commercially available. Topas, a Polish variety registered in 1982, probably is the most extensively available today (14).

In Switzerland, the *H. perforatum* fields were attacked by the *Colletotrichum gloeosporioides* fungus, causing anthracnose starting in 1995 (7). Most of the 20 hectares of *H. perforatum* fields planted in this country are currently managed with a biological specification. Anthracnose can destroy those perennial cultures from the first year, mainly in heavy soil and damp regions (1). Since the specification does not allow the use of fungicides, those cultures usually are irretrievably lost. Hollow lesions circling the stems are noted in infected plants (9,21) (Figure 1); these rapidly turn red as though they were burned, then wither and die (Figure 2).

Médiplant terms of reference. A cooperation protocol was set up in 1996 between Médiplant and Bioforce, a Swiss company in Roggwil/TG, to select a St. John's wort variety that is both productive and *C. gloeosporioides* tolerant. The main selection criterion for this new variety is a *pathogen fungus toler ance*. The *H. perforatum* plots often are in the mountains and an early bloom-ing genotype would thus be quite suitable for growing in altitude. To be quite cost effective, this variety must also be high-yielding in flowery tops and be easy to harvest. Finally, since the antidepressive molecules are not yet known, a plant with a high secondary metabolites content and a chemical profile simi-lar to that of Topas must be sought.

FIGURE 1. *Hypericum perforatum* stem necrosis. Lesions typical of anthracnose caused by *Colletotrichum gloeosporioides*.

FIGURE 2. *Hypericum perforatum* anthracnose symptoms in the second growth year.

MATERIALS AND METHODS

Seed origin. Twenty-four batches of seed of various origins were compared during a test undertaken in 1997-1998.

Seed origin		Code
3 commercial varieties	Topas	P1
	Hyperimed	P2
	Elixir	P3
21 wild varieties	from Switzerland, Germany, Italy, Australia and Canada	P4 to P24

Topas, the *H. perforatum* variety most extensively available on the market, served as a reference·for this test.

Culture sites. Three experimental plots were selected to acquire as much information as possible regarding the behavior of those 24 varieties when cultured.

		Plot 1 Fougères	Plot 2 Epines	Plot 3 Bruson
Altitude	(in m)	480	480	1060
Soil type	(17)	Silt	Sand	Sandy silt
pH		8	8	7

Experimental design. The experimental design was composed of Fisher blocks with 3 replications in each test station.

Basic plots		Fougères	Epines	Bruson
Number of plants		10	10	10
Surface	(in m^2)	3.2	3.2	2.4
Density	(plants/m^2)	3.1	3.1	4.2

Only 18 accessions out of 24 were cultured in Fougères for lack of available space. Ten plants from the other 6 accessions (P4, P6, P8, P10, P17 and P23) were grown outside the experimental design on the same site.

Culture schedule. The culture schedule is shown below.

Greenhouse seeds	Seed trays	Early March
Greenhouse rooting	Compressed root balls	Early April
Planting	Field	Mid May
Harvesting	Full bloom	June-September

The plants were harvested while they were in full bloom. The stems were cut with shears 15 cm above the inflorescence and folded over by 10 cm after harvest. The experimental surfaces were weeded manually and regularly irrigated as long as the cultures lasted, i.e., 2 years.

Observations and measurements. Plant development and the plots' sanitary conditions were monitored throughout the season. Once harvested, the flowery tops were dried at 35°C for approximately 10 days and weighed. The yields by weight were expressed per plant on a 10-plant per plot basis. Samples were collected and powdered to analyze secondary metabolites (1 analysis per accession and site). Ten flavonoids and two hypericins were quantified by HPLC in the Bioforce laboratory at Roggwil/TG. These measurements were done for every accession collected in 1997. They were repeated in 1998 for some interesting genotypes only.

RESULTS AND DISCUSSION

Out of 21 wild genotypes subjected to tests, P7 was the only one that met the requirements of the 5 initial selection criteria. The results were thus focused on the demonstration of the agronomical qualities of this variety compared to the three commercial ones, mainly Topas.

Anthracnose tolerance. Figure 3 reports the sanitary conditions of the *H. perforatum* plants tested after two years' culture. Anthracnose is quite virulent in the plains–94% of the plants growing on the Epines plot and 89% of those growing on the Fougères plot were dead or diseased after the second test year. The commercial varieties' rate of attack was 64 to 100%; it was only 17% (Fougères) and 50% (Epines) for genotype P7. An analysis of variance (ANOVA) followed by a Newman-Keuls test ($\alpha = 5\%$) indicated that genotype P7 was, on the Fougères plot, significantly more dieback tolerant than the *H. perforatum* varieties currently available on the market. On the Epines site it was statistically comparable to Topas, the reference variety, as well as genotype P17. These three plant types were more dieback resistant than the other *Hypericum* accessions grown in this experimental device.

Anthracnose did not seem so virulent on the Bruson site in high altitude. During the 1998 harvest, 49% of the plants were still healthy. Furthermore, 6 of the 24 genotypes were totally symptom free, the Topas and P7 varieties amongst them; 30% of the Hypermed plants and 17% of the Elixir ones were diseased.

Phenotypes. Highly different growing modes were noted between batches throughout the first year. The accessions differ as to how they hold themselves up, how their main stems are branched and how homogenous their flower horizon is. These are classified into five phenotype categories:

1. Erect plant, no base branching, compact flower horizon
2. Erect plant, little base branching, compact flower horizon
3. Irregular plant, heavy base branching, large flower horizon
4. Irregular plant, heavy base branching, vague flower horizon
5. Creeper

We did not observe any significant morphological variations between plants of a same variety. The specific mode of sexual reproduction of this spe-

FIGURE 3. Dieback level in the second year for 24 *Hypericum perforatum* accessions on three growth sites (notes made upon harvest in 1998).

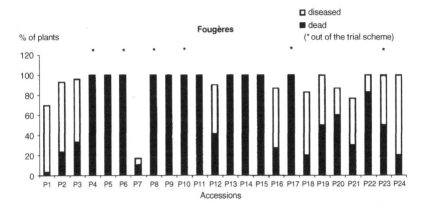

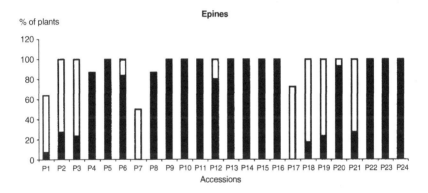

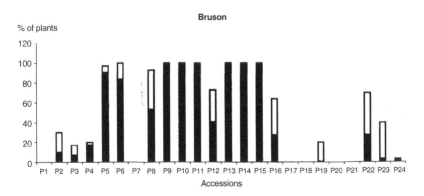

cies, apomixis, is probably responsible for this feature (3,16). P7 belongs to the first of the five phenotype categories. It is easy to harvest, thanks to its erect stand, highly homogenous flower horizon and flowers laid out in a same plane. Categories 3, 4 and 5 are more difficult to harvest because the corymbs are at different heights. The commercial varieties, Topas and Hyperimed, are in the third category. Elixir belongs to the fourth one. The morphological differences between plants were reduced during the second year. Every plant had an erect stand and a varying number of vertical stems. P7's advantage over Topas was a very compact flower horizon (Figure 4).

Blooming. In the first year, *H. perforatum* was harvested in the plains from July 8 to August 20. The next year, the plants bloomed one month earlier and were harvested from June 9 to July 13. In the mountains, the harvest was one month late and the difference between the first year (August 13-September 4) and the second one (July 6-August 4) remained. Early, intermediate or late genotypes were determined according to the harvest dates (Figure 5). The Topas variety probably had, amongst all those considered, the longest growing period before blooming. In the mountains, this late genotype, as 6 others, did not bloom early enough to be harvested during the first year; it was the last to be harvested in Bruson in 1998. P7, on the other hand, is in the early blooming category. In 1997 in the plains, it was blooming approximately one month be-

FIGURE 4. Selected genotype (P7) blooming during the second year. Its erect stand and homogenous flower horizon is to be noted.

FIGURE 5. Early blooming differences between genotypes P7 (on the left) and Topas (on the right).

fore Topas, the reference variety. In 1998, it was blooming two weeks before Topas. It can thus bloom at high altitude from the first year. Hyperimed also blooms early. Elixir blooms intermediate to late and could not be harvested in Bruson in 1997.

Yields by weight. Their dieback sensitivity excepted, the soil type, the plot's altitude and the morphology of the plants also have an influence on the flower yield of the *H. perforatum* accessions (Figure 6). The results demonstrated that flowery tops production is improved on a properly irrigated sand rather than silt soil. The plants developed three times more flowers the first year on the Epines site compared to the Fougères one. Epines' sandy soil is particularly suited for Topas and its cumulative yield over two years (176 g) could not be exceeded by the other accessions tested on this site (21 to 111 g). Stems without base branches and very early blooming proved detrimental to genotype P7 during the first year in the plain sites. Its bloom production was 1.5 to 3 times lower than that of the three commercial varieties. However, in the second year, this yield (31 g in Fougères and 80 g in Epines) was comparable to that of Topas, the best of the three commercial varieties (76 g in Fougères and 93 g in Epines). This early blooming gives P7 an incomparable advantage in high altitude. In Bruson, its cumulative yield over two years (113 g) exceeded that of Topas (63 g) and Elixir (76 g) that did not bloom until the second season. It

FIGURE 6. Flowery top yields for 24 *Hypericum perforatum* accessions on three sites over two years.

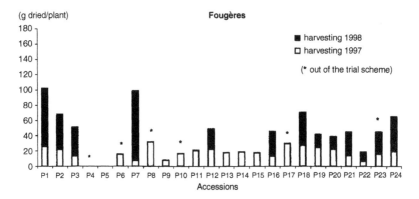

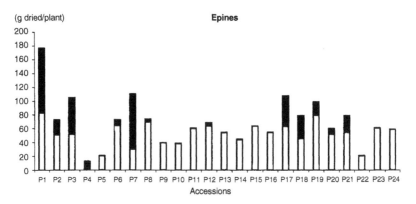

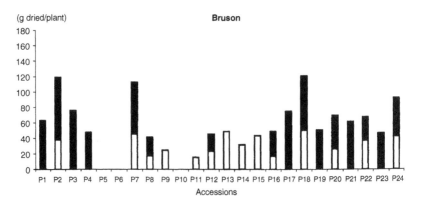

provided roughly the same quantity as Hyperimed (119 g). Even if the cumulative yields of P7 in the plains were not significantly higher than those of Topas, this genotype was still one of the most productive plants analyzed during those tests (ANOVA; Newman-Keuls test, $\alpha = 5\%$).

Dosed substances contents. The flavonoid and hypericin (hypericin and pseudo-hypericin) contents quantified in 1997 for the Epines plants are presented in Table 1. Amongst flavonoids are included rutin, hyperoside, isoquercitrin, quercitrin, quercetin and biapigenin; four of these remain unidentified. Globally, a very interesting variation was evidenced for the substances dosed as a whole, with, for example, extreme values ranging from 1 to 7.5 for hypericin. It must be noted that Topas apparently had the least flavonoids and hypericins. Genotype P7 appears promising for the preparation of pharmaceutical extracts; it includes the same range of measured substances as Topas and also contains 26% more flavonoids and 79% more hypericins compared to the latter.

The chemical profile defining the quality of the *H. perforatum* flowery tops is dependent upon the plant development stage upon harvest (14) but it does not seem to be influenced by the soil type, the altitude or the culture's age (Figure 7). This principal component analysis demonstrated, for each of the four accessions analyzed, a very low scatter of the chemical profiles whatever the site or the harvesting year may have been.

CONCLUSION

The high variation detected for the five selection criteria initially retained provided interesting vegetable material and allowed for satisfactory selection.

TABLE 1. Flavonoid and hypericin contents (in mg/100 g of dried flowery tops) for 24 *Hypericum perforatum* accessions harvested in full bloom during the first year (Epines site, 1997).

Accessions	rutin	hyperoside + isoquercitrin	flavonoid X	flavonoids Y(1)	quercitrin	flavonoid U	quercetin	biapigenin	pseudohypericin	hypericin	Total Sum of Flavonoids content	index	Total Sum of Hypericins content	index
P1 (Topas)	910	1343	104	39	97	0	119	82	125	34	2693	*100*	159	*100*
P2 (Hyperimed)	411	1209	380	408	227	0	169	322	236	108	3125	*116*	343	*216*
P3 (Elixir)	500	1559	482	408	189	56	200	256	128	58	3649	*135*	185	*116*
P7	1024	1016	453	176	411	0	93	213	204	80	3384	*126*	284	*179*
Mean	711	1282	355	258	231	14	145	218	173	70	3213	-	243	-

(1) 2 peaks

FIGURE 7. Principal component analysis of the chemical composition of the dried flowery tops of four *Hypericum perforatum* varieties grown on three sites over two years.

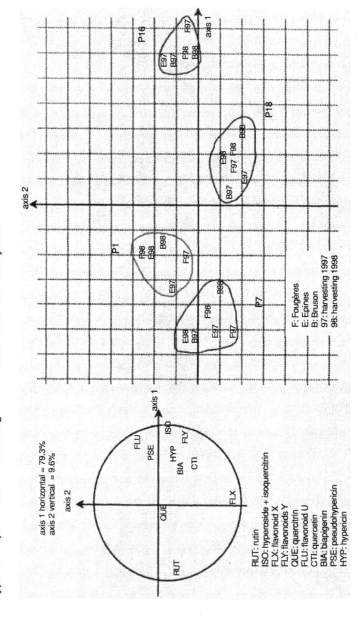

A *H. perforatum* variety tolerating the severe anthracnose problem and well suited for mountain growth was selected. The flowery top yield of this new genotype is competitive compared to that of the *H. perforatum* varieties currently available on the market. In addition, its secondary metabolite chemical profile meets industrial requirements (Table 2). This variety is now being registered. Tests are currently in progress to optimize its growth.

TABLE 2. Comparing three commercial *Hypericum perforatum* varieties with the genotype selected.

Selection criteria		Topas	Hyperimed	Elixir	P7
1. anthracnose tolerance		average	low	low	**good**
2. blooming	1st year	late	intermediate	late	**early**
	2nd year	late	intermediate	intermediate	intermediate
3. flower horizon		large	large	vague	**compact**
4. flower yield		**good to very good**	**good**	average	**good to very good**
5. contents	flavonoids	low	low	**high**	**good**
	hypericins	low	**high**	low	**good**
Remarks			No longer sold	Heterogenous, does not appear to be a true variety	

Boldface indicates advantageous traits.

REFERENCES

1. Bomme U. 1997. Produktionstechnologie von Johanniskraut (*Hypericum perforatum* L.) (German) (Production technology of St. John's-Wort (*Hypericum perforatum* L.)). *Z. Arzn. Gew.pfl.* 2:127-134.

2. Bruneton J. 1993. *Pharmacognosie–Phytochimie des plantes médicinales* (French) (Pharmacognosy–Phytochemistry of Medicinal Herbs). Lavoisier TecDoc, Londre, Paris et New York. 915 pp.

3. Cellarova E., R. Brutovska, Z. Daxnerova, K. Brunakova, and R.C. Weigel. 1997. Correlation between hypericin content and the ploidy of somaclones of *Hypericum perforatum* L. *Acta Biotechnol.* 1:83-90.

4. Chatterjee S.S., M. Nöldner, E. Koch, and C. Erdelmeier. 1998. Antidepressant activity of *Hypericum perforatum* and hyperforin: the neglected possibility. *Pharmacopsychiat.* 31(special issue):7-15.

5. Czygan F. 1993. Kulturgeschichte und Mystik des Johanniskrautes (German) (History and Mysticism of St. John's-Wort). *Zeitsch. Phytoth.* 14:272-278.

6. Debrunner N., and X. Simonnet. 1998. Le millepertuis: une plante médicinale antidépressive d'un très grand intérêt (French) (St.-John's-wort: an antidepressive medicinal plant of interest). *Revue Suisse Vitic. Arboric. Horic.* 30(4):271-273.

7. Debrunner N., A.L. Rauber, A. Schwarz, and V. Michel. 2000. First report of St. John's-Wort anthracnose caused by *Colletotrichum gloeosporioides* in Switzerland. *Plant Dis.* 84(2):203.

8. Erdelmeier C.A.J. 1998. Hyperforin, possibly the major non-nitrogenous secondary metabolite of *Hypericum perforatum* L. *Pharmacopsychiat.* 31(special issue): 1-6.

9. Hildebrand P.D., and K.I.N. Jensen. 1991. Potential for the biological control of St. John's-Wort (*Hypericum perforatum*) with endemic strain of *Colletotrichum gloeosporioides. Can. J. Plant Path.* 13:60-70.

10. Hobbs C. 1989. St. John's Wort (*Hypericum perforatum* L.)–A review. *Herbal Gram* 18/19:24-33.

11. Hölzl J. 1993. Inhaltstoffe und Wirkmechanismen des Johanniskrautes (German) (Content and Efficiency of St. John's-Wort). *Zeit. Phytother.* 14(5):255-264.

12. Hostettmann K. 1997. *Tout savoir sur le pouvoir des plantes sources de médicaments* (French) (All about the Power of Plants Sources of Medicines). Favre, Lausanne. 239 pp.

13. Hostettmann K., O. Potterat, and J.L. Wolfender 1994. Millepertuis et gentianes: nouveaux antidépresseurs? (French) (St.-John's-Wort and Gentians: New Antidepressants?). *In Ressources et potentiels de la flore médicinale des Alpes.* actes du troisième colloque Médiplant, 20 Octobre, 1994, Bruson, Suisse. pp. 51-58.

14. Kartnig T., B. Heydel, L. Lässer, and N. Debrunner. 1997. Johanniskraut aus schweizer Arzneipflanzenkultur (German) (St.-John's-Wort cultivated as a medicinal herb in Switzerland). *AgraForsch.* 4(7):299-302.

15. Laakmann G., C. Schüle, T. Baghai, and M. Kieser. 1998. St. John's-Wort in mild to moderate depression: the relevance of hyperforin for the clinical efficacy. *Pharmacopsychiat.* 31(special issue):54-59.

16. Martonfi P., R. Brutovska, and E. Cellarova. 1996. Apomixis and hybridity in *Hypericum perforatum. Folia Geobot. Phytotax.* 31(3):37-44.

17. Rod P. 1980. Granulométrie–Nouvelles limites, nouvelles définitions (French) (Granulation–New limits, new definitions). *Revue suisse Vitic. Arbor. Hortic.* 12(2): 1-8.

18. Roth L. 1990. *Hypericum–Hypericin* (German). Ecomed, Landsberg. 158 pp.

19. Schaffner W., B. Häfelfinger, and B. Ernst. 1992. *Compendium de phytothérapie* (French) (Compendium of Phytotherapy). Arboris-Verlag, Hinterkappelen, Bern. 336 pp.

20. Schauenberg P., and F. Paris. 1977. Delachaux et Niestlé, eds. *Guide des plantes médicinales* (French) (Medicinal and Aromatic Plants Guide). Neuchâtel. 386 pp.

21. Schwarczinger I., and L. Vajna. 1998. First report of St. John's-Wort anthracnose caused by *Colletotrichum gloeosporioides* in Hungary. *Plant Dis.* 82:711.

Evaluation of Spontaneous Oleander (*Nerium oleander* L.) as a Medicinal Plant

Maurizio Mulas
Barbara Perinu
Ana Helena Dias Francesconi

SUMMARY. *Nerium oleander* L. (Apocynaceae) is a spontaneous plant widely diffused throughout the Mediterranean region. In Sardinia, it grows mainly near water flows, preferring humid and sunny environments. Oleander is used mainly as an ornamental. Since ancient times, both the toxicity and the medicinal properties of this species were well known. In fact, all plant parts are toxic if ingested and they can cause death by heart paralysis. This paper consists of a review on the medicinal characteristics of *Nerium oleander* and on its possible applications. In addition, a survey conducted on oleander types growing in Sardinia, aiming at evaluating their aptitude for medicinal uses, is discussed. *[Article copies available for a fee from The Haworth Document Delivery Service: 1-800-HAWORTH. E-mail address: <getinfo@haworthpressinc.com> Website: <http://www.HaworthPress. com> © 2002 by The Haworth Press, Inc. All rights reserved.]*

Maurizio Mulas, Barbara Perinu, and Ana Helena Dias Francesconi are affiliated with the Dipartimento di Economia e Sistemi Arborei, Università di Sassari, Via E. De Nicola, 07100 Sassari, Italy (E-mail: mmulas@ssmain.uniss.it).

The authors equally contributed to this study and are grateful to M.O.C.-Arasolè (Sardinia, Italy) for financial support.

[Haworth co-indexing entry note]: "Evaluation of Spontaneous Oleander (*Nerium oleander* L.) as a Medicinal Plant." Mulas, Maurizio, Barbara Perinu, and Ana Helena Dias Francesconi. Co-published simultaneously in *Journal of Herbs, Spices & Medicinal Plants* (The Haworth Herbal Press, an imprint of The Haworth Press, Inc.) Vol. 9, No. 2/3, 2002, pp. 121-125; and: *Breeding Research on Aromatic and Medicinal Plants* (ed: Christopher B. Johnson, and Chlodwig Franz) The Haworth Herbal Press, an imprint of The Haworth Press, Inc., 2002, pp. 121-125. Single or multiple copies of this article are available for a fee from The Haworth Document Delivery Service [1-800-HAWORTH 9:00 a.m. - 5:00 p.m. (EST). E-mail address: getinfo@haworthpressinc.com].

KEYWORDS. Oleander, germplasm, cultivation, medicinal use, biomass yield

INTRODUCTION

Oleander (*Nerium oleander* L., Apocynaceae) is an evergreen shrub which grows spontaneously throughout the Mediterranean region. In Sardinia, it grows mainly near water flows, preferring humid and sunny environments (6). All plant parts of oleander are toxic if ingested, because it contains extremely toxic glycosides and alkaloids which can cause death by heart paralysis (6). Even though this species is mainly used as an ornamental, many studies have shown its multiple medicinal properties, especially in veterinary and human pharmacology.

The most well-known effects of oleander are due to two glycosides, neriin and nerianthin, and an alkaloid, oleandrin, which have a cardiostimulatory action (14), and to the glycosides gentiobiosyl-oleandrin, gentiobiosyl-nerigoside and gentiobiosyl-beaumontoside extracted from the leaves (1). Oleander is also diuretic and lenitive on dermatosis and contusion (19). In addition, its lymph is rich of minerals (11) and α-tocopherol, an important antioxidant (13).

Using modern analysis techniques, more than 40 compounds have been isolated from oleander roots and leaves, such as glycosides and steroids that are antibacterial against many species harmful to humans (2,8,10). Moreover, some substances found in its leaves, such as neridiginoside, nerizoside, neritaloside and odoroside-H, are analgesic and sedative on animals (17). Extracts obtained from its leaves also have an anti-tumoral action (3,5).

Oleander preparations have herbicidal (18) and insecticidal action (7,12). Its lymph also has nematicidal action against *Meloidogyne* (20). Leaf extracts inhibit mycelium growth and spore germination of many seed-borne fungi of sorghum (15), affect mycoplams on rice (16), and act as a slug repellent on beans (9).

In spite of the numerous medicinal properties of oleander, this species is cultivated only for ornamental purposes. The main objectives of our research were to individuate and characterize oleander selections, with emphasis on aspects related to the production of biomass containing active principles for medicinal purposes. The great diffusion of oleander in Sardinia allowed the identification of a large number of ecotypes, which were observed during bloom and submitted to morphological and biometrical analyses of shoot samples.

MATERIALS AND METHODS

Seventeen oleander ecotypes were identified in Sardinia in 1999. All plants were identified during bloom, being differentiated mainly on the basis of flower characteristics. *In situ* observations on the morphological and vegetative (size, habitus and vigour) characteristics of each plant were taken. Successively, several biometric analyses of flowering shoots were performed, to evaluate the aptitude of each ecotype for biomass production for medicinal uses. The following parameters were evaluated: shoot length, percentage of leaves and woody parts of shoots (on a fresh and dry weight basis), leaf number per shoot, leaf length and width, leaf fresh and dry weight, and percentage of dry matter of leaves. Biometric data were submitted to analysis of variance using MSTAT-C software and mean separation was done using Duncan's Multiple Range Test.

RESULTS AND DISCUSSION

Oleander ecotypes differed considerably regarding tree height, which varied from 1.5 to 6.0 m, and width, varying between 1.4 and 5.0 m.

Flowering-shoot length varied from 21.8 to 38.4 cm. The percentage of leaves or woody parts in relation to total shoot fresh weight varied between 39.4% and 61.9%, and between 11.3% and 20.9%, respectively. The percentage of leaves or woody parts in relation to total shoot dry weight varied from 46.3% to 68.8%, and from 13.6% to 22.4%, respectively.

The most important biometric characters of flowering shoots for the medicinal use of oleander are those regarding the leaves, from which most active principles are extracted. Therefore, it is important to evaluate the following characters during clone selection: leaf number per shoot, and leaf dimension, weight and percentage of dry weight. Those parameter values ranged as follows: from 10.6 to 27.6 (leaf number per shoot), from 8.9 to 13.3 cm (leaf length), from 1.4 to 2.2 cm (leaf width), from 0.35 to 0.82 g (leaf fresh weight), from 46.3% to 68.8% (leaves in relation to total shoot dry weight), and from 34.2% to 44.1% (leaf dry matter) (Table 1).

Among the studied oleander ecotypes, a high variability regarding the parameters considered important for medicinal uses was found. Considering that the leaves are the main product of this plant, the optimal plant should have a quite compact vegetative habitus, with predominance of leaves over woody parts, large leaves and a high content of dry matter in the leaves. Because of the large variability available in the oleander germplasm for those parameters, and considering the easy agamic propagation, obtaining selections suitable for pharmaceutical uses will be possible.

TABLE 1. Leaf characteristics of seventeen oleander ecotypes.[z]

Ecotype	N. of leaves per shoot	Leaf length (cm)	Leaf width (cm)	Leaf fresh weight (g)	Leaves per shoot (% d. w.)	Leaf dry matter (%)
ALG 1	18.7 bc	13.2 a	1.99 ab	0.72 a-d	50.9 cde	37.9 def
ALG 2	19.7 bc	11.7 ab	1.85 a-d	0.61 a-d	57.8 bcd	39.6 cde
ALG 3	18.6 bc	12.9 a	1.76 bcd	0.63 a-d	59.6 abc	39.9 bcd
ALG 4	17.9 bcd	11.6 ab	1.66 b-e	0.68 a-d	52.1 b-e	38.4 def
MA 1	10.6 d	12.9 a	2.13 a	0.82 a	61.2 ab	35.6 fg
OLM 1	19.7 bc	11.7 ab	1.67 b-e	0.50 cde	57.5 bcd	36.2 efg
OLM 2	18.0 bcd	13.3 a	1.72 b-e	0.69 a-d	50.3 cde	39.7 cde
OLM 3	16.8 bcd	12.3 ab	1.93 abc	0.72 a-d	46.6 e	35.7 fg
OLM 4	27.6 a	12.1 ab	1.69 b-e	0.73 abc	68.8 a	34.4 g
OLM 5	19.8 bc	11.9 ab	1.69 b-e	0.67 a-d	61.5 ab	34.2 g
OLM 6	19.0 bc	12.0 ab	1.75 bcd	0.71 a-d	58.3 bcd	36.4 d-g
CR 1	11.9 cd	11.7 ab	1.51 de	0.50 cde	53.2 b-e	36.6 d-g
ALG 5	22.7 ab	13.3 a	1.53 de	0.74 abc	53.9 b-e	39.9 bcd
ALG 6	18.3 bcd	12.5 a	2.17 a	0.79 ab	50.9 cde	37.1 d-g
CAP 1	14.1 cd	8.87 c	1.36 e	0.35 e	50.2 cde	43.2 ab
CAP 2	17.8 bcd	11.3 ab	1.66 b-e	0.55 b-e	46.3 e	42.5 abc
MUR 1	15.2 bcd	10.0 bc	1.59 cde	0.47 de	49.4 de	44.1 a

[z] Data on the same column having different letters are statistically different at P ≤ 0.01 by Duncan's Multiple Range Test.

REFERENCES

1. Abe, F. and T. Yamauchi. 1992. Cardenolide triosides of oleander leaves. *Phytochemistry* 31 (7): 2459-2463.

2. Ahmed, B.A., K.D. Sulayman, A.A. Aziz, A.A. Abd, and J.J. Rashan. 1993. Antibacterial activity of the leaves of *Nerium oleander. Fitoterapia* 64 (3): 273-274.

3. Alkofahi, A.S., A. Abdelaziz, I. Mahmoud, M. Abuirjie, A. Hunaiti, and A. El-Oqla. 1990. Cytotoxicity, mutagenicity and antimicrobial activity of forty Jordanian medicinal plants. *International Journal of Crude Drug Research* 28 (2): 139-144.

4. Zia, A., B.S. Siddiqui, S. Begum, S. Siddiqui, and A. Suria. 1995. Studies on the constituents of the leaves of *Nerium oleander* on behavior pattern in mice. *Journal of Ethnopharmacology* 49 (1): 33-39.

5. Basoglu, A., M. Sevinc, M. Ortatath, F. Birdane, and I. Camkerten. 1998. Effect of an aqueous extract of *Nerium oleander* plant in cattle with cutaneous fibropapilloma. *Indian Veterinary Journal* 75 (10): 915-917.

6. Camarda, I. and F. Valsecchi. 1983. *Nerium L. In* I. Camarda and F. Valsecchi, eds. *Alberi e arbusti spontanei della Sardegna* (in Italian). (Spontaneous trees and shrub of Sardinia). Ed. Giovanni Gallizzi, Sassari, Italy: 423-426.

7. El-Shazly, M.M., M.I. Nassar, and H.A. El-Sherief. 1996. Toxic effect of ethanolic extract of *Nerium oleander* (Apocynaceae) leaves against different developmental stages of *Muscina stabulans* (Diptera-Muscidae). *Journal of the Egyptian Society of Parasitology* 26 (2): 461-473.

8. Hanada, R., F. Abe, and T. Yamauchi. 1992. Steroid glycosides from the roots of *Nerium odorum. Phytochemistry* 31(9): 3183-3187.

9. Howard, A.S., K.L. Andrews, R. Caballero, and T. Madrid. 1991. Use of botanical extracts to prevent damage by the slug *Sarasinula plebeia* (Fisher) on common bean\ *Phaseolus vulgaris. CEIBA* 32 (2): 187-200.

10. Huq, M.M., A. Jabbar, M.A. Rashid, and C.M. Hasan. 1999. A novel antibacterial and cardiac steroid from the roots of *Nerium oleander. Fitoterapia* 70 (1): 5-9.

11. Jayabalan, M., K. Rjarathinam, G.D.P.S. Augustus, T. Sekar, and S. Veerasamy. 1995. Analysis of minerals by EDAX in the latex of *Apocynaceae. Journal of Ecotoxicology and Environmental Monitoring* 5 (1): 45-49.

12. Khan, S.M. and M.N. Siddiqui. 1994. Potential of some indigenous plants as pesticides against the larvae of cabbage butterfly *Pieris brassicae* L. *Sarhad Journal of Agriculture* 10 (3): 291-301.

13. Mallet, J.F., C. Cerrati, E. Ucciani, J. Gamisans, and M. Gruber. 1994. Antioxidant activity of plant leaves in relation to their α-tocopherol content. *Food Chemistry* 49 (1): 61-65.

14. Marchioni, A.R. and F. Caliò Distefano. 1989. *Nerium oleander* L. *Le piante medicinali della Sardegna–Guida pratica per il riconoscimento di 102 specie* (in Italian). (Medicinal plants of Sardinia–Practical guide-book for 102 species recognition). Ed.della Torre: 156-157.

15. Meena, S.S. and V. Mariappan. 1993. Effect of plant products on seed borne mycoflora of sorghum. *Madras Agricultural Journal* 80 (7): 383-387.

16. Rao, G.N. and P. Narayanasamy. 1990. Effect of plant extracts and oils on rice yellow dwarf infection. *Madras Agricultural Journal* 77 (5-6): 197-201.

17. Begum, S., B.S. Siddiqui, R. Sultana, A. Zia, and A. Suria. 1999. Bio-active cardenolides from the leaves of *Nerium oleander. Phytochemistry* 50 (3): 435-438.

18. Uygur, F.N. and S.N. Iskenderoglu. 1997. Allelopathic and bioherbicide effects of plant extracts on germination of some weed species. *Turkish Journal of Agriculture and Forestry* 21 (2): 177-180.

19. Valnet, J. 1976. Oleandro. *Fitoterapia–cura delle malattie con le piante* (in Italian). (Oleander. Phytotherapy–diseases cure with plants). Aldo Martello-Giunti, Firenze, Italy: 332-333.

20. Zureen, S. and M.I. Khan. 1985. Nematicidal activity in some plant latices. *Horticultural Abstract* 55: 3806.

Myrtle (*Myrtus communis* L.)
as a New Aromatic Crop:
Cultivar Selection

Maurizio Mulas
Ana Helena Dias Francesconi
Barbara Perinu

SUMMARY. *Myrtus communis* L. (Myrtaceae) is a typical shrub of the Mediterranean maquis which grows spontaneously in Sardinia. It is used in the drug, perfume and food industries. Intensive myrtle cultivation systems should be developed, in order to assure both a constant supply of good quality material for the liqueur industry and the preservation of natural myrtle populations. Cultivar selection is essential for the successful cultivation of this new aromatic crop and has been the main goal of our research. The main morphological and phenological characters of 16 cultivars selected for fruit and biomass production are presented here. *[Article copies available for a fee from The Haworth Document Delivery Service: 1-800-HAWORTH. E-mail address: <getinfo@haworthpressinc.com> Website: <http://www.HaworthPress.com> © 2002 by The Haworth Press, Inc. All rights reserved.]*

Maurizio Mulas, Ana Helena Dias Francesconi, and Barbara Perinu are affiliated with the Dipartimento di Economia e Sistemi Arborei, Università di Sassari, Via E. De Nicola, 07100 Sassari, Italy (E-mail: mmulas@ssmain.uniss.it).

The authors equally contributed to this study and are grateful to the Ministry of Agriculture Policies of Italy, Special Grant I.P.P.O.

[Haworth co-indexing entry note]: "Myrtle (*Myrtus communis* L.) as a New Aromatic Crop: Cultivar Selection." Mulas, Maurizio, Ana Helena Dias Francesconi, and Barbara Perinu. Co-published simultaneously in *Journal of Herbs, Spices & Medicinal Plants* (The Haworth Herbal Press, an imprint of The Haworth Press, Inc.) Vol. 9, No. 2/3, 2002, pp. 127-131; and: *Breeding Research on Aromatic and Medicinal Plants* (ed: Christopher B. Johnson, and Chlodwig Franz) The Haworth Herbal Press, an imprint of The Haworth Press, Inc., 2002, pp. 127-131. Single or multiple copies of this article are available for a fee from The Haworth Document Delivery Service [1-800-HAWORTH 9:00 a.m. - 5:00 p.m. (EST). E-mail address: getinfo@haworthpressinc.com].

KEYWORDS. Myrtle, aromatic crop, germplasm, cultivar, selection

INTRODUCTION

Both the demand of the liqueur industry for a constant raw-material supply (berries and shoots) and the need to protect spontaneous myrtle (*Myrtus communis* L.) plants against indiscriminate harvest (2,3) led to the promotion of the cultivation of this species. As a consequence, research has been conducted on cultural techniques suitable for its cultivation and on the selection of productive cultivars well-adapted to cultivation conditions (1,2).

A research program, which has been developed for the domestication of myrtle, has consisted of harvesting, studying, propagating and characterizing the spontaneous germplasm of the Mediterranean maquis (1). During the initial phases of the program, mass selection, accession evaluation and cultivar characterization were performed. In a subsequent phase, cultivars were analyzed in detail. In this paper, a synthetic description of morphological, biometric and phenological aspects of some selected cultivars is presented.

MATERIALS AND METHODS

In 1995 and 1996, more than 70 ecotypes were selected from the spontaneous myrtle germplasm of Sardinia. Mother plants were identified and ecological, environmental and plant characteristics were recorded *in situ*. Successively, branch samples of each ecotype were brought to a laboratory for biometric and qualitative analysis. From 1996 to 1999, cuttings of selected ecotypes were planted in a repository located at the Experimental Station of the University of Sassari, in Oristano (39° 55″ lat. North), central-western Sardinia. Planting distance was 1 m × 3 m. Plants were irrigated by a drop system in summer.

Cultivars were classified into three groups related to their use: fruit (for red myrtle liqueur production), vegetative biomass (for white myrtle liqueur production) and fruit and biomass (double purpose) producing cultivars. The following biometric and morphological parameters were determined on three replicates of ten spring shoots per cultivar: fruit size, peduncle length, pulp/seed ratio, leaf size, fruit shape, and fruit peel colour.

Biometric data were submitted to analysis of variance using MSTAT-C software and mean separation was done using Duncan's Multiple Range Test. Plant vigour and yield were also evaluated. Phenological observations of myrtle cultivars were performed with regard to sprouting, bloom and fruit ripening periods.

RESULTS AND DISCUSSION

The observed sixteen cultivars showed high variability for most studied characters (Table 1) as well as phenology (Figure 1). Fruit weight varied from 0.28 ('Piera') to 0.69 g ('Grazia'). Peduncle length ranged from 1.20 cm in 'Giovanna' to 2.64 cm in 'Angela'. Pulp/seed ratio ranged between 2.13 ('Erika') and 6.64 ('Angela'). Length/width ratio of leaves ranged from 1.98 ('Grazia' and 'Marta') to 3.37 ('Piera'), indicating variability in leaf shape: elliptical, ovate, and acute.

Fruit shape was round (e.g., 'Giovanna'), round-oval (e.g., 'Simona'), oval-elliptical ('Maria Elisa'), oval-pyriform (e.g., 'Ilaria'), or pyriform ('Maria Rita'); fruit peel colour was generally dark blue, except for 'Angela' (white-green), 'Giovanna' (light purple), and 'Grazia' (white-yellow).

TABLE 1. Some fruit and leaf characters of 16 myrtle cultivars selected in Sardinia.[z]

Cultivar	Fruit weight (g)	Pulp/seed ratio	Leaf l/w ratio	Peduncle length (cm)	Fruit shape	Fruit peel color
'Nadia'	0.44 efg	4.30 bc	2.42 bcd	1.96 b	round-oval	dark blue
'Marta'	0.43 fg	5.03 b	1.98 f	1.64 cde	oval	dark blue
'Maria Rita'	0.46 def	3.32 de	2.39 b-e	1.36 fg	pyriform	dark blue
'Barbara'	0.37 gh	3.20 def	2.36 cde	1.36 fg	oval	dark blue
'Daniela'	0.66 a	2.54 fg	2.04 f	1.71 bcd	round	dark blue
'M. Antonietta'	0.58 b	3.81 cd	2.55 bc	1.46 ef	round	dark blue
'M. Elisa'	0.33 hi	3.02 def	2.00 f	1.40 efg	oval-elliptical	dark blue
'Angela'	0.52 bcd	6.64 a	2.44 bcd	2.64 a	round	white-green
'Simona'	0.42 fg	2.51 fg	2.20 def	1.25 fg	round-oval	dark blue
'Giovanna'	0.57 bc	2.74 efg	2.07 f	1.20 g	round	light purple
'Ilaria'	0.44 efg	4.61 b	2.39 b-e	1.88 bc	oval-pyriform	dark blue
'Grazia'	0.69 a	3.72 cd	1.98 f	1.90 b	oval	white-yellow
'Erika'	0.51 cde	2.13 g	2.63 b	1.61 de	oval	dark blue
'Piera'	0.28 i	2.49 fg	3.37 a	1.79 bcd	oval-pyriform	dark blue
'Carla'	0.43 efg	2.20 g	2.11 f	1.85 bcd	round	dark blue
'Ana'	0.37 gh	3.57 cd	2.16 ef	1.86 bcd	round-oval	dark blue

[z] Data on the same column having different letters are statistically different at $P \leq 0.01$ by Duncan's Multiple Range Test.

FIGURE 1. Phenology of 16 myrtle cultivars grown in Oristano, Sardinia (1999/2000).

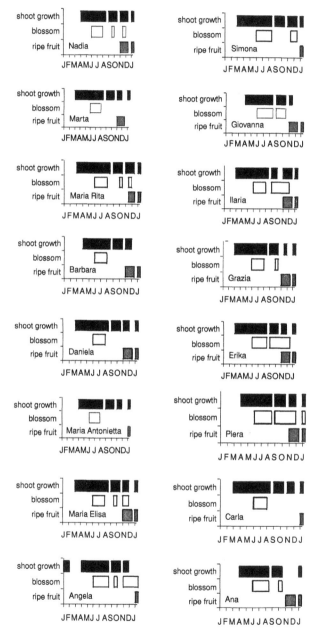

Fruit yield and plant vigour ranged from low to high, depending on the cultivar. Low fruit yield was observed in 'Piera', while 'Barbara', 'Maria Rita', 'Angela', 'Giovanna', and 'Grazia' were characterized by high fruit yield. Fruit producing cultivars should have preferably larger fruit and longer peduncle to ease harvest, and higher pulp/seed ratio to release lower tannin content in liqueur. Low vegetative vigour was observed in 'Piera', while 'Angela', 'Giovanna', and 'Grazia' showed high vigour and may be considered as promising cultivars for biomass production.

As a synthesis of our first evaluation of cultivars, 'Maria Rita' and 'Barbara' show high yields and could be used especially for fruit production. 'Daniela', 'Maria Antonietta', 'Giovanna' and 'Ilaria' are quite vigorous and have medium/high or high fruit yield, being suitable for fruit and biomass production (double purpose). 'Grazia' and 'Angela' are vigorous and show high production of non-pigmented fruit, being suited for biomass production and, in the case of 'Grazia', possibly for liqueur production from non-pigmented fruit, if that practice becomes reality.

Regarding the phenology of the selected cultivars (Figure 1), spring vegetative flush is the most important one, even though almost all cultivars show secondary vegetative flushes, more or less important, in summer and autumn. Autumnal reflowering occurs only in those cultivars that show intense secondary vegetative flushes and scarse fruit-set during the main vegetative flush. This character, which is particularly accentuated in cultivars 'Angela', 'Ilaria', 'Erika' and 'Piera', is considered detrimental for fruit production for the liqueur industry due to the non-uniform maturation of fruit originated in the various bloom periods. On the other hand, reflowering is basically absent in 'Barbara' and 'Maria Rita', which show a quite early fruit-ripening period.

REFERENCES

1. Mulas, M. and M.R. Cani. 1999. Germplasm evaluation of spontaneous myrtle (*Myrtus communis* L.) for cultivar selection and crop development. *Journal of Herbs, Spices & Medicinal Plants* 6(3): 31-49.

2. Mulas, M., M.R. Cani, and N. Brigaglia. 1998a. Characters useful to cultivation in spontaneous populations of *Myrtus communis* L. *Acta Horticulturae* 457: 271-278.

3. Mulas, M., M.R. Cani, N. Brigaglia, and P. Deidda. 1998b. Study of myrtle (*Myrtus communis* L.) genetic resources to promote extensive crop as integration of spontaneous harvests. *Acta Horticulturae* 502: 85-88.

Selection of Rosemary
(*Rosmarinus officinalis* L.) Cultivars
to Optimize Biomass Yield

Maurizio Mulas
Ana Helena Dias Francesconi
Barbara Perinu
Erika Del Vais

SUMMARY. *Rosmarinus officinalis* L. (Labiatae) is a typical shrub of the Mediterranean maquis. In spite of its multiple ornamental and aromatic uses, and the great interest in its cultivation, only a few cultivars or clones have been well characterized. Our breeding program has focused on optimizing biomass yield and improving the quality of cultivated rosemary. In a preliminary phase (1996-1997), 31 mother plants were identified and described *in situ* throughout Sardinia; spring-shoot samples were then evaluated in laboratory. In a second phase (1996-2000), cuttings of selected clones were planted in a repository at the Experimental Station of the University of Sassari, in central-western Sardinia. In this paper, morphological, biometric, qualitative and phenological characters of fifteen rosemary selections of the repository are reported. Selections tended to produce flowers, fruit and new shoots in various periods of the year. Selections showed a high variability for spring-shoot

Maurizio Mulas, Ana Helena Dias Francesconi, Barbara Perinu, and Erika Del Vais are affiliated with the Dipartimento di Economia e Sistemi Arborei, Università di Sassari, Via E. De Nicola, 07100 Sassari, Italy (E-mail: mmulas@ssmain.uniss.it).

The authors equally contributed to this study and are grateful to the Ministry of Agriculture Policies of Italy, Special Grant I.P.P.O.

[Haworth co-indexing entry note]: "Selection of Rosemary (*Rosmarinus officinalis* L.) Cultivars to Optimize Biomass Yield." Mulas, Maurizio et al. Co-published simultaneously in *Journal of Herbs, Spices & Medicinal Plants* (The Haworth Herbal Press, an imprint of The Haworth Press, Inc.) Vol. 9, No. 2/3, 2002, pp. 133-138; and: *Breeding Research on Aromatic and Medicinal Plants* (ed: Christopher B. Johnson, and Chlodwig Franz) The Haworth Herbal Press, an imprint of The Haworth Press, Inc., 2002, pp. 133-138. Single or multiple copies of this article are available for a fee from The Haworth Document Delivery Service [1-800-HAWORTH 9:00 a.m. - 5:00 p.m. (EST). E-mail address: getinfo@haworthpressinc.com].

133

length, leaf length and width, essential oil content of leaves, and percentage of leaf weight in relation to total shoot fresh weight. Clones with higher leaf/wood ratio and higher essential oil content were considered more suitable for the optimization of biomass yield. *[Article copies available for a fee from The Haworth Document Delivery Service: 1-800-HAWORTH. E-mail address: <getinfo@haworthpressinc.com> Website: <http://www. HaworthPress.com> © 2002 by The Haworth Press, Inc. All rights reserved.]*

KEYWORDS. Rosemary, germplasm, selection, biomass, essential oil

INTRODUCTION

Rosmarinus officinalis L. (Labiatae), a typical shrub of the Mediterranean maquis, is widely distributed in Sardinia (1). It has multiple aromatic and ornamental uses (5,6). However, in spite of the great interest in rosemary cultivation, only a few cultivars or clones have been well characterized so far.

The wide distribution and diversity of rosemary germplasm in Sardinia has provided a unique opportunity for our breeding program, which aims at characterizing its phenotypical variability and selecting cultivars for various potential uses (2,3,4). In the first phase of the program (1996-1997), 31 mother plants were identified and described *in situ* in thirteen areas of Sardinia. Shoot samples collected and brought to a laboratory for evaluation showed a high variability for many traits (2,3). In a second phase (1996-2000), cuttings of selected clones were planted in a repository at the Experimental Station of the University of Sassari in central-western Sardinia. Observations of those clones were taken in the field and shoot samples were analyzed in laboratory (4). In this paper, fifteen rosemary selections of our breeding program are characterized for some leaf, shoot and phenological traits, with emphasis on the optimization of biomass yield.

MATERIALS AND METHODS

The rosemary (*R. officinalis* L.) plants used in our study were obtained by cuttings from mother plants collected throughout Sardinia in a preliminary phase of the breeding program (1996-1997) (2,3) and planted in a repository located at the Experimental Station of the University of Sassari, located in Oristano, central-western Sardinia. In 1999, phenological observations of fifteen rosemary selections were taken periodically. In July 1999, spring-shoot samples were taken from each selection and brought to a laboratory for biometric, morphological, and qualitative analysis, and for photographic records.

Three replicates of ten spring-shoots per clone were evaluated for: shoot length and weight, leaf length and width, and percentage of leaf weight in relation to total shoot weight (Table 1). Preliminary qualitative analysis consisted of extracting essential oil by distillation, using a vapour stream from a sample of fresh leaves (ca. 300 g) of each ecotype.

Data were submitted to analysis of variance using MSTAT-C software. Mean separation was done using Duncan's Multiple Range Test at 1% significance level.

TABLE 1. Biometric characters of spring-shoots and leaves of rosemary clones (July 1999, Oristano, Sardinia).[z]

Selection	Shoot length (cm)	Shoot fresh weight (g)	Leaf/Total shoot weight (g/g)	Leaf length (cm)	Leaf width (cm)
CAG 5	34.40 ab	3.82 c-f	63.89 c-f	1.30 h	0.14 def
ORS 1	31.90 ab	2.63 fg	67.34 a-d	1.76 a	0.10 g
ORS 3	30.27 b	3.52 d-g	69.72 ab	1.62 b	0.13 efg
SAT 11	34.90 ab	5.33 a	70.17 a	1.47 def	0.18 a
SAT 13	31.47 ab	4.03 b-e	61.26 fg	1.38 g	0.15 bcd
SAT 14	36.07 a	4.95 abc	66.93 a-e	1.48 def	0.17 ab
VIG 10	30.80 ab	2.58 fg	62.61 d-g	1.52 cde	0.12 efg
VIG 11	23.57 c	3.03 efg	68.55 a-c	1.45 efg	0.13 efg
VIG 12	33.00 ab	5.11 ab	64.49 b-f	1.58 bc	0.16 bcd
VIG 13	34.10 ab	5.40 a	61.53 efg	1.43 fg	0.15 bcd
VIG 14A	33.97 ab	3.48 d-g	58.27 g	1.41 fg	0.11 fg
VIG 14B	33.17 ab	2.32 g	69.10 abc	1.20 i	0.14 cde
SAT 10	31.63 ab	4.51 a-d	60.85 fg	1.29 h	0.17 ab
CAG 1	29.83 b	3.34 d-g	59.55 fg	1.52 cde	0.16 abc
CAG 4	30.83 ab	3.31 d-g	62.40 d-g	1.54 cd	0.17 ab

z: Data in the same column having different letters are statistically different at P < 0.01 by Duncan's Multiple Range Test.

RESULTS AND DISCUSSION

The selected clones presented a high variability for many of the characters evaluated. Spring-shoot length varied from 23.6 to 36.1 cm. Spring-shoot fresh weight ranged from 2.6 to 5.4 g. Leaf length values varied from 1.20 to 1.76 cm, while leaf width varied from 0.10 to 0.18 cm (Table 1). A positive correlation was found between leaf width and shoot fresh weight ($r = 0.66$, $p \leq 0.01$), suggesting that the first character could be used to estimate leaf biomass. The percentage of leaf weight in relation to total shoot fresh weight varied from 58 to 70% (Table 1). This character is particularly important because essential oil content is much higher in leaves than in woody parts of the shoot. In addition, based on our preliminary analysis, essential oil in fresh leaves was also variable (from 0.33 to 1.17%) (Figure 1).

Under the Mediterranean environmental conditions of Sardinia, rosemary plants had bloom, fruiting and shoot growth in various periods of the year (Figure 2). This type of information helps in defining the cultural practices suitable for each clone, depending on its potential use. For instance, since the best moment to harvest leaves for oil extraction is before bloom, techniques that could induce fewer and shorter bloom periods could be helpful.

Finally, clones with higher leaf/wood ratio and higher essential oil content are more suitable for the optimization of biomass yield of rosemary.

FIGURE 1. Essential oil content of leaves of fifteen rosemary selections.

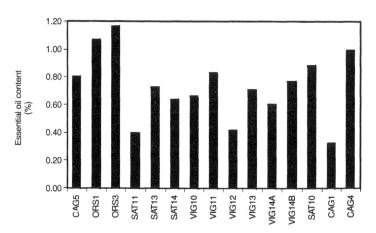

FIGURE 2. Phenology of fifteen rosemary selections, in Oristano, Sardinia (1999/2000).

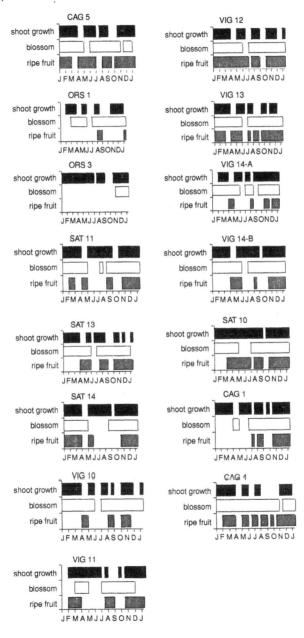

REFERENCES

1. Camarda, I. and F. Valsecchi. 1983. Labiatae. *In* I. Camarda and F. Valsecchi, eds. *Alberi e arbusti spontanei della Sardegna* (in Italian). (Spontaneous trees and shrubs of Sardinia). Edizioni Gallizzi, Sassari, Italy. pp. 427-430.

2. Mulas, M., N. Brigaglia, and M.R. Cani. 1998. Clone selection from spontaneous germplasm to improve *Rosmarinus officinalis* L. crop. *Acta Horticulturae* 457: 287-294.

3. Mulas, M. and P. Deidda. 1998. Domestication of woody plants from Mediterranean maquis to promote new crops for mountain lands. *Acta Horticulturae* 457:294-301.

4. Mulas, M., B. Perinu, and A.H.D. Francesconi. 2000. Selezioni provenienti dalle popolazioni spontanee per la coltivazione intensiva del rosmarino (*Rosmarinus officinalis* L.) (in Italian). (Selections from spontaneous populations of rosemary (*Rosmarinus officinalis* L.) for intensive cultivation). *Atti V Giornate Scientifiche S.O.I.*, Sirmione, 28-30 Marzo 2000. pp. 169-170.

5. Prakash, V. 1990. Leafy spices. CRC Press, Inc., Boca Raton, Florida (USA). 114 pp.

6. Tesi, R. 1994. Rosmarino (*Rosmarinus officinalis* L.). In: *Principi di orticoltura e ortaggi d'Italia* (in Italian) (Principles of horticulture and horticultural plants of Italy). Edagricole, Bologna, Italy. pp. 290-292.

BIOCHEMISTRY, BIOTECHNOLOGY, MOLECULAR GENETICS AND PHYSIOLOGY

Manipulation of Natural Product Accumulation in Plants Through Genetic Engineering

Barry V. Charlwood
Marcia Pletsch

SUMMARY. The efficient strategies now available through which the stable incorporation of artificial (chimeric) genes into a plant genome may be brought about provides the possibility of manipulating biosynthetic pathways in commercially important plants. This technology

Barry V. Charlwood is affiliated with the Division of Life Sciences, King's College London, 150 Stamford Street, London SE1 8WA, UK, and the Department of Chemistry, Universidade Federal de Alagoas, 57072-970 Maceió-AL, Brazil.

Marcia Pletsch is affiliated with the Department of Chemistry, Universidade Federal de Alagoas, 57072-970 Maceió-AL, Brazil.

[Haworth co-indexing entry note]: "Manipulation of Natural Product Accumulation in Plants Through Genetic Engineering." Charlwood, Barry V., and Marcia Pletsch. Co-published simultaneously in *Journal of Herbs, Spices & Medicinal Plants* (The Haworth Herbal Press, an imprint of The Haworth Press, Inc.) Vol. 9, No. 2/3, 2002, pp. 139-151; and: *Breeding Research on Aromatic and Medicinal Plants* (ed: Christopher B. Johnson, and Chlodwig Franz) The Haworth Herbal Press, an imprint of The Haworth Press, Inc., 2002, pp. 139-151. Single or multiple copies of this article are available for a fee from The Haworth Document Delivery Service [1-800-HAWORTH 9:00 a.m. - 5:00 p.m. (EST). E-mail address: getinfo@haworthpressinc. com].

can provide the plant breeder with a much wider range of traits for use in a classical breeding programme than is normally available within natural biodiversity. Whilst the main focus of this form of plant biotechnology has been centred on the genetic modification of crop plants, many opportunities exist for the improvement of medicinal and ornamental plants. Using genetic modification it has been shown to be possible to alter, amongst other attributes, flower colour, lignin levels, and the accumulation of biologically active principles. Examples of such studies are the accumulation of alizarin, shikonin, solasodine, artemisinin, scopolamine which are examined in this paper, along with a consideration of some of the possible future developments in this type of technology. *[Article copies available for a fee from The Haworth Document Delivery Service: 1-800-HAWORTH. E-mail address: <getinfo@haworthpressinc.com> Website: <http://www.HaworthPress.com> © 2002 by The Haworth Press, Inc. All rights reserved.]*

KEYWORDS. Genetic engineering, medicinal plants, alizarin, shikonin, solasodine, scopolamine, artemisinin

INTRODUCTION

Over the last two decades much research has been focused on the production of secondary metabolites by unorganized plant cells as a potential substitute for traditional cropping. Such studies have increased our understanding of the pathways leading to secondary products and of their regulation. However, limited success has been achieved regarding commercial application of cell cultures because of the very low, and often negligible, yields of target compounds obtained. As gene cloning and genetic transformation of plant cells has become feasible, and increasingly easier, so has the possibility of manipulating metabolic flux through pathways in order to enhance the production of those metabolites which are of commercial value. However, the question that must be addressed is whether plant cells will actually allow us to alter the load of compounds which may offer them no advantage in the environment in which they are growing and multiplying.

It is well known that many bioactive secondary compounds are phytotoxic and, unless stored in specific compartments, can reduce growth and viability of the producing cells (3). In an undifferentiated cell system, therefore, cells which have been genetically manipulated to produce more of a specific secondary compound will, in all probability, multiply slower than those which are less productive. In the absence of "positive selection," callus or suspension cells are picked every two weeks solely on the basis of rapid multiplication.

Ipso facto, cells which accumulate high levels of secondary compounds could eventually be eliminated. An example of this may be shown by the studies of Verpoorte and co-workers (19) who overexpressed tryptophan decarboxylase and strictosidine synthase genes in *Catharanthus roseus* cells which initially yielded increased levels of alkaloids; following one to two years of subculture, however, alkaloid levels in these cells fell back to those found in non-transgenic cells, even though expression of the transgenes remained high. Presumably, those transgenic cells which had developed a mechanism for reducing (by whatever means) the excess of alkaloid were more suited for growth in the culture environment and hence were favoured.

In the field, of course, selection is for plants which accumulate advantageous secondary metabolites in their natural environment and, furthermore, differentiated cell systems do not undergo the rapid genetic variation as seen in unorganized cells. It would appear, therefore, that organized cell systems should provide the best models in which to investigate the effects of genetic manipulation on secondary compound accumulation, and more importantly for future commercial applications.

GENETIC MANIPULATION
OF SECONDARY COMPOUND ACCUMULATION

The main strategies that have been followed in attempts to alter the accumulation of specific secondary compounds so far may be listed as: (i) reduction of the gene product of an extant gene in order to block or partially block a pathway; (ii) insertion of a novel gene to introduce a new pathway; (iii) amplification of an existing gene to augment carbon flux through an existing pathway.

With respect to the first of these strategies, the most studied approach has involved the introduction of anti-sense constructs utilizing cDNAs coding for enzymes associated with lignin production (e.g., cinnamyl alcohol dehydrogenase and cinnamyl CoA reductase) and with flower colour (e.g., chalcone synthase and dihyroflavonol 4-reductase). Results have been varied: despite considerable research activity it has still not been possible to reduce lignin levels significantly in crop plants using the anti-sense approach, although the composition of the lignin formed by transgenic plants can be altered so as to render the plant material more suitable as forage or feed-stock for the paper industry (4). On the other hand, the use of various anti-sense constructs has given rise to a number of uniform white or patterned white flowered varieties of petunia, torenia, carnation and lisianthus (17).

An alternative approach for modulating levels of enzymes of the flavonoid pathway has recently been described (13): in this study antibodies that recognize chalcone synthase and chalcone isomerase were overexpressed in *Arabidopsis thaliana* and a significantly altered flavonoid content was demonstrated in the transgenic plants so derived. The general applicability of this method of pathway blocking is, however, yet to be assessed.

Studies concerning the upregulation of existing routes to secondary compounds have mainly focussed on the introduction or amplification of a single gene in the pathway to the target compound. Clearly, in order that the accumulation of product be enhanced, the biosynthetic step being manipulated should be one that is (or is likely to be) associated with the regulation of flux through the appropriate pathway. An example of this strategy is demonstrated by the studies made in our laboratory on the biosynthesis of alizarin in *Rubia* spp.

The roots of *Rubia peregrina* L., a plant commonly known as wild madder, have been used in popular medicine and as a dye since ancient times. The medicinal properties of the roots (diuretic, anti-spasmodic and laxative) are due to the presence of a number of anthraquinones of which alizarin is the major component. As shown in Figure 1, alizarin is derived from shikimic acid via the intermediacy of chorismic acid, isochorismic acid and *O*-succinylbenzoic acid. A number of enzymes compete for chorismic acid in higher plants, including anthranilate synthase (AS), chorismate mutase (CM) and isochorismate synthase (ICS), although it appears that the alteration of flux through one branch of the competing pathways does not automatically imply that another branch will be directly regulated in order to compensate (12). One strategy for increasing flux through the *O*-succinylbenzoic acid branch of the pathway would thus be to overexpress ICS (6). Although a cDNA for a plant-derived ICS is not available, the *ent C* gene (which is associated with the biosynthesis of the powerful iron chelating agent enterobactin) from *E. coli* codes for ICS and has been fully sequenced (11). We amplified bacterial *ics* from *E. coli* by polymerase chain reaction, fused the gene with CaMV 35S promoter and octopine synthase terminator sequences, and cloned the product into a binary vector which was then mobilized into *Agrobacterium rhizogenes*. Excised stems of *R. peregrina* were infected with *A. rhizogenes* harbouring this plasmid (5) and hairy root clones which contained the plant transformation marker (PTM) gene were selected. Transgenic roots which expressed bacterial *ics* were darker purple than were control hairy roots. The ICS activity in hairy roots transgenic for *ics* was typically more than twice that found in roots transformed with either wild type *A. rhizogenes* or with the bacterium harbouring a binary vector containing only the PTM. Furthermore, the amount of free alizarin increased from 0.13 mg/g dry weight in the control hairy roots to 0.24 mg/g dry weight in the roots transgenic for *ics*. Interestingly, the roots of the parent plant (field-grown and cultured) accumulated only ca. 0.04 mg/g dry weight of free alizarin although the amount of bound alizarin (measured as ruberythric acid) was around 0.11 mg/g dry weight, a value some 27% higher than found in the transgenic roots. This shows that the incorporation of T_L from *A. rhizogenes* into the root genome itself has a marked effect on secondary compound accumulation, a result that appears general for a wide range of metabolites. With respect to the overexpression of *ics,* clearly the extra flux through the *O*-suc-cinylbenzoic acid pathway did manifest itself as an increase in accumulation of product alizarin, along with other anthraquinones, even though the inserted bacterial gene was not targeted either spatially or temporally within the trans-formed tissue.

A particularly interesting possibility to enhance carbon flux through one section of a pathway deriving from shikimic acid has recently been reported by Sommer and co-workers (16). The naphthoquinone pigment shikonin, an anti-inflammatory, anti-tumour and wound-healing component of *Lythospermum erythrorhizon,* is formed from geranyl diphosphate and 4-hydroxybenzoic acid (4-HB), the latter being derived from chorismic acid through a ten step se-quence involving phenylalanine and cinnamic acid (Figure 1). In contrast, chorismate pyruvate-lyase (CPL), an enzyme not present in *L. erythrorhizon,* is able to convert chorismate directly into 4-HB; in *E. coli*, CPL is encoded by the *ubiC* gene. Previously it had been shown (15) that when the *ubiC* gene (fused to a chloroplast transit peptide sequence and placed under the control of a constitutive plant promoter) was expressed in tobacco, the transformants were characterized by high CPL activity and they accumulated 4-HB in a bound form. Clearly if 4-HB could be more efficiently synthesized in *L. erythrorhizon* by the shortened pathway, and if the product was available for further metabolism *in situ,* then this strategy might provide an elegant means of upregulating shikonin accumulation. Hairy root clones of *L. erythrorhizon* transgenic for the *ubiC* gene, again fused to a chloroplast transit peptide se-quence but with two different promoter/terminator combinations, were pro-duced (16), each containing 1-5 copies of the integrated *ubiC* construct; 4-HB production in transgenic roots was enhanced, and radiofeeding experiments showed that some 20% of the 4-HB glucoside present in these roots had been formed via the novel pathway. Studies with the phenylalanine ammonia-lyase inhibitor 2-aminoindan-2-phosphonic acid demonstrated that when the natural pathway to 4-HB was blocked in the transgenic roots, the alternative pathway could partially substitute. However, because of a large variation in the accu-mulation of shikonin between the hairy root clones, the presence of the transgene and the enhanced 4-HB production could not be demonstrated to al-ter the yield of shikonin in a statistically significant manner. It should be pointed out that such a large variation in secondary compound accumulation in transgenic roots has certainly not been detected, for example, in the case of alizarin accumulation in *R. peregrina* (5). In roots of *L. erythrorhizon* trans-genic for *ubiC,* an enhanced product of the novel pathway was the nitrile glucoside menisdaurin. It must be presumed that this storage compound was synthesized in response to the overproduction of 4-HB in the chloroplasts and that this metabolite was either not available, or not required, for further elabo-

FIGURE 1. Biosynthetic pathways from chorismate.

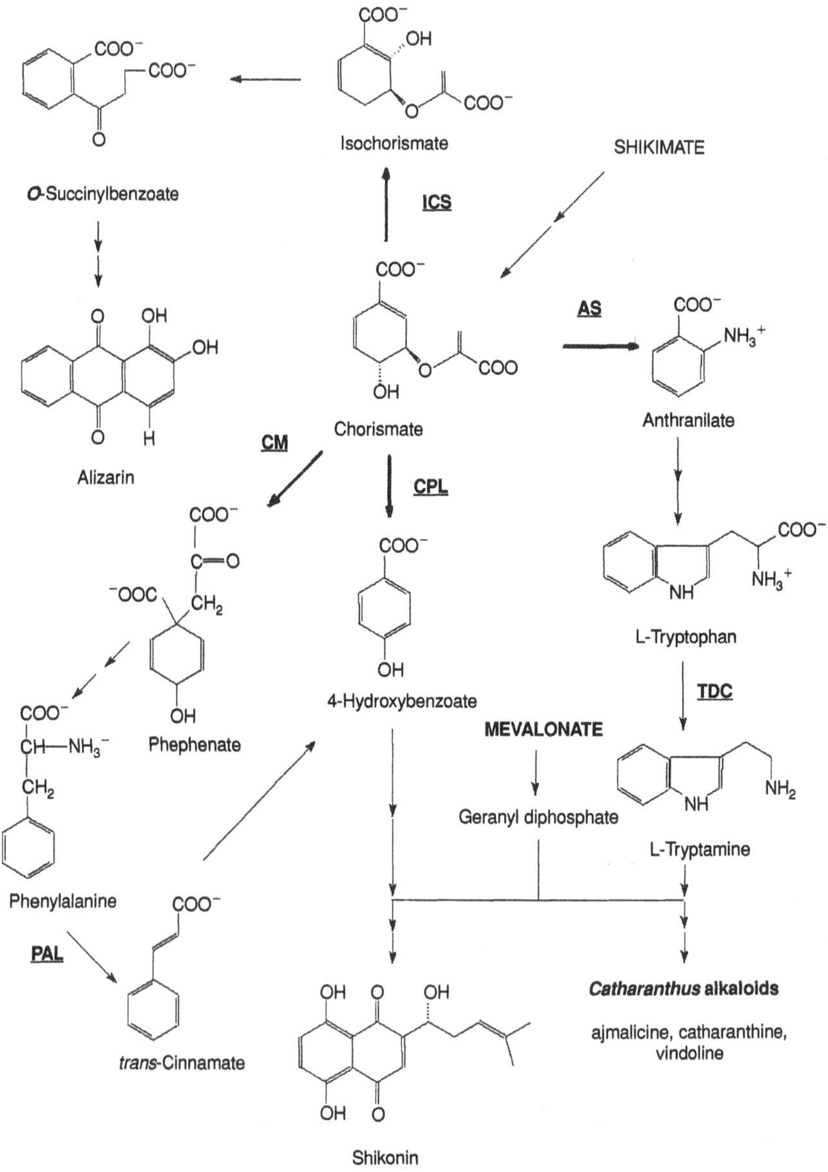

ration to shikonin. Indeed, 4-HB and shikonin formation are normally located in the cytosol and the endoplasmic reticulum, and there is little evidence to suggest that free 4-HB is able to be transported through the plastid membrane.

It is interesting to note that the shikonin pathway and the route to the *Catharanthus* alkaloids are both highly compartmentalized and involve the coupling of a precursor from the isoprenoid pathway to one from the shikimate pathway (Figure 1). In each case, regulation of accumulation of the end products of the pathway depends to a greater extent on the availability of the isoprenoid precursor and the activity of the coupling enzyme than on an increased flux through the shikimate pathway.

The synthesis of isoprenoid units is now known to be more complex than was originally realized following the demonstration (10) of a mevalonate-independent pathway in higher plant chloroplasts (leading to the mono- and diterpenoids) which operates alongside the classical acetate-mevalonate pathway involved in the synthesis of cytoplasmic-derived products such as sesqui- and tri-terpenoids and steroids. The latter pathway is believed to be regulated at the level of 3-hydroxy-3-methylglutaryl CoA reductase (HMGR), the enzyme which catalyses the irreversible conversion of HMG to mevalonate. In most higher plants, HMGR is encoded by a family of genes, each presumed to be involved in the biosynthesis of a set of unique products at various stages of the life-cycle of the plant. Schaller and co-workers (14) introduced one such gene (*hmgr1*) from *Hevea brasiliensis*, driven by a CaMV CoA promoter, into *Nicotiana tabacum* and analyzed HMGR activity in membranes isolated from leaves of the transgenic lines. They studied in detail three of the five transgenic lines produced and showed high *hmgr1* mRNA level with a concomitant large increase (4- to 8-fold) in HMGR activity compared with wild-type plants or control plants transgenic for PTM. Furthermore, although the transgenic lines with high HMGR were morphologically identical to the wild-type plants, they overproduced by some 6-fold total sterols including steroid intermediates such as cycloartenol and obtusifoliol.

The steroidal alkaloid solasodine is also synthesized via the isoprenoid pathway through the intermediacy of mevalonate (Figure 2) and occurs in glycosidic form (as solasonine and solamargine) in a number of *Solanum* species. Because solasodine has potential as a substitute for diosgenin as a feedstock for the pharmaceutical industry in the production of semi-synthetic contraceptive and anti-inflammatory drugs, we have been interested to study whether its accumulation can be enhanced through the up-regulation of HMGR. To this end we have expressed an *hmgr* gene, derived from the medicinal plant *Artemisia annua* and controlled by the CaMV 35S promoter, in hairy roots of *Solanum aviculare* and *S. paludosum* (1). Roots of *S. paludosum* transgenic for *hmgr* could not be maintained in culture, but those of *S. aviculare* grew well, typically faster than transformed control roots. When incubated under light

FIGURE 2. Possible biosynthetic routes to solasodine.

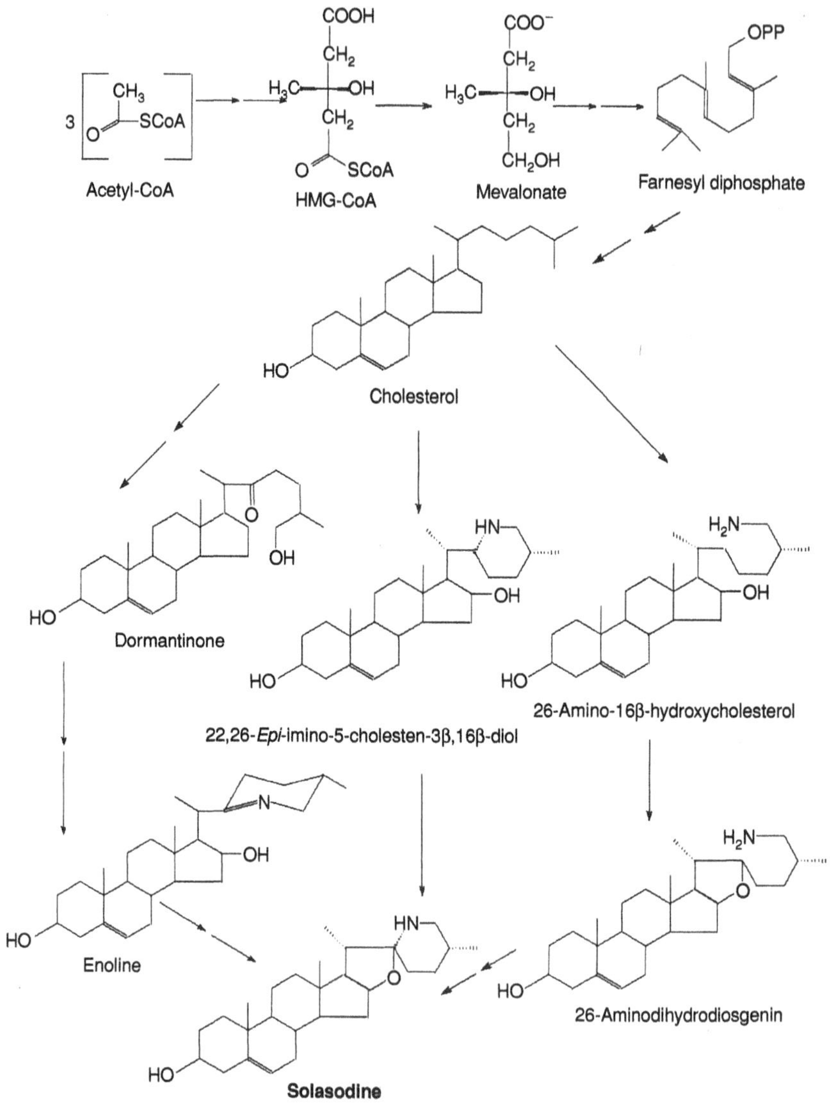

conditions, root lines transgenic for *hmgr*, transgenic controls and non-transgenic roots of the parent plant all accumulated amounts of solasodine (ca. 2.5 mg/g dry weight) which were not statistically significantly different from one another. On the other hand, roots transgenic for *hmgr* accumulated between 4.8 to 7.1 mg solasodine/g dry weight when incubated in the dark whilst steroidal alkaloid accumulation in the transgenic and non-transgenic controls remained statistically unchanged. In terms of solasodine accumulation in total root biomass per flask, the dark grown cultures accumulated between 3.3 and 4.2 times more aglycone than the controls (Figure 3). Since the introduced *hmgr* gene was under the control of a CaMV 35S promoter, which is not itself light regulated, it must be presumed that the enzyme is inactivated by light or that steps in the pathway later than steroid nucleus formation are rate-limiting and light regulated.

Attempts to introduce extra copies of the *hmgr* gene into plants are not without their difficulties. For example, when the gene was introduced into *A. annua* using a transformation system that had been validated using PTM genes (18), the resulting transformed callus could no longer be regenerated. Unfortunately, callus of *A. annua* accumulates only extremely low levels of the anti-malarial sesquiterpene artemisinin and none of the derivative artemisitene (which can also be readily converted into artemisinin). Callus lines transgenic for the homologous *hmgr* gene (driven by a CaMV 35S promoter) have been

FIGURE 3. Accumulation of solasodine in *hmgr*-transformed and non-transformed root clones of *Solanum aviculare* grown under dark conditions.

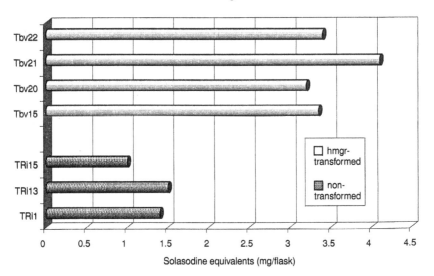

produced in a collaborative project (unpublished) between our laboratory and that of Prof. E. Van Den Eeckhout (University of Gent, Belgium) and have been shown to accumulate some 3- to 4-fold more artemisinin than non-transgenic callus together with significant amounts of artemisitene. However, in the absence of transgenic plants, such results are of limited value in the quest to upregulate these commercially important medicinal compounds.

The examples of pathway upregulation provided so far have all involved attempts to "push" carbon flux towards a target product by manipulation of enzyme steps located early in the pathway. It is clear that an increase of flux derived in this manner is not necessarily channelled in the expected direction as the natural regulatory processes of the plant attempt to neutralize the effects of the introduced gene. An alternative strategy would be to "pull" flux towards the desired target by upregulating a key step at the end of a pathway. Such a strategy has been employed with considerable success in order to increase the amounts of the commercially important anti-cholinergic drug scopolamine in relationship to the co-occurring, but much less valuable, hyoscyamine. Most solanaceous plants that produce these two alkaloids accumulate hyoscyamine in larger quantities than scopolamine, but it appears that only one enzyme, hyoscyamine-6 β-hydroxylase (H6H), is responsible for both the hydroxylation of substrate and the formation of the epoxide ring. In 1993, Yamada's group (8) produced hairy roots of *Atropa belladonna* transformed with cDNA for H6H derived from *Hyoscyamus niger,* and were able to select a line which showed 5-times higher H6H protein and H6H activity compared with control hairy roots. The latter accumulated levels of hyoscyamine between 0.05-0.3% dry weight, with scopolamine being less than 0.1% dry weight. The hairy roots transgenic for *h6h,* however, accumulated some 0.45% scopolamine and only 0.02% of hyoscyamine. In very recent work (9), cDNA for H6H derived from *H. niger* was inserted into hairy roots of *H. muticus* and of the 43 clones that were transgenic for *h6h* only 22 showed elevated levels of scopolamine compared with the control roots. The best of these clones, KB7, accumulated 100-times more scopolamine than did the control clones, but this still represented only ca. 10% of the hyoscyamine which remained as the major alkaloid produced in these roots. On the other hand, the scopolamine accumulated by clone KB7 was 4-fold higher then the level found in the leaves of the mother plant (0.2 mg/g dry weight).

In order to produce whole plants transgenic for *h6h,* Yamada and co-workers (20) transformed leaf explants of *A. belladona* with disarmed *Agrobacterium tumefaciens* harbouring a binary vector containing cDNA for H6H controlled by the CaMV 35S promoter. Of the regenerated transformants, the one showing the highest activity of H6H contained three copies of the *h6h* gene: this plant was selfed and the progeny showed uniformly high levels of scopolamine in the aerial parts compared with the wild plant or the transformed controls. Indeed, one progeny line accumulated more than 1.2% dry weight of scopolamine, a level that cannot be attained by classical plant breeding.

FURTHER CONSIDERATIONS AND FUTURE PERSPECTIVES

Whilst there have been some advances in the area of upregulation of secondary compounds, more problems have been exposed than have been solved. For instance, the complexity of compartmentalisation of biosynthetic route to even the simplest of products is only now emerging. Perhaps part of the solution to this problem may involve moving genes for an entire block of a pathway into a single compartment. An excellent candidate would be the plastid (2), since here related genes may be transcribed together under the same control in the form of an operon. Various types of plastids (chloroplasts, chromoplasts, elaioplasts, etioplasts, etc.) have different roles in the plant and could very well stand as storage sites in their own right for the target product. Genetic manipulation of plastids is a reality and yields a number of advantages, perhaps the most important of which is that genes inserted into the plastid of the parent are not present in pollen and cannot possibly be transmitted to close relatives of the plant growing nearby; furthermore, since transgene products will also not be produced in pollen, there will be no harmful effect on pollinating insects.

Another concern that derives from the results of genetic manipulation so far is that biosynthetic routes to secondary compounds exist as a labyrinth of interconnected highways and byways. It has become obvious that most products cannot be upregulated by manipulating only one, or even a few, isolated steps. The alleviation of one bottleneck only brings us to the next, or worse still to a metabolic black-hole through which hard-won carbon flux is siphoned off into obscurity. Again a solution may be in sight through the study of natural transcriptional factors. For example, Grotewald and co-workers (7) have shown that in maize and other cereals two complete flavonoid pathways (one leading to the 3-hydroxy flavonoids and the other to the 3-deoxy flavonoids) are independently regulated by transcriptional regulators C1/R and P, respectively. Transformation of Black Mexican Sweet maize, which does not express R, C1 or P factors, led to suspension cell lines which were transgenic for either C1/R or P; the former cells were red due to the presence of anthocyanidins, whilst the latter remained colourless and were shown to accumulate luteoforol, a flavan-4-ol. Furthermore, there was a direct correlation between the degree of overexpression of the transcriptional factors and the amounts of compounds accumulated.

It is known that, in some circumstances, plants control their accumulation of end-product not only by increasing the levels of bottleneck enzymes but by diminishing those associated with competing pathways: for example, when stressed, tobacco cells form sesquiterpene phytoalexins through up-regulation of HMGR and sesquiterpene cyclases, and by down-regulation of enzymes responsible for the formation of steroids from squalene. If transcriptional factors could be found that control such events *in planta,* they might be harnessed to regulate entire pathways in transgenic medicinal plants.

REFERENCES

1. Argôlo, A.C.C., B.V. Charlwood and M. Pletsch. 2000. The regulation of solasodine production by *Agrobacterium rhizogenes*-transformed roots of *Solanum aviculare. Planta Medica* (in press).

2. Bogorad, L. 2000. Engineering chloroplasts: An alternative site for foreign genes, proteins, reactions and products. *TIBTECH* 18: 257-263.

3. Brown, J.T., P.K. Hegarty and B.V. Charlwood. 1987. The toxicity of monoterpenes to plant cell cultures. *Plant Science* 48: 195-201.

4. Dixon, R.A., C.J. Lamb, S. Masoud, V.J.H. Sewalt and N.L. Paiva. 1996. Metabolic engineering: Prospects for crop improvement through the genetic manipulation of phenylpropanoid biosynthesis and defence responses: A review. *Gene* 179: 61-71.

5. Lodhi, A.H. and B.V. Charlwood. 1996. *Agrobacterium rhizogenes*-mediated transformation of *Rubia peregrina* L.: *In vitro* accumulation of anthraquinones. *Plant Cell Tissue Organ Culture* 46: 103-108.

6. Lodhi, A.H., R.J.M. Bongaerts, R. Verpoorte, S.A. Coomber and B.V. Charlwood. 1996. Expression of bacterial isochorismate synthase (EC 5.4.99.6) in transgenic root cultures of *Rubia peregrina. Plant Cell Reports* 16: 54-57.

7. Grotewold, E., M. Chamberlin, M. Snook, B. Siame, L. Butler, J. Swenson, S. Maddock, G. St. Clair and B. Bowen. 1998. Engineering secondary metabolism in maize cells by ectopic expression of transcription factor. *Plant Cell* 10: 721-740.

8. Hashimoto, T., D.-J. Yun and Y. Yamada. 1993. Production of tropane alkaloids in genetically engineered root cultures. *Phytochemistry* 32: 713-718.

9. Jouhikainen, K., L. Lindgren, T. Jokelainen, R. Hiltunen, T.H. Teeri and K.-M. Oksman-Caldentey. 1999. Enhancement of scopolamine production in *Hyoscyamus muticus* L. hairy root cultures by genetic engineering. *Planta* 208: 545-551.

10. Lichtenthaler, H.K., J. Schwender, A. Disch and M. Rohmer. 1997. Biosynthesis of isoprenoids in higher plant chloroplasts proceeds via a mevalonate-independent pathway. *FEBS Letters* 400: 271-274.

11. Ozenberger, B.A., T.J. Brickmann and M.A. McIntosh. 1989. Nucleotide sequence of *Escherichia coli* isochorismate synthase gene *ent* C and evolutionary relationship of isochorismate synthase and other chorismate utilizing enzymes. *Journal of Bacteriology* 171: 775-783.

12. Poulsen, C., O.J.M. Goddijn, J.H.C. Hoge and R. Verpoorte. 1994. Anthranilate synthase and chorismate mutase activities in transgenic tobacco plants overexpressing tryptophan decarboxylase from *Catharanthus roseus. Transgenic Research* 3: 43-49.

13. Santos, M.C.O. and B.S.J. Winkel. 2000. Modulating flavonoid biosynthesis in transgenic *Arabidopsis*. Abstracts of 6th International Congress of Plant Molecular Biology, June 18-24, 2000, Quebec, Canada. S10-14.

14. Schaller, H., B. Grausem, P. Benveniste, M.-L. Chye, C.-T. Tan, Y.-H. Song and N.-H. Chua. 1995. Expression of the *Hevea brasiliensis* (H.B.K.) Müll. Arg. 3-hydroxy-3-methylglutaryl-coenzyme A reductase 1 in tobacco results in sterol over-production. *Plant Physiology* 109: 761-770.

15. Siebert, M., S. Sommer, S.-M. Li, Z.-X. Wang, K. Severin and I. Heide. 1996. Genetic engineering of plant secondary metabolism. Accumulation of 4-hydroxy-benzoate glucosides as a result of the expression of the bacterial *ubiC* gene in tobacco. *Plant Physiology* 112: 811-819.

16. Sommer, S., A. Köhle, K. Yazaki, K. Shimomura, A. Bechthold and L. Heide. 1999. Genetic engineering of shikonin biosynthesis in hairy root cultures of *Lithospermum erythrorhizon* transformed with the bacterial *ubiC* gene. *Plant Molecular Biology* 39: 683-693.

17. Tanaka, Y., S. Tsuda and T. Kusumi. 1998. Metabolic engineering to modify flower colour. *Plant Cell Physiology* 39: 1119-1126.

18. Vergauwe, A., R. Cammaert, D. Vandenberghe, C. Genetello, D. Inze, M. Van Montagu and E. Van Den Eeckhout. 1996. *Agrobacterium tumefaciens*-mediated transformation of *Artemisia annua* L. and regeneration of transgenic plants. *Plant Cell Reports* 15: 929-933.

19. Verpoorte, R., R. van der Heiden, H.J.G. ten Hoopen and J. Memelink. 1999. Metabolic engineering of plant secondary metabolite pathways for the production of fine chemicals. *Biotechnology Letters* 21: 467-479.

20. Yun, D.-J., T. Hashimoto and Y. Yamada. 1992. Metabolic engineering of medicinal plants: Transgenic *Atropa belladonna* with an improved alkaloid composition. *Proceedings of the National Academy of Science USA* 89: 11799-11803.

Biochemical and Molecular Regulation of Monoterpene Accumulation in Peppermint (*Mentha ×piperita*)

Jonathan Gershenzon
Marie McConkey
Rodney Croteau

SUMMARY. Recent advances in the biochemistry and molecular biology of monoterpene metabolism in peppermint suggest that increased accumulation of the menthol-rich essential oil in this species might be achieved by a general enhancement of the rate of biosynthesis and, more specifically, by increasing the expression of the gene encoding (−)-limonene synthase, the first committed step in monoterpene biosynthesis. *[Article copies available for a fee from The Haworth Document Delivery Service: 1-800-HAWORTH. E-mail address: <getinfo@haworthpressinc. com> Website: <http://www.HaworthPress. com>* © *2002 by The Haworth Press, Inc. All rights reserved.]*

KEYWORDS. Monoterpenes, menthol, limonene, essential oil, peppermint, Lamiaceae, biosynthesis, limonene synthase

Jonathan Gershenzon, Marie McConkey, and Rodney Croteau are affiliated with the Institute of Biological Chemistry, Washington State University, Pullman, WA 99164-6340 USA.

Address correspondence to: Jonathan Gershenzon, Max Planck Institute for Chemical Ecology, Carl-Zeiss Promenade 10, D-07745 Jena, Germany (E-mail: gershenzon@ ice.mpg.de).

[Haworth co-indexing entry note]: "Biochemical and Molecular Regulation of Monoterpene Accumulation in Peppermint (*Mentha ×piperita*)." Gershenzon, Jonathan, Marie McConkey, and Rodney Croteau. Co-published simultaneously in *Journal of Herbs, Spices & Medicinal Plants* (The Haworth Herbal Press, an imprint of The Haworth Press, Inc.) Vol. 9, No. 2/3, 2002, pp. 153-156; and: *Breeding Research on Aromatic and Medicinal Plants* (ed: Christopher B. Johnson, and Chlodwig Franz) The Haworth Herbal Press, an imprint of The Haworth Press, Inc., 2002, pp. 153-156. Single or multiple copies of this article are available for a fee from The Haworth Document Delivery Service [1-800-HAWORTH 9:00 a.m. - 5:00 p.m. (EST). E-mail address: getinfo@haworthpressinc.com].

153

One of the most widely cultivated essential oil crops in the world is peppermint (*Mentha* ×*piperita* L.) (5). This perennial herb of the family Lamiaceae produces a monoterpene-rich oil, stored in glandular trichomes, whose major constituent is (−)-menthol. Given the importance of peppermint, the development of cultivars with increased essential oil content could have significant economic benefit. However, such an effort requires precise knowledge of the factors regulating monoterpene synthesis and accumulation. While the pathway of monoterpene biosynthesis in peppermint was established over 15 years ago and nearly all of the relevant enzymes have been described (3,6), the chief factors regulating monoterpene accumulation in this species are still unknown.

The accumulation of any type of metabolite is controlled by the balance between the rate of its formation and the rate of its loss, with loss being attributable either to metabolic breakdown or direct release into the environment. Hence, the rates of monoterpene biosynthesis, catabolism and volatilization in peppermint leaves were assessed at different stages of development. The rate of monoterpene biosynthesis was measured by exposing rooted plants to a five minute pulse of $^{14}CO_2$, and determining the incorporation of ^{14}C into monoterpenes in leaves of various ages after a six hour chase period. There was a high level of monoterpene biosynthesis in expanding leaves (12-20 days old), but only very low rates of monoterpene biosynthesis in both younger and older leaves (4). At the same time, the monoterpene content of leaves rose steadily with leaf growth, until it leveled off at about 20 days. The close correspondence of the decline in biosynthetic rate with the leveling off of monoterpene content indicates that the rate of biosynthesis is likely a principal factor controlling monoterpene accumulation. In contrast, no significant degradation of monoterpenes was detected throughout leaf development, and monoterpene volatilization was found to occur at a very low rate, which on a monthly basis represented less than 1% of the total pool of stored monoterpenes (4).

Having demonstrated that biosynthesis is the chief process that determines the rate of monoterpene accumulation in peppermint, subsequent studies were focused on the first committed step of monoterpene biosynthesis in this species, the cyclization of geranyl diphosphate to (−)-limonene (Figure 1). The enzyme responsible for this conversion, (−)-limonene synthase, has already been purified to homogeneity (1) and the corresponding cDNA isolated (2). We measured the activity of (−)-limonene synthase in cell-free extracts prepared from leaves of different ages. The level of this enzyme activity was high only in expanding leaves (12-20 days old), exhibiting a close correlation with the rate of monoterpene biosynthesis (7). Hence, (−)-limonene synthase may represent an important control point in monoterpene formation.

If (−)-limonene synthase is an important regulatory enzyme in monoterpene biosynthesis, then it is important to learn what regulates its activity. Western and northern blot analyses carried out on leaves of varying age re-

FIGURE 1. The first committed step of monoterpene biosynthesis in peppermint is (−)-limonene synthase, which converts geranyl diphosphate, a ubiquitous intermediate of isoprenoid metabolism, to (−)-limonene. Subsequent reactions convert (−)-limonene to (−)-menthone and (−)-menthol, the principal monoterpenoid consitutuents of the essential oil.

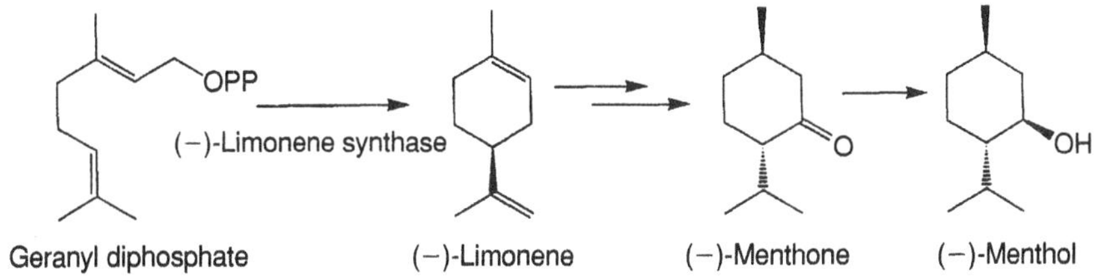

Geranyl diphosphate   (−)-Limonene   (−)-Menthone   (−)-Menthol

vealed that (−)-limonene synthase protein and limonene synthase mRNA levels both showed peaks in leaves of 12-20 days old (7). The close correspondence of enzyme activity, enzyme protein and steady-state-A levels in this developmental time course suggests that (−)-limonene synthase activity is controlled at the level of gene expression. Thus, to increase monoterpene accumulation in peppermint, it may prove fruitful to increase the transcription of the (−)-limonene synthase gene.

An interesting feature of many monoterpene synthases is their ability to produce multiple products from geranyl diphosphate (10). (−)-Limonene synthase, for example, has been shown to produce low levels of α-pinene, β-pinene and myrcene in addition to limonene (2,8). The existence of such multi-product enzymes may be one reason why monoterpenes are usually present as complex mixtures in essential oils. Overexpression of genes encoding such enzymes may be problematic if increased accumulation of only a single product is desired. However, the recent publication of the three-dimensional structure of a plant terpene synthase (9) heralds the time when such enzymes might be re-engineered to produce a custom blend of single or multiple products, as desired.

REFERENCES

1. Alonso W.R., J.I.M. Rajaonarivony, J. Gershenzon, and R. Croteau. 1992. Purification of 4S-limonene synthase, a monoterpene cyclase from the glandular trichomes of peppermint (*Mentha ×piperita*) and spearmint (*Mentha spicata*). *J Biol. Chem.* 267:7582-7587.

2. Colby S.M., W.R. Alonso, E.J. Katahira, D.J. McGarvey, and R. Croteau. 1993. 4S-Limonene synthase from the oil glands of spearmint (*Mentha spicata*). *J. Biol. Chem.* 268:23016-23024.

3. Croteau R., and K.V. Venkatachalam. 1986. Metabolism of monoterpenes: Demonstration that (+)-*cis*-isopulegone, not piperitenone, is the key intermediate in the conversion of (−)-isopiperitenone to (+)-pulegone in peppermint (*Mentha piperita*). *Arch. Biochem. Biophys.* 249:306-315.

4. Gershenzon J., M.E. McConkey, and R.B. Croteau. 2000. Regulation of monoterpene accumulation in leaves of peppermint. *Plant Physiol.* 122:205-213.

5. Hay R.K.M., and P.G. Waterman, eds. 1993. *Volatile Oil Crops: Their Biology, Biochemistry and Production.* Longman Scientific and Technical, Essex. 185 pp.

6. Kjonaas R., and R. Croteau. 1983. Demonstration that limonene is the first cyclic intermediate in the biosynthesis of oxygenated *p*-menthane monoterpenes in *Mentha piperita* and other *Mentha* species. *Arch. Biochem. Biophys.* 220:79-89.

7. McConkey M.E., J. Gershenzon, and R.B. Croteau. 2000. Developmental regulation of monoterpene biosynthesis in the glandular trichomes of peppermint. *Plant Physiol.* 122:215-223.

8. Rajaonarivony J.I.M., J. Gershenzon, and R. Croteau. 1992. Characterization and mechanism of (4*S*)-limonene synthase, a monoterpene cyclase from the glandular trichomes of peppermint (*Mentha ×piperita*). *Arch. Biochem. Biophys.* 296:49-57.

9. Starks C.M., K.W. Back, J. Chappell, and J.P. Noel. 1997. Structural basis for cyclic terpene biosynthesis by tobacco 5-epi-aristolochene synthase. *Science* 277: 1815-1820.

10. Wise M.L., and R. Croteau. 1999. Monoterpene biosynthesis. *In* D. Barton, K. Nakanishi, and O. Meth-Cohn, eds. *Comprehensive Natural Products Chemistry, Vol. 2, Isoprenoids Including Carotenoids and Steroids.* Elsevier Science, Amsterdam. pp. 97-152.

Influence of Environmental Factors (Including UV-B Radiation) on the Composition of the Essential Oil of *Ocimum basilicum*–Sweet Basil

Zivojin Rakic
Christopher B. Johnson

SUMMARY. *Ocimum basilicum* L. (Lamiaceae)–sweet basil–is a fragrant herb with a minty, pleasantly sweet flavour, mainly used for production of essential oil and as a condiment. Three different commercial varieties of basil, one broad-leaved and two different small-leaved varieties, were grown in a greenhouse at MAICh under controlled conditions and used for a period of one year. Seasonal variation in each variety was examined and, in addition, each variety was treated with supplementary UV light in order to assess the effects of UV-B, following previous work showing a substantial UV-B mediated induction of essential oil. *[Article copies available for a fee from The Haworth Document Delivery Service: 1-800-HAWORTH. E-mail address: <getinfo@haworthpressinc.com> Website: <http://www.HaworthPress.com> © 2002 by The Haworth Press, Inc. All rights reserved.]*

Zivojin Rakic and Christopher B. Johnson are affiliated with the Department of Natural Products, Mediterranean Agronomic Institute of Chania, Greece.

Address correspondence to: Christopher B. Johnson, Department of Natural Products, Mediterranean Agronomic Institute of Chania, Alsyllion Agrokepion, P.O. Box 85, 73100 Chania, Greece.

The authors thank George Naxakis for excellent technical assistance and Melpo Skoula for helpful discussions.

[Haworth co-indexing entry note]: "Influence of Environmental Factors (Including UV-B Radiation) on the Composition of the Essential Oil of *Ocimum basilicum*–Sweet Basil." Rakic, Zivojin, and Christopher B. Johnson. Co-published simultaneously in *Journal of Herbs, Spices & Medicinal Plants* (The Haworth Herbal Press, an imprint of The Haworth Press, Inc.) Vol. 9, No. 2/3, 2002, pp. 157-162; and: *Breeding Research on Aromatic and Medicinal Plants* (ed: Christopher B. Johnson, and Chlodwig Franz) The Haworth Herbal Press, an imprint of The Haworth Press, Inc., 2002, pp. 157-162. Single or multiple copies of this article are available for a fee from The Haworth Document Delivery Service [1-800-HAWORTH 9:00 a.m. - 5:00 p.m. (EST). E-mail address: getinfo@haworthpressinc.com].

KEYWORDS. *Ocimum basilicum* L., volatile oil, UV-B treatment, seasonal variation

INTRODUCTION

Ocimum basilicum L. (Lamiaceae)–sweet basil–is a fragrant herb with a minty, pleasantly sweet flavour. Traditionally it has been used as a medicinal plant, but nowadays its leaves, fresh or dried, are used as flavourings or spices, and its essential oil, which is made up of mono- and sesquiterpenes and phenyl-propanoids, is used in all kinds of flavours and in the perfume industry.

Variations in the yield and composition of the essential oil are affected by several factors. There is considerable genetic variation in the spectrum of compounds found in the essential oil (1). The stage of development of the plant, its nutritional status, temperature and light all affect yield and composition, and recent work has shown a substantial UV-B mediated induction of essential oil (2).

In the experiments reported here, three different commercial varieties of basil (one broad-leaved and two different small-leaved varieties) were grown in a greenhouse at MAICh under controlled conditions and used for a period of one year. Seasonal variation in each variety was examined and, in addition, each variety was treated with supplementary UV light in order to assess the effects of UV-B.

MATERIALS AND METHODS

Plant material. From July 1998 to July 1999, three varieties of sweet basil (*Ocimum basilicum* L.) were grown in a greenhouse at MAICh. The varieties were:

Variety 1: Broad-leaved sweet basil (Vilmorin, La Verpilliere Cedex, France)
Variety 2: Small-leaved sweet bush basil (Megastar)
Variety 3: Small-leaved Greek variety (obtained locally)

Seeds were sown in trays with a mixture of peat and perlite (3:1). Seven to 10 days after the appearance of the first leaves, the seedlings were moved to pots with the same substrate, ready for a two week treatment. The procedure was repeated at intervals of three weeks, except during the period December-February, when plant growth was extremely slow.

Light treatments. For each series of plants the experimental period was of two weeks duration. Control plants and those to be treated with supplementary

UV-B were placed on parallel benches in the glasshouse, separated by a plastic screen opaque to UV-B light. Plants were treated with UV-B for one hour each day before dawn from two Philips 20 W/12 tubes placed 1.0 m above the bench and 1.0 m apart.

Analysis of volatile oil. After plant harvesting the leaves were collected and dried to constant weight at ambient temperature. For each sample, 50 mg of dry leaf material was used for headspace analysis as described in Johnson et al. (2). Compounds were identified as reported in Johnson et al. (2).

RESULTS AND DISCUSSION

Essential oil composition. The principal volatile constituents of the three varieties are shown in Table 1. The broad-leaved variety (Variety 1) and the second small-leaved variety (Variety 3) had much higher total oil content than the other small-leaved variety (Variety 2). In Variety 1, linalool and 1,8-cineole were the major constituents, with eugenol the predominant phenylpropanoid. In Variety 2, linalool predominated, with significant amounts of 1,8-cineole and β-*trans*-bergamotene. The major phenyl-propanoids were eugenol and methyl eugenol. In Variety 3, 1,8-cineole was the predominant terpene. This variety had a much greater phenylpropanoid content: eugenol and methyl eugenol representing 24% of the total volatiles.

Seasonal variation. In Figure 1, amounts of the constituents of the basil essential oil have been grouped as monoterpenes, sesquiterpenes and phenylpropanoids. All three varieties show some variation during the year. This variation differs from one variety to another but, in general, the monoterpenes are greater in summer and autumn, and reduced in winter and spring as a proportion of the total. This may reflect the consequences of changes in photosynthetic rate on the mainly chloroplastic route of synthesis of the precursor IPP (3). The sesquiterpenes, which are reported to be synthesized mainly via the cytoplasmic mevalonic acid pathway (4), remain rather constant throughout. The phenylpropanoids (eugenol and methyl eugenol) show no consistent seasonal pattern. This seasonal effect was less noticeable in the narrow-leaved Variety 2.

Effects of UV-B. Neither of the small-leaved varieties showed much response to UV-B, whereas the broad-leaved (Variety 1) displayed a significant response to added UV-B (Figure 2). Both monoterpenes and phenylpropanoids were stimulated by UV-B, with negligible effect on sesquiterpenes. The UV-B effect on phenylpropanoids was much less noticeable in spring than in summer.

TABLE 1

Peak	Compound	Ret Time	Variety 1	Variety 2	Variety 3
1	α-pinene	8,629	2.32	2.72	2.68
2	sabinene	10,775	2.31	1.84	2.64
3	β-pinene	10,871	3.94	2.75	4.62
4	β-myrcene	11,985	4.25	2.30	3.02
5	limonene	14,262	1.51	2.49	1.48
6	1,8-cineole	14,406	26.03	16.31	30.41
7	*trans*-β-ocimene	15,908	3.42		0.57
8	*cis*-sabinene hydrate	17,114	0.90	2.36	0.88
9	α-terpinolene	18,707	0.33	0.57	0.73
10	linalool	19,967	34.16	33.86	15.09
11	camphor	23,011	1.50	1.09	1.97
12	borneol	24,965	0.21	0.38	
13	4-terpineol	26,001		5.04	
14	α-terpineol	27,219	1.64	1.40	1.78
15	*iso*-bornyl-acetate	35,623	0.40	0.76	0.60
16	eugenol	41,874	8.36	5.20	5.44
17	methyl eugenol	46,323	1.95	4.01	18.29
18	α-*trans*-bergamotene	48,416	4.04	12.07	6.43
19	α-humulene	49,421	0.39	0.44	0.52
20	*trans*-β-farnesene	50,571	0.39	0.70	1.25
21	germacrene D	51,795	1.22	1.96	0.85
22	bicyclogermacrene	53,115	0.33	0.66	0.23
23	γ-cadinene	54,603	0.72	1.17	0.53
	Total area		825020	513481	1052579

FIGURE 1. Seasonal variation in the volatile oil content of the three basil varieties. Symbols: (♦) monoterpenes; (▲) sesquiterpenes; (■) phenylpropanoids. Units of oil content are the integrated peak areas as detected by FID in gas chromatography.

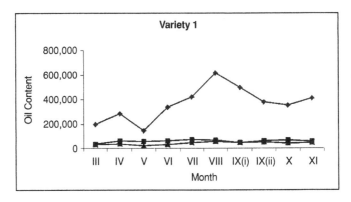

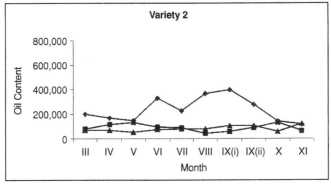

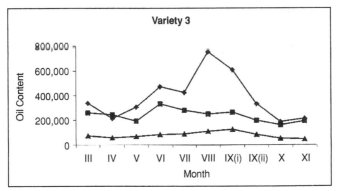

FIGURE 2. Effect of two weeks of supplementary UV-B radiation on volatile oil content in basil in May (above) and August (below). Monoterpenes are indicated by solid bars, sesquiterpenes by open bars, and phenylpropanoids by hatched bars. In August the residual effect of UV-B was followed for a further two weeks.

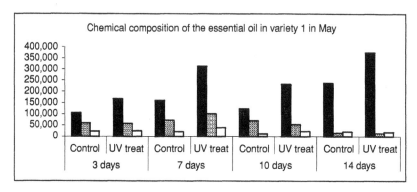

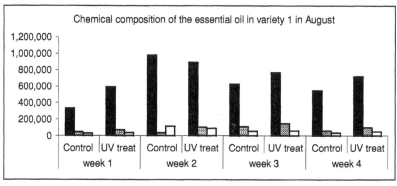

REFERENCES

1. Grayer RJ, Kite CG, Goldstone JF, Byan ES, Paton A, Putievsky E (1996). Infraspecific taxonomy and essential oil chemotypes in sweet basil, *Ocimum basilicum*. Phytochemistry 43, 1033-1039.

2. Johnson CB, Kirby JD, Naxakis G, Pearson S (1999). Substantial induction by UV-B of essential oils in sweet basil (*Ocimum basilicum* L.). Phytochemistry 51, 507-510.

3. Rohmer M (1999). The discovery of a mevalonate-independent pathway for isoprenoid biosynthesis in bacteria, algae and higher plants. Natural Product Reports 16, 565-574.

4. Lichtenthaler HK (1999). The 1-deoxy-D-xylulose-5-phosphate pathway of isoprenoid biosynthesis in plants. Annual Review of Plant Physiology & Molecular Biology 50, 47-65.

Characterization
of *Hypericum perforatum* L. Plants
from Various Accessions
by RAPD Fingerprinting

Birgit Arnholdt-Schmitt

SUMMARY. RAPD fingerprints of individual plants of *Hypericum perforatum* from different accessions are discussed in this paper. From a wild population, 10 offspring were investigated by genome analyses. The results obtained revealed that the collection included significantly polymorphic genotypes. Whereas 7 of the individuals displayed identical fingerprints, obviously representing the dominant genotype of this accession, a deviating but also identical genotype was shown by 2 plants and a third genotype was represented by another plant. Additionally, genome characterization of progenies from individual plants of 4 further

Birgit Arnholdt-Schmitt is affiliated with the Departamento de Bioquímica e Biologia Molecular, Universidade Federal do Ceará, Caixa Postal 1065, 60001-970 Fortaleza, Ceará, Brasil (E-mail: birgit@ufc.br).

The author is grateful to Prof. Dr. J. Hölzl from the University of Marburg (Germany) and Bionorica/Plantamed Arzneimittel for making plant material from *Hypericum perforatum* collections available, and especially to Dr. M. Popp for granting financial support for research on this plant species. Without laboratory space and general support provided by Prof. K.-H. Neumann from the University of Giessen (Germany), this work could not have been performed. He also would like to thank Dr. Shikha Roy of the University of Rajasthan, Jaipur, India, a visitor through DAAD cooperation, for help with DNA preparation.

[Haworth co-indexing entry note]: "Characterization of *Hypericum perforatum* L. Plants from Various Accessions by RAPD Fingerprinting." Arnholdt-Schmitt, Birgit. Co-published simultaneously in *Journal of Herbs, Spices & Medicinal Plants* (The Haworth Herbal Press, an imprint of The Haworth Press, Inc.) Vol. 9, No. 2/3, 2002, pp. 163-169; and: *Breeding Research on Aromatic and Medicinal Plants* (ed: Christopher B. Johnson, and Chlodwig Franz) The Haworth Herbal Press, an imprint of The Haworth Press, Inc., 2002, pp. 163-169. Single or multiple copies of this article are available for a fee from The Haworth Document Delivery Service [1-800-HAWORTH 9:00 a.m. - 5:00 p.m. (EST). E-mail address: getinfo@haworthpressinc.com].

accessions were performed. The results show that most of the fingerprints from each accession indicated an identical mode of reproduction for *H. perforatum*. Nevertheless, non-identical progenies could be discovered as a minor event by RAPD analyses. *[Article copies available for a fee from The Haworth Document Delivery Service: 1-800-HAWORTH. E-mail address: <getinfo@haworthpressinc.com> Website: <http://www.HaworthPress. com> © 2002 by The Haworth Press, Inc. All rights reserved.]*

KEYWORDS. Accession, apomixis, *Hypericum perforatum*, RAPD fingerprints

INTRODUCTION

Successful use of St. John's wort in phytotherapy as a natural antidepressant (5) has caused an augmentation of interest in the species *Hypericum perforatum* during recent years. Up to now a great part of the plant material was collected from wild populations as an established practice. However, industrial producers of drugs and extracts are required nowadays to guarantee the homogeneous quality of their products. To achieve stability in yield production and quality characteristics, an increase of efforts in cultivation as well as in breeding is required.

Efficient strategies in breeding depend on a good knowledge of the reproductive biology of the species of interest. For *H. perforatum*, apomixis was described as an important mode of reproduction (3), and Noack (4) reported in his early cytological studies an extent of apomixis of about 97%. RFLP analysis with an rDNA probe was performed by Halusková and Cellárová (2) on somaclones and their progenies. The authors interpreted the obtained variations in RFLP patterns as an indication of cross pollination. In the present paper, results on RAPD fingerprints of individual offsprings from different accessions of *H. perforatum* will be discussed.

MATERIALS AND METHODS

Plant material. Plants of *Hypericum perforatum* were available from collections at the botanical garden in Marburg (Germany) and Neumarkt (Bionorica/Plantamed Arzneimittel, Germany). For RAPD analysis only young leaves were taken. At harvest, the leaves were washed twice in sterile aqua dest., shock frozen with liquid nitrogen and stored at −80°C until DNA isolation.

DNA isolation. Genomic DNA was isolated with the help of the "DNeasy Plant Mini Kit" from Qiagen (Germany). The quantity and quality of the extracted DNA was checked in comparison to high molecular weight lambda DNA in a 1% agarose gel by using a video densitometer.

RAPD analysis. Analysis with random primers was performed with a RAPD analysis kit ("RAPD analysis beads," Amersham-Pharmacia). Operon Technologies 10mers obtained by Roth (Germany) were applied as primers using the following conditions for the PCR reaction: 5 min 95°C; 45 cycles: 1 min 95°C, 1 min 36°C, 2 min 72°C; and 5 min 72°C. Amplifications were carried out by using a thermocycler from Techne (model "Progene"). The PCR fragments were separated in a 1.5% agarose gel and evaluated with the help of a video documentation system.

RESULTS

In a pre-screen performed with 44 primers, 6 primers were selected for further analyses of individual plants from different accessions. These primers were chosen because of their ability to discriminate all accessions included in the pre-screen and, additionally, to indicate quantitative or qualitative polymorphism between two plants from one accession. In Figure 1 the results of the pre-screen of one of the primers (026) is given as an example. It can be seen that the accessions could be discriminated by this primer, and, also, the two plants of accession 5 displayed a polymorphic band (see arrow).

FIGURE 1. RAPD fingerprints of 9 individual plants from 8 accessions, obtained with primer 026 (M: marker, bp: basepairs). Adapted from Arnholdt-Schmitt (1).

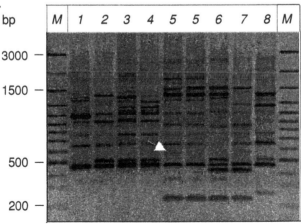

With the help of the selected primers, seedlings of a plant collection of a wild population of *H. perforatum* (accession A) were characterized by RAPD fingerprinting. The results obtained demonstrate that the collection contains highly polymorphic genomes. Nevertheless, a high degree of putative identity between plants was also found. In Figure 2 it becomes clear by lanes 1, 2 and 9 that in fact three different genotypes could be revealed amongst the 10 seedlings which were investigated. Additionally, it is obvious by lanes 1 and 6, as well as by lanes 2, 3, 4, 5, 7, 8 and 10, that at least very similar if not identical genomes were present. This observation was confirmed by the other five selected primers (data not shown, see (1)). To follow up the question of whether, in general, offspring of *H. perforatum* plants show identical genomes or display variation, 10 progenies of individual plants from 4 different accessions were analyzed. The genotype analyses of offspring of three accessions indicated a high degree of identity. Figure 3 illustrates RAPD fingerprints of 10 individual plants from one accession (E) as an example. From another accession (B), 10 progeny plants of two individual plants (B/1 and B/2) were analyzed. Figure 4 shows, as clearly confirmed by the RAPD fingerprints, that the randomly chosen two parent plants originated from the same accession. Additionally, it can be seen in Figure 4a that 10 progenies of plant B/2 again demonstrated identity. However, two offspring of plant B/1 (Figure 4b) show polymorphism, which is due in each case to only one polymorphic band (see arrows). Minor deviations in the genomes of these two progeny plants were also confirmed by the additional 5 primers (data not shown).

DISCUSSION

RAPD fingerprints of 10 seedlings from a plant collection of *H. perforatum* demonstrated typical signs of a wild population that proliferates by vegetative reproduction or by self-pollination. This became evident by the complex fingerprint patterns which indicate a mixture of highly distinctive groups of identical individual plants. Further, RAPD analyses of progenies from individual plants displayed predominantly identical reproduction. However, polymorphic fingerprints due to single bands were also evaluated as a minor event among progenies from individual *H. perforatum* plants. The results are also in good agreement with the findings of Noack (4), who reported a high degree of apomixis for *H. perforatum,* as well as with RFLP analyses, which also revealed the occurrence of non-identical reproduction (2). Nevertheless, further cytological and genomic studies are required in diverse populations to elucidate the putative occurrence and extent of apomixis and intra- as well as interspecies cross-fertilization in *H. perforatum.*

FIGURE 2. RAPD fingerprints of 10 seedlings from accession A, obtained with primer 026 (M: marker, bp: basepairs). Adapted from Arnholdt-Schmitt (1).

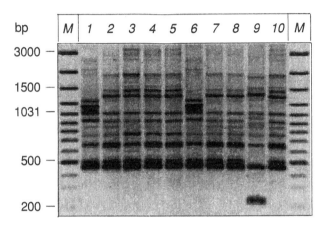

FIGURE 3. RAPD fingerprints of 10 individual plants from accession E, obtained with primer 026 (M: marker, bp: basepairs). Adapted from Arnholdt-Schmitt (1).

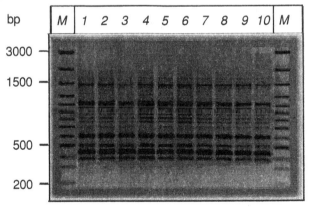

FIGURE 4. RAPD fingerprints of 10 progenies of plant B/2 (a) and plant B/1 (b) from accession B, obtained with primer 029 (M: marker, bp: basepairs). Adapted from Arnholdt-Schmitt (1).

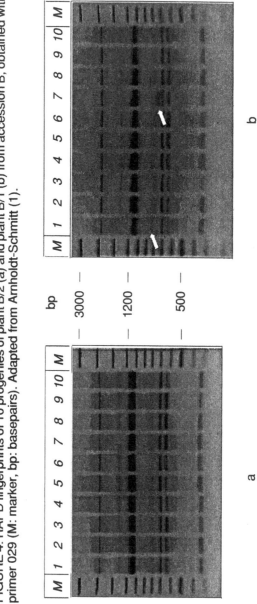

REFERENCES

1. Arnholdt-Schmitt B (2000). RAPD analysis: A method to investigate aspects of the reproduction biology of *Hypericum perforatum* L. *Theor Appl Genet* 100(6): 906-911.

2. Halusková J, Cellárová E (1997). RFLP analysis of *Hypericum perforatum* L. somaclones and their progenies. *Euphytica* 95:229-235.

3. Mártonfi P, Brutovská R, Cellárová E, Repcák M (1996). Apomixis and hybridity in *Hypericum perforatum*. *Folia Geobot Phytotax* 31:389-396.

4. Noack, KL (1939). Über Hypericum–Kreuzungen VI. Fortpflanzungsverhältnisse und Bastarde von *Hypericum perforatum* L. *Z Indukt Abstamm Vererbungslehre* 76:569-601.

5. Stevinson C, Ernst E (1999). Hypericum for depression–An update of the clinical evidence. *Europ. Neuropsychopharmacology* 9(6):501-505.

DNA Fingerprinting by RAPD on *Origanum majorana* L.

Evelyn Klocke
Jan Langbehn
Claudia Grewe
Friedrich Pank

SUMMARY. Breeding research for improvement of the homogeneity of marjoram pollinator lines requires reliable information on the genetic diversity. Three progenies of marjoram pollinator lines were examined by means of RAPD assays. For this reason a suitable protocol for DNA isolation and the conditions for the PCR reactions were worked out. A single plant analysis as well as an analysis with bulked DNA samples were used. The RAPD fingerprints of single plants illustrate the heterogeneity inside the accessions in the first year and the enhancement of the homogeneity in the next progenies causing by selfing. For revealing the genetic distances of the accessions, the pooling strategy is more suitable than the single plant analysis. The simplicity of the RAPDs makes it possible to realize a high number of analyses, but the results should be evaluated with caution, with knowledge about some drawbacks of the

Evelyn Klocke, Jan Langbehn, Claudia Grewe, and Friedrich Pank are affiliated with the Federal Centre for Breeding Research on Cultivated Plants, Institut of Horticultural Crops, Neuer Weg 22/23, 06484 Quedlinburg, Germany.

The authors wish to thank Elisabeth Fleck for her excellent technical assistance.

The authors thank the European Union for financial support (FAIR3-CT96-1914).

[Haworth co-indexing entry note]: "DNA Fingerprinting by RAPD on *Origanum majorana* L." Klocke, Evelyn et al. Co-published simultaneously in *Journal of Herbs, Spices & Medicinal Plants* (The Haworth Herbal Press, an imprint of The Haworth Press, Inc.) Vol. 9, No. 2/3, 2002, pp. 171-176; and: *Breeding Research on Aromatic and Medicinal Plants* (ed: Christopher B. Johnson, and Chlodwig Franz) The Haworth Herbal Press, an imprint of The Haworth Press, Inc., 2002, pp. 171-176. Single or multiple copies of this article are available for a fee from The Haworth Document Delivery Service [1-800-HAWORTH 9:00 a.m. - 5:00 p.m. (EST). E-mail address: getinfo@haworthpressinc.com].

171

method. Nevertheless, there is the conclusion that RAPDs are a useful tool in marjoram breeding programs. *[Article copies available for a fee from The Haworth Document Delivery Service: 1-800-HAWORTH. E-mail address: <getinfo@haworthpressinc.com> Website: <http://www.HaworthPress. com> © 2002 by The Haworth Press, Inc. All rights reserved.]*

KEYWORDS. *Origanum majorana* L., DNA extraction, RAPD, inbred progenies

INTRODUCTION

Marjoram is an important spice plant in Central Europe. One of the tasks in marjoram breeding is the development of a hybrid system. The formation of pollinator lines with a high homogeneity by using selection of suitable plants and inbreeding by selfpollination is a necessarily step in this process.The introduction of RAPD markers to marjoram breeding programs could have a significant impact on the success of breeding. The RAPD technique is a very fast and relatively easy PCR procedure. The main focus of this work was the establishment of a rapid method for isolation of total marjoram DNA, the optimization and standardization of reaction conditions for RAPD-PCR for marjoram, and the use of RAPD investigations for the characterization of various marjoram strains over three years.

MATERIALS AND METHODS

Plant material. In 1997, the first year of testing, 20 marjoram accessions from various sources (breeders, gene bank, etc.) were analyzed. In the second and third years, a few single plants were selected for the development of inbred progenies by selfpollination. For this reason, the numbers of accessions in each year have another genetic background but the level of inbreeding increased from year to year.

Isolation of total DNA. Total genomic DNA was individually extracted from fresh or frozen marjoram leaf material with a rapid method according to Dorokhov and Klocke (1).

RAPD-PCR analysis. The 12.5 µl reaction mixture contained the following components: 1X incubation buffer, 100 µM dNTPs, 0.2 µM primer, 0.25 U Taq polymerase (Appligene) and 20 ng DNA. The PCR programme began with a denaturation step for 3 min at 94°C, then 45 cycles with primer annealing (36°C, 0.5 min), extension (72°C, 1 min) and denaturation (94°C, 0.3 min). Dekamer primers from kit A (Operon Technologies, USA) were used. Amplified products were electrophoresed in a 2% agarose gel stained with ethidium bromide and the gel image was recorded on a computer using a videocamera and the image software package "Kontron KS 300." For the analysis of the digital images of the gels, the software RFLPscan Plus version 3.0 (1994, Scanalytics, USA) was used. The data from single plants and pooled samples were processed separately.

RESULTS

DNA extraction. The DNA extraction method should be rapid since it was necessary to extract DNA from a lot of individual marjoram plants. The isolation of DNA is an important requirement for the following PCR investigations. About 50 mg of leaf material yielded an average of 1 µg DNA in 2-3 hours, sufficient for at least 50 RAPD-PCR reactions. There was no difference between fresh or frozen material.

PCR conditions. Under the described conditions, depending on the primer, there were 5 to 20 amplified fragments produced. A large majority of the strong, reproducible products fall in the range of 250-2000 bp.

Analysis of single plant samples in 1997, 1998 and 1999. In 1997, for the analyses of single plants, four plots of 12 plants from each accession were used. In the next two years, only two plots from each progeny were analyzed by RAPD-PCR. In summary, 454 gels each with 18 DNA samples were electrophoresed. The Operon primers A-02, A-03, A-04 and A-20 were used for each DNA sample.

For the statistical analysis, 38 user-defined match bins were used for characterizing polymorphic RAPD markers of 4 primers and all gel images of single plant samples from 1997-1999. The user defined bins comprise 65.2% of the detected bands. The cluster analysis shows an increase in the differentiation of the targeted strains over three years. The complexity of the dendrograms from 1997, 1998 and 1999 rises continuously (Figure 1a, b). The growth of genetic homogeneity of the single plants inside the strains due to self-pollination causes the changes in the dendrograms as well as in the genetic distances.

Analysis of bulked DNA samples in 1997. Bulked DNA samples were used to detect DNA polymorphisms in accessions of 1997. For this reason the isolated DNA of 4 single plants from each accession was pooled. This simple modification levels out differences between the individual plants, simultaneously intensifying accession-typical DNA differences. Each pooled sample was tested with 16 different primers with one replication. For matching the database, 128 user-defined bins of 16 primers were taken into account. Despite the high number of considered bins, only 58.7% of the revealed RAPD bands

FIGURE 1. Dendrogram of 20 accessions in 1997(a) and for 20 progenies in 1999 (b) (4 primers, single plant samples).

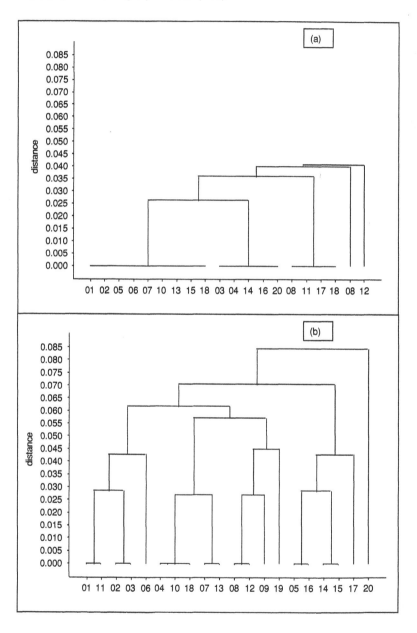

were used in the analysis using the RFLPscan software. Nevertheless, the pooled samples together with the higher number of primers gave a better discrimination of the accessions in comparison to the analyses of single plants (Figure 2).

DISCUSSION

Marjoram, as a spice plant, contains a lot of secondary metabolites which can inhibit the PCR reaction. Despite this, it was found that a rapid extraction method gives enough DNA from marjoram leaf material of a suitable quality for PCR. The rapid isolation method and the storage possibility of the plant material at $-80°C$ were prerequisites for the handling of the high number of individual plants.

RAPDs are a kind of PCR reaction with only one short (10 bp) arbitrary primer. Depending on the primer used, DNA fragments will be enzymatically amplified. The number and size of the fragments are characteristic for each genome (3,5), but the reproducibility is sometimes affected by inadequate

FIGURE 2. Dendrogram of 18 accessions in 1997 (16 primers, pooled samples).

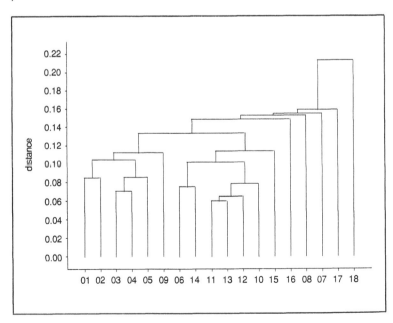

amplification caused by the relatively low annealing temperature (34-36°C) (2,4). For this reason, the optimization of reaction conditions and the selection of suitable 10bp primers were important. The sometimes poor reproducibility of the RAPD reaction as well as the insufficient resolution of an agarose gel and an image quality loss by the digitalizing of the gel could cause ignored amplicons for the statistical analysis. Some problems could be resolved by the application of an automatic DNA sequencer with a high resolution capacity and further developments in software for gel image analysis.

In spite of some problems with reproducibility, RAPD analysis could be a useful tool in the marjoram breeding process. Depending on the breeding task, either the single plant or bulked samples analysis should be applied.

REFERENCES

1. Dorokhov, B. D. and E. Klocke. 1997. A rapid and economic technique for RAPD analysis of plant genomes. *Russ J. of Genetics* 33(4): 358-365.

2. Meunier J. R. and P. A. Grimont. 1993. Factors affecting reproducibility of random amplified polymorphic DNA fingerprinting. *Res. Microbiol.* 144: 373-379.

3. Rafalski, J. A., S. V. Tingey, and J. G. K. Williams. 1991. RAPD markers–a new technology for genetic mapping and plant breeding. *AgBiotech News and Information* 3: 645-648.

4. Skroch, P. and J. Nienhuis. 1995. Impact of scoring error and reproducibility of RAPD data on RAPD based estimates of genetic distance. *Theor. Appl. Genet.* 91: 1086-1091.

5. Williams, J. G. K., A. R. Kubelik, K. J. Livak, J. A. Rafalski, and S. V. Tingey. 1990. DNA polymorphisms amplified by arbitrary primers are useful as genetic markers. *Nucl. Acids Res.* 18: 6531-6535.

Determination of the Progenitors and the Genetic Stability of the Artichoke Cultivar SALUSCHOCKE® Using Molecular Markers

Monika Messmer
Ernst Scheider
Günter Stekly
Bernd Büter

SUMMARY. The registered artichoke (*Cynara cardunculus* L. subsp. *flavescens* Wilk.; synonym to *Cynara scolymus* L.) cultivar Blanc amélioré/ SALUSCHOCKE® of Salus-Haus was developed by open pollination of four selected populations (M1-M4) in a polycross design. Objectives of this study were to determine the progenitors, to evaluate the genetic stability of SALUSCHOCKE in advanced generations and to assess the genetic similarity within and between artichoke populations with the aid of RAPD (random amplified polymorphic DNA) markers. DNA of 31 single plants was extracted and tested with 57 RAPD primers. The synthetic variety SALUSCHOCKE was most similar to the populations M1 and

Monika Messmer and Bernd Büter are affiliated with VitaPlant AG, Benkenstr. 254, CH-4108 Witterswil, Switzerland.

Ernst Scheider and Günter Stekly are affiliated with Salus-Haus, Dr. med. Greither Nachf. GmbH & Co. KG, Natur-Arzneimittel, Bahnhofstr. 24, D-83052 Bruckmühl, Germany.

[Haworth co-indexing entry note]: "Determination of the Progenitors and the Genetic Stability of the Artichoke Cultivar SALUSCHOCKE® Using Molecular Markers." Messmer, Monika et al. Co-published simultaneously in *Journal of Herbs, Spices & Medicinal Plants* (The Haworth Herbal Press, an imprint of The Haworth Press, Inc.) Vol. 9, No. 2/3, 2002, pp. 177-182; and: *Breeding Research on Aromatic and Medicinal Plants* (ed: Christopher B. Johnson, and Chlodwig Franz) The Haworth Herbal Press, an imprint of The Haworth Press, Inc., 2002, pp. 177-182. Single or multiple copies of this article are available for a fee from The Haworth Document Delivery Service [1-800-HAWORTH 9:00 a.m. - 5:00 p.m. (EST). E-mail address: getinfo@haworthpressinc.com].

M3 and very different from population M4. Little genetic variation was found between single plants of the F_4 generation and between the F_1 and F_4 generation of the variety SALUSCHOCKE, indicating homogeneity and genetic stability across generations. *[Article copies available for a fee from The Haworth Document Delivery Service: 1-800-HAWORTH. E-mail address: <getinfo@haworthpressinc.com> Website: <http://www.HaworthPress.com> © 2002 by The Haworth Press, Inc. All rights reserved.]*

KEYWORDS. *Cynara*, artichoke breeding, RAPD, genetic similarity

INTRODUCTION

Leaf extracts of artichoke are widely used in the treatment of dyspeptic disorders due to their strong choleretic effects (1). Leaf extracts also show antioxidative, hepatoprotective and anti-cholestatic effects as well as inhibiting actions on cholesterol biosynthesis and LDL oxidation (2). The essential ingredients of artichoke extracts are caffeic acid, chlorogenic acid, cynarin, luteolin and the glycosides scolymoside and cynaroside. Artichoke is an outcrossing species, thus special breeding effort is necessary to obtain a homogeneous, genetically stable cultivar. Objectives of this study were to determine the progenitors of the registered cultivar Blanc amélioré/SALUSCHOCKE® of Salus-Haus developed by polycross, to evaluate the genetic stability of SALUSCHOCKE in advanced generations and to assess the genetic similarity within and between artichoke cultivars.

MATERIALS AND METHODS

Four selected artichoke populations (M1-M4) were planted in a polycross design allowing open-pollination. The F_1 seed was selected for agronomically important traits, high content of pharmacologically relevant compounds and sensoric traits in order to develop the improved cultivar SALUSCHOCKE. For RAPD analysis, seeds of the original selection (F_1) and the fourth generation (F_4) of SALUSCHOCKE as well as seeds of the possible progenitors were planted. Leaf material of one F_1 and ten F_4 plants of SALUSCHOCKE and five plants of each potential progenitor were harvested. DNA extraction of the 31 plants was performed according to Siedler et al. (3). In order to compare the populations, DNA of single plants of each population was combined to one mixed sample (M1_mix, M2_mix, M3_mix, M4_mix, S2_mix). Altogether, 57 RAPD primers (10 base, Operon Technology, USA) were tested with the mixed DNA samples, whereas 27 RAPD primers were tested also on the DNA of the 31 single plants. PCR reactions contained 100 ng template DNA, 30 ng primer, 0.2 mMol of each dNTP (deoxynucleotides), and 2.5 units rTaq DNA polymerase (Amersham Pharmacia Biotech) in a buffer of 10 mMol Tris-HCl (pH 9.0), 1.5 mMol $MgCl_2$ and 50 mMol KCl in a volume of 50 µl. Amplification was performed in a thermocycler (GeneAmp PCR system 9700, Perkin Elmer) programmed for 6 min at 95°C, 40 cycles of 60 s at 95°C, 90 s at 36°C and 120 s at 72°C, and a final extension step at 72°C for 6 min. The amplification products were separated in 1.5% agarose gels and visualized by ethidium bromide staining. For statistical analysis, only those primers resulting in unambiguous bands were used. Genetic similarity values (GS) between each pair of genotypes were calculated according to the formula of Dice (4). Based on these similarity values a cluster analysis (UPGMA = unweighted pair group method average linkage) was performed using NTSYS-pc (5).

RESULTS

Of the 57 primers tested, only 26 revealed polymorphism between the different populations, whereas 5 were monomorphic and 19 showed no or only ambiguous bands. Altogether, 186 bands were used for the calculation of genetic similarity. Cluster analysis (Figure 1) revealed that the populations M1, M2 and M3 used in the polycross were related to each other, whereas M4 was very different. The synthetic variety SALUSCHOCKE was most similar to the populations M3 (GS = 0.97) and M1 (GS = 0.96) but genetically distinct from M4 (GS = 0.63).

The genetic relationship among the 31 single plants is represented in Figure 2. While the plants of M2 and M4 clustered together, single plants of M1, M3 and SALUSCHOCKE could not be classified to the respective populations by RAPD data. Mean, minimum and maximum values of GS within and between populations are presented in Table 1. Analysis of the single plant of the F_1 generation and the ten plants of the F_4 generation of the synthetic variety SALUSCHOCKE revealed no significant differences with respect to genetic similarity. Genetic similarity within populations was highest for M3 (GS = 0.95), M2 (GS = 0.94) and SALUSCHOCKE (GS = 0.94) and lowest for M4 (GS = 0.87).

DISCUSSION

Based on the RAPD data, the populations M3 and M1 contributed most to the genetic composition of SALUSCHOCKE. Since the M2 population proved to be related to M1 and M3 it cannot be decided if the genetic similarity found between M2 and S2 was due to the relatedness of M2 to the probable progeni-

FIGURE 1. Dendrogram of the 4 possible progenitor populations (M1-M4) and the synthetic variety SALUSCHOCKE® (S2) based on genetic similarity values calculated from RAPD data of the DNA mixture.

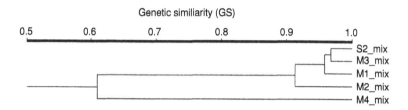

FIGURE 2. Dendrogram of the 31 single plants of the possible progenitor populations (M1-M4) and the synthetic variety SALUSCHOCKE® (S1, S2) based on genetic similarity values calculated from RAPD data.

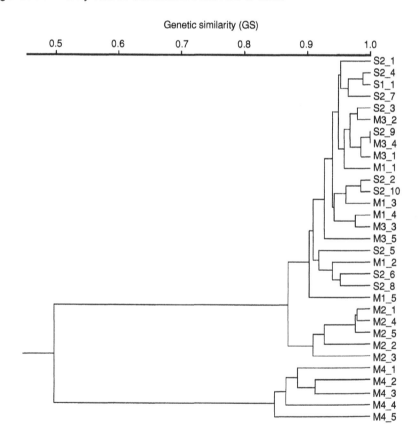

TABLE 1. Mean (minimum and maximum) of the pairwise genetic similarities (GS) of the 31 single plants within (**bold**) and between the four possible progenitor populations (M1, M2, M3, M4) and the synthetic variety SALUSCHOCKE® (S1 = F_1 generation, S2 = F_4 generation).

Population	S1	S2	M1	M2	M3	M4
Number of plants	1	10	5	5	5	5
S2	0.95 (0.91 - 0.99)	**0.93** **(0.89 - 0.99)**				
M1	0.94 (0.93 - 0.96)	0.92 (0.86 - 0.97)	**0.91** **(0.86 - 0.94)**			
M2	0.88 (0.83 - 0.92)	0.87 (0.81 - 0.95)	0.86 (0.79 - 0.94)	**0.94** **(0.87 - 0.98)**		
M3	0.94 (0.92 - 0.95)	0.94 (0.86 - 1.00)	0.93 (0.88 - 0.98)	0.88 (0.80 - 0.94)	**0.95** **(0.92 - 0.98)**	
M4	0.51 (0.47 - 0.58)	0.49 (0.40 - 0.55)	0.51 (0.45 - 0.60)	0.50 (0.46 - 0.58)	0.51 (0.45 - 0.60)	**0.87** **(0.83 - 0.91)**

tors of SALUSCHOCKE or due to direct contribution of gametes to the synthetic variety SALUSCHOCKE. In contrast, M4 can be excluded as a potential progenitor. Possibly M4 deviated in the flowering time and therefore could not participate in the development of SALUSCHOCKE. A small amount of genetic variation was found between single plants within population compared to the large variation found between unrelated populations (e.g., M3 and M4). As only one of the seeds of the F_1 generation germinated and thus only one plant could be analyzed, information on the genetic stability of the synthetic variety SALUSCHOCKE is limited. However, this plant was indistinguishable from plants of the F_4 generation, indicating no major changes in advanced generations. Since artichoke is an outbreeding species, it is possible that already in the F1 generation the Hardy-Weinberg equilibrium for gene frequencies was reached. In this case, SALUSCHOCKE is expected to be genetically stable across generations if the seed propagation is isolated from other artichoke populations.

REFERENCES

1. Fintelmann, V. 1996. Antidyspeptische und lipidsenkende Wirkung von Artischokenblätterextrakt. *Z. Allgemeine Medizin* 72:48-57.

2. Kraft K. 1997. Artichoke leaf extract–recent findings reflecting effects on lipid metabolism, liver and gastrointestinal tracts. *Phytomedicine* 4:369-378.

3. Siedler, H., M.M. Messmer, G.M. Schachermayr, H. Winzeler, M. Winzeler, and B. Keller. 1994. Genetic diversity in European wheat and spelt breeding material based on RFLP data. *Theoretical and Applied Genetics* 88:994-1003.

4. Dice, L.R. 1945. Measures of the amount of ecologic association between species. *Ecology* 26:297-302.

5. Rohlf, F.S. 1972. NTSYS-pc numerical taxonomy and multivariate analysis system. Exeter Publishing Ltd., Setauket, New York.

Investigations on Morphological, Biochemical and Molecular Variability of *Ocimum* L. Species

Sabine B. Wetzel
Hans Krüger
Karl Hammer
Konrad Bachmann

SUMMARY. The genus *Ocimum* L. contains several species which are important economic and medicinal plants. The most important are *O. basilicum* L., *O. americanum* L. and their hybrid *O. ×citriodorum* Vis., which are used for essential oil production and as potherbs. Until now, the complex relationships between these species and their high intraspecific variability in morphological and biochemical characters were not well understood. The morphological differentiation of these species is very difficult. For an unambiguous communication about cultivated basils it is necessary to analyze an extensive sample with a combination of morphological, karyotypic, chemical and molecular markers. *[Article copies available for a fee from The Haworth Document Delivery Service: 1-800-HAWORTH. E-mail address: <getinfo@haworthpressinc.com> Website: <http://www. HaworthPress.com> © 2002 by The Haworth Press, Inc. All rights reserved.]*

Sabine B. Wetzel and Konrad Bachmann are affiliated with the Institut für Pflanzengenetik und Kulturpflanzenforschung, IPK, Corrensstr. 3, 06466 Gatersleben, Germany.

Hans Krüger is affiliated with Bundesanstalt für Züchtungsforschung, Neuer Weg 22/23, 06484 Quedlinburg, Germany.

Karl Hammer is affiliated with the Universität Gesamthochschule Kassel, Steinweg 11, 37213 Witzenhausen, Germany.

[Haworth co-indexing entry note]: "Investigations on Morphological, Biochemical and Molecular Variability of *Ocimum* L. Species." Wetzel, Sabine B. et al. Co-published simultaneously in *Journal of Herbs, Spices & Medicinal Plants* (The Haworth Herbal Press, an imprint of The Haworth Press, Inc.) Vol. 9, No. 2/3, 2002, pp. 183-187; and: *Breeding Research on Aromatic and Medicinal Plants* (ed: Christopher B. Johnson, and Chlodwig Franz) The Haworth Herbal Press, an imprint of The Haworth Press, Inc., 2002, pp. 183-187. Single or multiple copies of this article are available for a fee from The Haworth Document Delivery Service [1-800-HAWORTH 9:00 a.m. - 5:00 p.m. (EST). E-mail address: getinfo@haworthpressinc.com].

KEYWORDS. *Ocimum*, basil, spice, medicinal plant, essential oil, chemotaxonomy, molecular biology, morphology

INTRODUCTION

Plants of the genus *Ocimum* have been used for at least 3000 years as culinary and medicinal herbs (1). Economically most relevant is *Ocimum basilicum*, well-known as a spice for tomato products and pesto sauce but also cultivated for essential oil production. Due to the antimicrobial activity of the essential oil against bacteria, molds and yeasts, basil is also used as a medicinal plant (2) and for biological plant protection (3). Next to basil, *Ocimum americanum* and their hybrid *Ocimum ×citriodorum* are the most intensively used species. These are very fragrant plants, with scents similar to those of cloves, cinnamon and lemon. During the second world war, *Ocimum kilimandscharicum* was cultivated to prepare camphor (4). *Ocimum gratissimum* has analgesic and spasmolytic activity (5) and is used in traditional African medicine. *Ocimum tenuiflorum,* the holy basil, is sacred to the Hindus and offered in special prayers and rituals (6). In phytotherapy it is considered as a protective agent against carcinogenesis (7). The potential of several other *Ocimum* species as medicinal herbs is still not exhausted.

The place of origin of the genus is unknown. Some experts suggest India or Asia (8), but tropical West Africa seems most likely at present, because most of the species are naturally distributed there (9).

Ocimum is not well understood taxonomically. A new revision of the genus (9) shows that *Ocimum* is not clearly delimited from related genera. The preliminary number of species is 65. The complex relationships among these species and their high infraspecific variability in morphological, chemical and molecular characters need to be characterized.

MATERIALS AND METHODS

In 1999, 257 accessions of the genebank in Gatersleben were compared in a field experiment. These originated from 39 countries and were obtained during several collecting missions and the international seed exchange.

The following species were included:

- *O. americanum* ssp. *americanum* var. *pilosum* (Willd.) Paton
- *O. americanum* var. *americanum* L.
- *O. basilicum* ssp. *basilicum* L.
- *O. basilicum* ssp. *basilicum* var. *purpurascens* Benth.

- *O. basilicum* ssp. *basilicum* var. *thyrsiflorum* (L.) Benth.
- *O.* ×*citriodorum* Vis.
- *O. campechianum* Mill.
- *O. gratissimum* ssp. *gratissimum* var. *gratissimum* L.
- *O. gratissimum* ssp. *gratissimum* var. *macrophyllum* Briq.
- *O. gratissimum* ssp. *iringense* Ayobangira ex Paton
- *O. kilimandscharicum* Baker ex Gürke
- *O. minimum* L.
- *O. selloi* Benth.
- *O. tenuiflorum* L.

At the beginning of June, 14 plants of each accession were transplanted from greenhouse to field at a 30 cm × 30 cm spacing. A black plastic mulch was used to conserve moisture, increase soil temperature and suppress weeds. In order to avoid evaluation mistakes, a standard with 20 repetitions spread over the whole field was used.

The measured and evaluated parameters can be divided into four groups:

1. Molecular characters: Using a DNA fingerprint method called AFLP (amplified fragment length polymorphism)
2. Vegetative parameters
3. Generative parameters
4. Agronomically important parameters

The essential oils of all *Ocimum* accessions were isolated and identified by coupled GC/MS analysis at the German Federal Centre for Breeding Research on Cultivated Plants (BAZ) in Quedlinburg.

RESULTS AND DISCUSSION

In order to clarify the status of the many varieties of cultivated *Ocimum* species and to provide an up-to-date intraspecific classification, the ongoing project in Gatersleben focuses on the description of the diversity of basil and its nearest relatives. Table 1 lists the ranges of variation of important quantitative characters.

The leaves may be the most variable organs of the plant. Within basil there is a continuous range of leaf length from three to eight centimetres, leaf shapes which can be flat to strongly convex with turned down margins, the surface of the leaves can be smooth to strongly puckered, leaf margins can be smooth to heavily serrated and, finally, leaf colours range from yellow-green to deep red-violet.

TABLE 1. Range of variation (min., max.), mean and median for some quantitative characters of *Ocimum* species.

Character	Ocimum basilicum					Ocimum ×citriodorum					Ocimum americanum				
	min.	mean	median	max.	N	min.	mean	median	max.	N	min.	mean	median	max.	N
Width of cotyledons	0.68	0.99	0.98	1.49	134	0.67	0.81	0.8	0.98	17	0.35	0.72	0.75	0.91	12
Length of cotyledons	0.5	0.7	0.69	1.03	134	0.51	0.61	0.62	0.69	17	0.26	0.46	0.47	0.58	12
Plant height	12.2	44.6	44.1	69.6	134	34	45.7	45.2	62.2	17	30.8	33.2	33.4	35.2	12
Plant diameter	11	26.8	26.9	37	134	26	29.7	30.2	36	17	26	30.9	30.8	37	12
Length of the main inflorescence	4.6	19.75	19.2	30.8	134	19.4	23.4	22.4	31.4	17	10.4	19.5	20.9	21.4	12
Number of whorls of main inflorescence	7	14.4	14.4	19	134	11.6	14.3	13.2	21.6	17	14	18.6	19.3	20	12
Calyx length	4.2	5.2	5.1	6.9	133	4	4.5	4.4	5.3	17	2.8	3.2	3.2	3.5	12
Corolla length	7.3	8.3	8.2	10.4	133	7	7.2	7.3	8.5	17	3.8	5.2	5.4	6	12
Length of the anterior stamen	6.3	8.1	8	9.9	133	5.9	6.7	6.7	8.3	17	4	5.2	5.4	6.1	12
Length of the posterior stamen	8.6	10.6	10.6	12.5	133	8.1	8.9	8.8	11.5	17	5.4	7.1	7.4	7.8	12
Leaf length	3.2	5.4	5.2	8.1	133	3.4	4.5	4.3	5.8	13	1.8	3.9	3.8	6.2	3
Leaf width	1.3	2.8	2.7	5.8	133	1.2	2	1.9	2.8	13	0.7	1.9	1.8	3.3	3
Essential oil content (ml/100 g dried leaves)	0.07	0.29	0.25	1.37	134	0.15	0.36	0.36	0.68	17	0.27	0.42	0.33	0.74	12

Character	Ocimum gratissimum					Ocimum tenuiflorum				
	min.	mean	median	max.	N	min.	mean	median	max.	N
Width of cotyledons	0.56	0.7	0.69	0.9	10	0.41	0.59	0.6	0.79	11
Length of cotyledons	0.34	0.44	0.44	0.56	10	0.27	0.41	0.43	0.53	11
Plant height	28.8	47.7	47.8	69	12	21	34.25	31.8	57	11
Plant diameter	15.2	28.4	25.1	47	12	18	26.4	25.4	38.4	11
Length of the main inflorescence	8.5	10.6	10.25	12.6	11	10	11.4	10	12.2	11
Number of whorls of main inflorescence	11	14.3	13.5	17	12	8.8	12.4	8.8	14	11
Calyx length	3.5	3.9	3.8	4.5	9	2.6	3	2.9	3.6	11
Corolla length	3.5	4.1	4.1	4.4	9	3.52	4	4.1	4.3	11
Length of the anterior stamen	3.5	4.1	4.1	5.3	9	3.12	4.2	4.3	5.1	11
Length of the posterior stamen	4.1	5.2	5.4	5.7	9	3.75	4.6	4.84	5.48	11
Leaf length	6.1	7.9	8.1	9.4	12	2.5	3.9	4	5.9	11
Leaf width	3.5	4.6	4.7	5.4	12	1.3	2	1.9	3.5	11
Essential oil content (ml/100 g dried leaves)	0.33	1.2	1.3	1.95	12	0.06	0.36	0.26	0.9	9

The onset and synchronicity of flowering are important parameters to determine the optimal time for extracting essential oil from leaves. In total, 21 essential oils could be found. In addition to possible uses suggested by the essential oil composition, the data can be used as infraspecific taxonomic characters in *O. gratissimum* and *O. americanum*. The quantitatively most important essential oils of *O. basilicum* are methylchavicol, linalool, eugenol and methylcinnamate. Using the first three major components, 16 essential oil profiles could be found. Though they cannot be used for infraspecific classification, they suggest possible uses of the various chemotypes as spices, insect repellents or for medicinal purposes. A principal component analysis showed that the most informative parameters differentiating various accessions of *O. basilicum*, highly correlated with the first component, are colours of different plant organs. The second component is positively correlated with methylcinnamate content, cotyledon length and width, total essential oil content and the degree of lignification, and negatively correlated with leaf length, width and surface.

For a detailed characterization of the various strains of basil and its relatives, it is necessary to combine morphologic characters together with chemical, karyotypic and molecular markers.

REFERENCES

1. Schultze-Motel, J. (ed.). 1986. Rudolf Mansfelds Verzeichnis landwirtschaftlicher und gärtnerischer Kulturpflanzen. Akademie Verlag, Berlin, 1998 pp.

2. Wan J., Wilcock A., Coventry M.J. 1998. The effect of essential oils of basil on the growth of *Aeromonas hydrophila* and *Pseudomonas fluorescens*. J. Appl. Microbiol. 84(2):152-158.

3. Montes-Belmont R., Carvajal M. 1998. Control of *Aspergillus flavus* in maize with plant essential oils and their components. J. Food Prot. 61(5):616-619.

4. Ryding, O. 1994. Notes on sweet basil and its wild relatives (Lamiaceae). Econ. Bot. 48:65-67.

5. Aziba, P.I., Bass D., Elegbe Y. 1999. Pharmacological investigation of *Ocimum gratissimum* in rodents. Phytother. Res. 13:427-429.

6. Darrah, H.H. 1972. The basils in folklore and biological science. The Herbarist 38:3-10.

7. Aruna K., Sivaramakrishnan V.M. 1990. Plant products as protective agents against cancer. Indian J. Exp. Biol. 28(11):1008-1011.

8. Morton, J.F. 1981. Atlas of medicinal plants of Middle America. Charles C. Thomas, Springfield, USA, 1420 pp.

9. Paton, A.H., Harley, R.M., Harley, M.M. 1999. *Ocimum*–an overview of relationships and classification, In: Holm, Y. and Hiltunen, R. (Eds.), Basil: the genus *Ocimum*. Amsterdam: Harwood Academic.

Peltate Glandular Trichomes of *Ocimum basilicum* L. (Sweet Basil) Contain High Levels of Enzymes Involved in the Biosynthesis of Phenylpropenes

David R. Gang
James Simon
Efraim Lewinsohn
Eran Pichersky

SUMMARY. The plant defense compounds of the phenylpropene class (chavicol, eugenol and their derivatives) have been recognized since antiquity as important spices for human consumption, and plants that contain high concentrations of these compounds (e.g., cloves) have high economic value. Several cultivars of basil (*Ocimum basilicum*) have been developed that produce essential oils rich in specific phenylpropene compounds. However, our understanding of the biosynthetic pathway

David R. Gang and Eran Pichersky are affiliated with the Department of Biology, University of Michigan, Ann Arbor, MI.

James Simon is affiliated with the Horticulture Department, Purdue University, West Lafayette, IN.

Efraim Lewinsohn is affiliated with the Aromatic, Medicinal and Spice Crops Unit, Newe Ya'ar Research Center, Agricultural Research Organization, P.O. Box 1021, Ramat Yishay, 30095, Israel.

Address correspondence to: David R. Gang, Department of Biology, University of Michigan, Ann Arbor, MI 48109-1048 (E-mail: dgang@umich.edu).

This work was funded by a USDA-BARD grant.

[Haworth co-indexing entry note]: "Peltate Glandular Trichomes of *Ocimum basilicum* L. (Sweet Basil) Contain High Levels of Enzymes Involved in the Biosynthesis of Phenylpropenes." Gang, David R. et al. Co-published simultaneously in *Journal of Herbs, Spices & Medicinal Plants* (The Haworth Herbal Press, an imprint of The Haworth Press, Inc.) Vol. 9, No. 2/3, 2002, pp. 189-195; and: *Breeding Research on Aromatic and Medicinal Plants* (ed: Christopher B. Johnson, and Chlodwig Franz) The Haworth Herbal Press, an imprint of The Haworth Press, Inc., 2002, pp. 189-195. Single or multiple copies of this article are available for a fee from The Haworth Document Delivery Service [1-800-HAWORTH 9:00 a.m. - 5:00 p.m. (EST). E-mail address: getinfo@haworthpressinc.com].

189

that produces these compounds in the plant has remained incomplete, with several enzymatic steps, as well as the site of synthesis and storage, still undetermined. Like other members of the Lamiaceae, basil leaves possess on their surface two types of glandular trichomes, termed peltate and capitate glands. Here we demonstrate that the essential oil constituents eugenol and methylchavicol accumulate in the peltate glands of, respectively, basil cultivars SW and EMX-1. Assays for phenylalanine ammonia lyase, the enzyme that catalyzes the first step in the biosynthesis of all phenylpropenes, and for chavicol *O*-methyltransferase, the enzyme that catalyzes the last step in the formation of methylchavicol, localized the corresponding enzyme activities almost exclusively to the peltate glands. *[Article copies available for a fee from The Haworth Document Delivery Service: 1-800-HAWORTH. E-mail address: <getinfo@haworthpressinc. com> Website: <http://www.HaworthPress.com> © 2002 by The Haworth Press, Inc. All rights reserved.]*

KEYWORDS. Essential oil, phenylpropanoid metabolism, eugenol, methylchavicol, terpenes

INTRODUCTION

The major constituent of cloves, one of the spices which led Columbus to sail from Spain, is the phenylpropene eugenol, which gives this spice its pungent, distinctive aroma (2). Eugenol and the related phenylpropene, methylchavicol (also called estragole), are major flavor compounds in certain varieties of basil (*Ocimum basilicum* L., Lamiaceae), which contain upwards of 40% eugenol or 90% methylchavicol in their essential oils (2,4). These and related phenylpropenes are present, in varying amounts, in many other spices and herbs (2). The adaptive value of the toxic (i.e., defensive) properties of the phenylpropenes are likely responsible for the widespread distribution of these chemicals among the angiosperms, and humans have made extensive use of these properties to further protect their plants and food stocks. Surprisingly little is known about the biosynthesis of eugenol, methylchavicol and related phenylpropenes. Eugenol and chavicol are the simplest phenylpropenes. Labeling experiments of leaf segments with radioactive phenylalanine indicated that phenylalanine is their initial precursor (6), but no further work has been carried out to identify the enzymes involved and the location in the plant of their synthesis. A characteristic of plants in the Lamiaceae, which includes basil and mint, is the possession of specialized glands known as glandular trichomes on the surface of their leaves. In these plants, two classes of secretory glandular trichomes (glands) can be found (1,10). In mint, it has been shown that the peltate glands, which are made of a stalk cell attached to the leaf, 8 se-cretory cells attached to the stalk cell, and an oil sac ("subcuticular space") above the secretory cells, contain the stored essential oil components (1,7). We have begun a project to identify the location of synthesis of the phenylpropenes in basil, and to isolate and study the enzymes responsible for their synthesis from phenylalanine. Here we describe evidence that the sacs of the peltate glands of basil contain phenylpropenes, and that the peltate glands themselves contain some of the enzymes involved in phenylpropene synthesis at much higher concentration that in the rest of the leaf tissue.

MATERIALS AND METHODS

Plant Material–Seeds for two cultivars of basil (*Ocimum basilicum* L.), designated EMX-1 and SW, were from stocks developed at Newe Ya'ar Research Center, Israel.

Gland Isolation–Peltate and capitate glandular trichomes were isolated from young leaves using a method modified from (1). Details of the modifications will be published elsewhere.

Enzyme Assays–PAL activity was determined using a method that measures the conversion of L-phenylalanine into cinnamic acid (3). Eugenol *O*-methyltransferase (EOMT) and chavicol *O*-methyltransferase (CVOMT) activities were determined by measuring the formation of radiolabeled methyleugenol or methylchavicol by protein extracts supplemented with the appropriate substrate and S-[methyl-^{14}C]-adenosyl-L-methionine, as previously described (5,9).

RESULTS

Basil Peltate Glandular Trichomes Contain Essential Oil. Although a previous report correlated gland density with essential oil content (10), a direct measurement of phenylpropene presence in the glands was not attempted. To investigate whether any of these glands (capitate or peltate) are the site of phenylpropene storage, we first used stretched glass pipettes to extract a small droplet of essential oil from sacs of individual peltate glands (observed on the leaf surface under a dissecting microscope) and then analyzed the oil on GC-MS. Although the yields of extracted oil were extremely small, the major essential oil components (methylchavicol for cultivar EMX-1 and eugenol and linalool for cultivar SW) were clearly identified in these extracts (data not shown). The sacs of the capitate glands are too small for this procedure, but extraction of isolated capitate glands (see below) did not reveal the presence of phenylpropenes, whereas a similar extraction did reveal the presence of these compounds in isolated peltate glands.

Isolation of Basil Peltate Glands and Assays of Key Enzymes Involved in Phenylpropene Synthesis. We optimized a procedure (1) for removing the peltate and capitate glands from the leaf surface without damaging them, and separating them from all other leaf material and from each other. Gland preparations from the EMX-1, SW and other cultivars yielded essentially identical results. The peltate glands were observed to be disks of four cells that are highly cytoplasmic and contain only small vacuoles (Figure 1 A and B). When

FIGURE 1. A and B: Isolated peltate glands, not stained with toluidine blue, with focus set at the interface between the gland disk cells and the overlying oil sac (A) and through the middle of the disk of secretory cells (B), scale = 20 μm. C and D: Isolated peltate glands stained with toluidine blue, showing the prominent nucleoli (C) and a comparison of the cross-sectional and transverse views of the disk of secretory cells (D), scale = 20 μm.

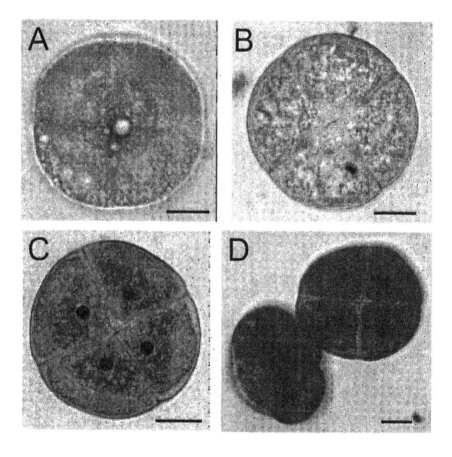

stained with toluidine blue (Figure 1 C and D), the nucleoli are very prominent, indicating that the secretory cells possess high metabolic activity.

Crude protein extracts obtained from isolated peltate glands, isolated capitate glands and from whole young leaves were assayed for activity of the first enzyme in the pathway leading to all the phenylpropenes, as well as for the last enzyme in the pathway leading to methylchavicol. These reactions are the well-known conversion of phenylalanine to cinnamic acid by the enzyme phenylalanine ammonia lyase (PAL), and the recently discovered conversions of chavicol to methylchavicol by the enzyme chavicol *O*-methyltransferase (CVOMT) (5). We also tested for the enzyme that converts eugenol to methyleugenol, EOMT (eugenol *O*-methyltransferase) (8). The peltate glands of both basil lines contained much higher specific activities for PAL than did the whole leaves (Figure 2). The peltate glands from the EMX-1 cultivar, which

FIGURE 2. The first and last enzymes in the phenylpropanoid pathway are active in the peltate glandular trichomes in basil varieties.

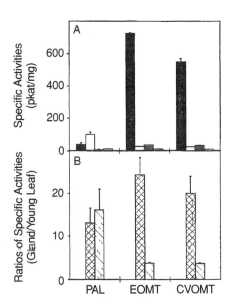

A. Comparison of enzymatic activities present in protein extracts from peltate glands and whole young leaves from the eugenol and methylchavicol producing basil lines. Black: methylchavicol line (EMX-1) peltate glands; white: eugenol line (SW) peltate glands; dark gray: methylchavicol line (EMX-1) young leaves; light gray: eugenol line (SW) young leaves.
B. Comparison of the ratios of specific enzymatic activities present in protein extracts from peltate glands and whole young leaves from the methylchavicol (EMX-1, cross-hatched filled) and eugenol (SW, slashed line filled) producing basil lines.

accumulates high levels of methylchavicol, also had high levels of CVOMT activities (and, interestingly, also high levels of EOMT). However, in extracts of peltate glands from the SW cultivar, which accumulates eugenol but not methyleugenol or methylchavicol, CVOMT and EOMT activities were greatly reduced in both glands and whole-leaf extracts (Figure 2). No activity for any of the enzymes was observed in extracts from capitate glands.

DISCUSSION

In basil, the leaf surface, as well as the stem, sepals and floral surfaces, are covered with several types of structures that include the four-celled peltate glands, as well as the smaller capitate glands (10). We have been able to separate and isolate the two types of glands, and to analyze them for phenylpropene content. Our results demonstrate that the capitate glands, which are more numerous on the leaf surface though much smaller, do not store these compounds nor do they appear to contain the enzymatic activities requisite for their formation. In contrast, we were able to show that the peltate glands contain the enzymes catalyzing the first and last steps in the synthesis of the methylated phenylpropene methylchavicol at much higher levels than elsewhere in the leaf. It remains now to identify the other enzymes in the phenylpropene pathway that operate in the peltate glands, and we have begun to do so by constructing an EST database from mRNAs isolated from these glands.

REFERENCES

1. Gershenzon, J., D. McCaskill, J.I.M. Rajaonarivony, C. Mihaliak, F. Karp, and R. Croteau. 1992. Isolation of secretory cells from plant glandular trichomes and their use in biosynthetic studies of monoterpenes and other gland products. *Anal. Biochem.* 200:130-138.

2. Guenther, E. 1949. *The Essential Oils.* Vol. 2. D. Van Nostrand Company, Inc., Princeton, NJ. 852 pp.

3. Lamb, C.J., T.K. Merritt, and V.S. Butt. 1979. Synthesis and removal of phenylalanine ammonia-lyase activity in illuminated discs of potato tuber parenchyme. *Biochim. Biophys. Acta* 582:196-212.

4. Lawrence, B.M. 1992. Chemical components of Labiate oils and their exploitation. *In* R.M. Harley and T. Reynolds, eds. *Advances in Labiate Science.* Royal Botanical Gardens, Kew.

5. Lewinsohn, E., I. Ziv-Raz, N. Dudai, Y. Tadmor, E. Lastochkin, O. Larkov, D. Chaimovitsh, U. Raid, E. Putievsky, E. Pichersky, and Y. Shoham. 2000. Biosynthesis of estragole and methyl-eugenol in sweet basil (*Ocimum basilicum* L.). Developmental and chemotypic association of allylphenol *O*-methyltransferase activities. *Plant Science* 160:27-35.

6. Manitto, P., D. Monti, and P. Gramatica. 1974. Biosynthesis of phenyl-propanoid compounds. I. biosynthesis of eugenol in *Ocimum basilicum* L. *J. Chem. Soc., Perkin Trans. I* 14:1727-1731.

7. McCaskill, D. and R. Croteau. 1995. Monoterpene and sesquiterpene bio-synthesis in gladular trichomes of peppermint (*Mentha* ×*piperita*) rely exclusively in plastid-derived isopentenyl diphosphate. *Planta* 197:49-56.

8. Wang, J., N. Dudareva, S. Bhakta, R. Raguso, and E. Pichersky. 1997. Floral scent production in *Clarkia breweri* (Onagraceae). II. Localization and developmental modulation of the enzyme S-adenosyl-L-methionine:(iso)eugenol O-methyltransferase and phenylpropanoid emission. *Plant Physiol.* 114(1):213-221.

9. Wang, J. and E. Pichersky. 1998. Characterization of *S*-adenosyl-L-methion-ine:(iso)eugenol *O*-methyltransferase involved in scent production in *Clarkia breweri*. *Arch. Biochem. Biophys.* 349:153-160.

10. Werker, E., E. Putievsky, U. Ravid, N. Dudai, and I. Katzir. 1993. Glandular hairs and essential oil in developing leaves of *Ocimum basilicum* L. (Lamiaceae). *Ann. Botany* 71(1):43-50.

Application
of Protoplast Fusion Technology
to Tansy (*Tanacetum vulgare* L.):
Biodiversity as a Source
to Enhance Biological Activity
of Secondary Compounds

Marjo Kristiina Keskitalo

SUMMARY. Tansy (*Tanacetum vulgare* L., Asteraceae) is an under-exploited species adapted to a northern climate. Its flower heads and leaves are characterized by their terpenoids which are moderately bioactive against insects. A related species, pyrethrum (*Tanacetum cinerariifolium* Vis.), produces highly bioactive terpenoids, pyrethrins, which are used in commercial biopesticides. The constraint of pyrethrum for further utilization is its cold sensitivity. In this study, tissue culture methods for tansy and pyrethrum were first developed to optimize the best conditions for protoplast isolation from donor tissues. Protoplast fusion between tansy and pyrethrum was developed using polyethyleneglycol. In addition, the genetic and chemical biodiversity of tansy was explored to find new germplasm for the further breeding experiments.

Marjo Kristiina Keskitalo is affiliated with the Agricultural Research Centre of Finland, Plant Production Research, 31600 Jokioinen, Finland (E-mail: marjo.keskitalo@mtt.fi).

The author gratefully acknowledges financial support from the Academy of Finland (grant 7798), Finnish Cultural Foundation, Kemira Foundation, Rotary Foundation of South-West Finland (district 1410).

[Haworth co-indexing entry note]: "Application of Protoplast Fusion Technology to Tansy (*Tanacetum vulgare* L.): Biodiversity as a Source to Enhance Biological Activity of Secondary Compounds." Keskitalo, Marjo Kristiina. Co-published simultaneously in *Journal of Herbs, Spices & Medicinal Plants* (The Haworth Herbal Press, an imprint of The Haworth Press, Inc.) Vol. 9, No. 2/3, 2002, pp. 197-203; and: *Breeding Research on Aromatic and Medicinal Plants* (ed: Christopher B. Johnson, and Chlodwig Franz) The Haworth Herbal Press, an imprint of The Haworth Press, Inc., 2002, pp. 197-203. Single or multiple copies of this article are available for a fee from The Haworth Document Delivery Service [1-800-HAWORTH 9:00 a.m. - 5:00 p.m. (EST). E-mail address: getinfo@haworthpressinc.com].

Up to 1×10^6 protoplasts per 1 g^{-1} FW of leaves could be obtained from tansy leaves. The production of protoplast-derived calli varied, but the best time for cell division was late winter or early spring. Protoplast-derived calli of tansy resulted in root and spontaneous shoot formation, although a protoplast-to-plant procedure still needs more studies. The hybridity of intraspecific (tansy × tansy) and interspecific (tansy × pyrethrum) protoplast fusion was successfully tested already at the callus stage. The nuclear DNA content was higher for calli derived from interspecific fusion (10.66 and 31.87 pg) than from intraspecific fusion (8.84, 15.96 and 19.59 pg). The nuclear DNA content of both types of fusion calli was more than the parental level in many fusion calli. The increased content of nuclear DNA in the fusion calli suggests that some degree of protoplast hybridization had occurred. The distance matrices calculated from the RAPD-PCR data with complete linkage cluster analysis showed that the calli derived from the intraspecific fusion were closer to tansy than to pyrethrum. In contrast, the calli derived from interspecific fusion were closer to pyrethrum than to tansy. Compounds such as syringaldehyde, coniferyl alcohol and artedouglasia oxide were only identified from protoplast-fusion derived calli, but no compounds common for parental tissues and fusion calli could be detected.

Tansy grown in Finland contains a wide range of genetic, chemical and morphological variation. Especially, among the larger number of volatile compounds detected from tansy, the chemotypes accumulating irregular monoterpenes such as artemisia ketone and davadone D, which are biochemically close to pyrethrins, are an interesting source for further biotechnological applications. In conclusion, biodiversity can be a valuable source of novel germplasm for breeding tansies that better accumulate more bioactive compounds. *[Article copies available for a fee from The Haworth Document Delivery Service: 1-800-HAWORTH. E-mail address: <getinfo@haworthpressinc.com> Website: <http://www.HaworthPress.com>*

KEYWORDS. Biopesticides, pyrethrum, *Tanacetum cinerariifolium* Vis., RAPD-PCR (random amplified polymorphic DNA-polymerase chain reaction), nuclear DNA content, flow cytometry, terpenoids, GC-MS (gas chromatography-mass spectrometry)

INTRODUCTION

Tansy (*Tanacetum vulgare* L., Asteraceae) is a wild and under exploited plant adapted to the northern climate and is characterized by its bioactivity against insects, mediated by isoprenoids. The disadvantage of tansy for further utilization is that isoprenoids exist at a low concentration and they are not as active against insects as the group of related isoprenoids, pyrethrins. Pyrethrins are produced by another *Tanacetum* species, pyrethrum (*Tanacetum cinerariifolium* (Trevir.) Schultz-bip.), which does not tolerate long and cold winters. Pyrethrins are environmentally benign insecticidal compounds with an increasing world marked demand (1). The goal of this research was to: (1) detect genetic, chemical and selected traits of morphological variation in tansy; (2) develop optimal *in vitro* conditions for tansy and pyrethrum to maximize the number of isolated protoplasts from the tissue cultured donor tissues; and (3) develop protoplast isolation and fusion methods for tansy and pyrethrum and to test the hybridity.

MATERIALS AND METHODS

Twenty tansy genotypes were collected from different parts of Finland and transplanted to the orchard at the Department of Plant Production, University of Helsinki (2). One accession of seeds from pyrethrum was sown in the greenhouse. Plant material derived from these plants was used in all experiments carried out in this study.

Genetic variation of tansy total DNA was tested with RAPD-PCR and the nuclear DNA content was analyzed with flow cytometry (2). Selected morphological traits from tansy were detected during three years from the genotypes grown in the orchard (2). Volatile compounds from tansy flower heads were extracted and identified with GC-MS (3).

Tissue culture procedures, including micropropagation, regeneration from leaf explants and suspension cultures, were developed for tansy (4,5) before protoplast fusion methods could be studied (4,6). A protoplast fusion procedure using polyethylene glycol was developed for tansy and pyrethrum. The hybridity of protoplast-fusion derived calli was detected with RAPD-PCR, flow cytometry and GC-MS (6).

Statistical analysis was carried out with SAS. The data from RAPD-PCR and GC-MS was recorded as present or absent and analyzed with complete linkage cluster analysis (2,3,5).

RESULTS AND DISCUSSION

Biodiversity of tansy. The data based on variation of total DNA and analyzed with RAPD-PCR divided the twenty tansy genotypes first into two groups, which were further divided into smaller clusters. Geographically, most of the genotypes in the first group originated south of latitude 60°30′N, whereas most of the genotypes in the second group originated north of this latitude (2). In total, 55 volatile compounds were recorded from the 20 tansy genotypes, of which 53 were identified. Most of the genotypes contained camphor as the main constituent, but also chemotypes containing thujone, 1-8-cineole, tricyclene, myrcene, artemisia ketone and davadone D could be identified, the latter two being irregular monoterpenes. The complete linkage analysis based on the volatiles identified from tansy divided the chemotypes first into two groups, which were further divided into smaller ones. Correlation between the distance matrices calculated from RAPD-PCR data and data on volatile compounds was 0.41, showing some analogy between these two matrices (Keskitalo et al. 2001).

Development of tissue culture methods for tansy. Proper tissue culture procedure was essential for the development of protoplast techniques (4,5). Especially, the occurrence of endophytic bacteria (Gram-negative) in *in vitro* cultured tansy plantlets was problematic. Antibiotics such as cefotaxime, gentamicin and rifampicin were the most effective against the isolated bacteria, although none of the antibiotics alone inhibited the growth of these bacteria totally. In contrast, the combination of gentamicin with cefotaxime or rifampicin inhibited bacteria growth effectively and was supplemented to the growth media. To avoid unwanted alteration in the growth of tansy, the shoot cultures were periodically transferred to antibiotic free medium.

Development of protoplast isolation for tansy and pyrethrum. In total more than 400 protoplast isolations were carried out from tansy and pyrethrum. *In vitro* growth conditions of tansy and pyrethrum used as donor tissues and the conditions used during the enzyme incubation had a significant effect on the yield of protoplast isolated from both of the species (Table 1). Number of successful protoplast isolations and the mean number of protoplasts isolated from the leaves varied according to the season (Figure 1). Optimization experiments resulted in cell differentiation, root and spontaneous shoot formation from protoplast-derived calli of tansy. However, no actual shoot regeneration could be obtained.

Development of protoplast fusion between tansy and pyrethrum. Totally, 46 chemical fusion studies between tansy and tansy (intraspecific) or between tansy and pyrethrum (interspecific) were carried out, from which 3 resulted in callus growth. More than 20 calli from these three successful fusion experiments could be obtained. Three different methods were used to characterize the fusion calli. The nuclear DNA content of hybrid calli was compared to pyrethrum, tansy and to protoplast-derived callus of tansy. The range of nuclear DNA content in the somatic hybrid calli was from 8.84 to 31.87 pg and was in most cases higher than the nuclear DNA content in the parental species (tansy 6.41 and 7.39 pg; pyrethrum 13.16-14.76 pg). The nuclear DNA content was higher for calli derived from interspecific fusion (10.66 and 31.87 pg) than from intraspecific fusion (8.84, 15.96 and 19.59 pg). Thus, the nuclear DNA

TABLE 1. The effect of culture conditions on the number of isolated protoplasts (per g^{-1} FW of leaves) from tansy and pyrethrum.

Intensity of illumination for donor tissue, mM m^{-2}s^{-1}	60-80		20-40	
Incubation temperature during enzyme digestion: 29°C				
Enzymes+sucrose[1,2]				
	Tansy	Pyrethrum	Tansy	Pyrethrum
Number of total protoplast isolations	62	68	76	26
Number of succeeded isolations	42	51	59	15
Quantiles of protoplast yield/g^{-1} FW				
75%	7,000,000	5,900,000	9,500,000	5,100,000
Median	3,290,000	1,800,000	4,200,000	2,600,000
25%	690,000	680,000	2,550,000	960,000

1. 0.375-0.5% Macerozyme, 0.5-1.0% Cellulase Onozuka, 0.5-1.0% Cellulysin Calbiochem, 0.25-0.5% Driselase, 0.03-0.04% Pectolyase
2. 0.5 M Sucrose + 5 mM $CaCl_2 \cdot 2H_2O$

content of the two different types of fusion varied, but the content was more than the parental level in many fusion calli. The increased content of nuclear DNA in the fusion calli suggests that some degree of protoplast hybridization had occurred (6). RAPD-PCR was another method to test the hybridity of fusion calli. The distance matrices calculated with complete linkage cluster analysis showed that the calli derived from the intraspecific fusion were closer to tansy than to pyrethrum. In contrast, the calli derived from interspecific fusion were closer to pyrethrum than to tansy. As a result of RAPD-PCR analysis, it was found that fusion between tansy and pyrethrum had occurred because new DNA arrangements were formed. The third method to test the hybridity of fusion-derived calli was to detect the volatile compounds with GC-MS from the parental tissue and from the fusion products. Compounds such as syringaldehyde, coniferyl alcohol and artedouglasia oxide were only identified from protoplast-fusion derived calli. No compounds common for parental tissues and fusion calli could be detected.

CONCLUSIONS

Root and spontaneous shoot formation show that cell differentiation in protoplast-derived calli of tansy is possible, although a protoplast-to-plant pro-

FIGURE 1. Seasonal changes in the mean yield of protoplasts (g⁻¹ fresh weight) and in the percent of successful protoplast isolations of tansy (a) and pyrethrum (b). The four seasons are: 1 = summer (June-August), 2 = fall (September-November), 3 = winter (December-February), and 4 = spring (March-May). Symbols: squares: successful isolations; diamonds: protoplast yield.

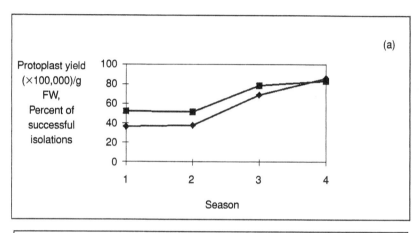

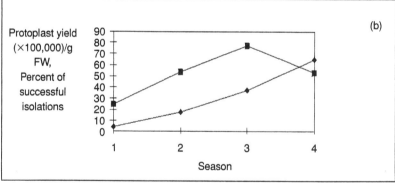

cedure still needs more studies. One reason for the limited regeneration may be incomplete cell partitioning, which inhibits the signalling of hormones from cell to cell. Protoplast fusion between tansy and pyrethrum was possible and the hybridity was tested successfully at the callus stage. The low number of compounds found from fusion-derived calli may not indicate absence of hybridization, but rather the inability of non-organized callus to produce volatile compounds.

Tansy grown in Finland contains a wide range of genetic, chemical and morphological variation. Especially, among the larger number of volatile compounds detected from tansy, the chemotypes accumulating irregular mono-terpenes such as artemisia ketone and davadone D, which are biochemically close to pyrethrins, are an interesting source for further biotechnological applications (6). These chemotypes also had the highest number of flowerheads, which is the site of isoprenoid accumulation. In contrast, chemotypes containing high concentration of camphor usually had a low number of flowerheads. In conclusion, biodiversity can be a valuable source for novel germplasm for breeding tansies that better accumulate more bioactive compounds.

REFERENCES

1. Keskitalo, M. 1999. Exploring biodiversity to enhance bioactivity in the genus *Tanacetum* through protoplast fusion. *Academic Dissertation, University of Helsinki, Department of Plant Production, Section Crop Husbandry*. Publication No. 53, 112 pp.

2. Keskitalo, M., Linden, A. and Valkonen, J.P.T. 1998a. Genetic and morphological diversity of Finnish tansy (*Tanacetum vulgare* L., Asteraceae). *Theor Appl Genet* 96: 1141-1150.

3. Keskitalo, M., Pehu, E. and Simon, J.E. 2001. Variation in volatile compounds from tansy (*Tanacetum vulgare* L.) related to genetic and morphological differences of genotypes. *Biochemical Systematics and Ecology* 29: 267-285.

4. Keskitalo, M., Kanerva, T. and Pehu, E. 1995. Development of *in vitro* procedures for regeneration of petiole and leaf explants and production of protoplast-derived callus of *Tanacetum vulgare* L. (Tansy). *Plant Cell Reports* 14: 281-296.

5. Keskitalo, M., Pohto, A., Savela, M.-L., Valkonen, J.P.T., Simon, J. and Pehu, E. 1998. Alterations in growth of tissue-cultured tansy (*Tanacetum vulgare* L.) treated with antibiotics. *Ann Appl Biol* 133: 281-296.

6. Keskitalo, M., Angers, P., Earle, E. and Pehu, E. 1999. Chemical and genetic characterization of calli from somatic hybridization between tansy (*Tanacetum vulgare* L.) and pyrethrum (*Tanacetum cinerariifolium* (Trevir.) Schultz-Bip.). *Theor Appl Genet* 98: 1335-1343.

Rapid SPME-GC Analysis
of Volatile Secondary Metabolites
in Various Wild Species
of the Genus *Allium*

Hartwig Schulz
Hans Krüger

SUMMARY. The individual volatile sulphur components formed in the presence of alliinase were analyzed in nine different *Allium* wild genotypes by SPME-GC and SPME-GC-MS, respectively. Relating to the individual profiles, two chemotypes can be discriminated which correspond more or less to a "garlic type" and an "onion type." Since the described method is very rapid and reproducible and needs only small sample amounts, it can be efficiently used to evaluate large *Allium* gene bank accessions with special regard to valuable flavour and healthful properties. *[Article copies available for a fee from The Haworth Document Delivery Service: 1-800-HAWORTH. E-mail address: <getinfo@haworthpressinc.com> Website: <http://www.HaworthPress.com> © 2002 by The Haworth Press, Inc. All rights reserved.]*

Hartwig Schulz and Hans Krüger are affiliated with the Federal Centre of Breeding Research on Cultivated Plants, Institute for Quality Analysis, Neuer Weg 22-23, D-06484 Quedlinburg, Germany.

The authors are grateful to Mrs. B. Zeiger and Mrs. C. Langanke for their technical support. The authors also express special thanks to Dr. E. R. J. Keller of the Institute of Plant Genetics and Crop Plant Research in Gatersleben for making available the gene bank material.

KEYWORDS. *Allium*, SPME-GC, alk(en)yl cysteine sulphoxides, dialk(en)yl (poly)sulphides, gene bank evaluation

INTRODUCTION

Since ancient times, several *Allium* species have been widely used because of their flavouring and healthful properties. Onions, garlic, leeks, chives and shallots are vegetables of the genus *Allium* which are predominantly used to flavour foods. In addition to their flavouring application, medicinal properties of garlic and onions have been known for centuries. Based on the first systematic studies of antibiotic properties of garlic, allicin was found to be effective in the range of 1:125,000 against a number of Gram-positive and Gram-negative microorganisms (1,2). In addition to these antibacterial activities, onion and garlic preparations were also investigated for their antifungal potential. Recently published studies have demonstrated that there are some inhibitory effects of various *Allium* oils against dermatophytic fungi. The best antifungal results were observed for *Microsporum cani, M. gypseum*, and *Trichophyton simii* at a concentration of 200 ppm onion oil. Juice prepared from onions was also active against several yeast species (3,4). Furthermore, the antithrombotic potential of aqueous *Allium* extracts has been evaluated (5). The aroma and many of the described health benefits can be attributed to organosulphur compounds, representing 1-5% of the dry weight of bulbs (6,7). The most important active principles are the S-substituted cysteine sulphoxides which are converted into the related volatile dialk(en)yl(poly)sulphides by the enzyme alliinase, when the plants are crushed. Lipid lowering effects, which were intensively studied for garlic, are related predominantly to the occurrence of (+)-S-methyl-L-cysteine sulphoxide (8). The active principle (+)-S-allyl-L-cysteine sulphoxide (alliin), which is typical for garlic, is mainly responsible for the inhibition of platelet aggregation in man; but other sulphur components such as volatile trisulphides and other alk(en)ylpolysulphides also contribute to this effectiveness (9). In this study, a new efficient solid phase micro extraction (SPME) method is described which allows the classification of *Allium* species in a very short time on the basis of their specific profiles built by the volatile, S-containing artefacts (Table 1). Relating these analytical data to a rough prediction of the individual precursor substances should be possible.

MATERIALS AND METHODS

Approximately 1-2 g of the freshly minced sample, combined with 5 mL of distilled water, was transferred into a 20 mL headspace vial. The sample suspension was thermostated at 32°C for 30 min until most of the enzymatic reac-

TABLE 1. Variation (GC %) of the aroma profiles of some selected *Allium* gene bank accessions. Compound identification according to Figure 2.

	Propane-thiol	Allyl methyl sulphide	Dimethyl disulphide	Methyl propyl disulphide	Dimethyl thiophene	Allyl methyl disulphide	Dipropyl disulphide	Allyl propyl disulphide	(E)-1-Propenyl propyl disulphide	Diallyl disulphide	Allyl (E)-1-propenyl disulphide	Dipropyl trisulphide
A. tuberosum	0	2.9	10.39	0.27	0	43.90	0	0	0	33.13	2.43	0.36
A. altaicum	4.63	0	0	2.85	0.45	0	59.38	0.81	21.96	0	0	5.56
A. hymenorrhizum	4.63	0	1.88	20.73	0.52	0.58	36.37	2.25	15.14	0	0.17	6.10
A. ramosum	0	4.59	32.46	1.44	0	41.88	0	0.48	0	10.68	1.62	0.44
A. lineare	0.35	1.79	0.97	1.70	0.44	18.60	0.32	7.54	1.42	49.61	11.03	0.17
A. albidum	27.98	0	0	3.94	4.73	0	33.27	0.41	18.85	0	0.40	2.47
A. rubens	33.88	0	0	1.81	0.73	0	48.68	0	8.25	0	0	3.47
A. nutans	18.21	0	0	4.67	21.67	0	10.66	0	31.74	0	0	2.67
A. carolinianum	0	17.87	3.42	0	0	32.29	0	0	0	30.77	5.46	0.69

tions were finished. Headspace sampling was done by piercing the septa with the SPME syringe and exposing the poly(dimethylsiloxane) coated fibre (film thickness of 100 μ) to the sample for 10 min (Figure 1). The fibre was then retracted, immediately inserted into the injection port of the gas chromatograph and thermally desorbed at 230°C (Figure 2). The identification of the detected GC signals was confirmed by co-chromatography and mass spectra of authentic standards.

GC parameters: fused silica capillary column HP INNOWAX 60 m × 0.25 mm i.d., film thickness 0.5 μm; split 1:10; linear temperature programme: 35°C → 10°C/min → 220°C; carrier gas (H_2) flow rate: 1 mL/min.

MS parameters: carrier gas (He) velocity: 1 mL/min; ionization voltage: 70 eV; source temperature: 195°C; scan range: 27-300 m/z.

MS libraries: Wiley 138, NBS 75 K.

RESULTS AND DISCUSSION

As demonstrated in Figure 1, the applied poly(dimethylsiloxane) fibre material seems to be very suitable to adsorb the volatile sulphur analyt molecules from the aqueous *Allium* matrix. In comparison to the usually performed

FIGURE 1. Adsorption of volatile sulphur components on a SPME fibre

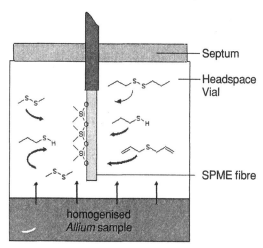

aroma isolation techniques, such as hydrodistillation, solvent extraction or simultaneous distillation/extraction (SDE) (10,11), a surprisingly high reproducibility, sensitivity and selectivity of all relevant S-containing substances is achieved. In spite of the fact that several competing consecutive reactions prevent the establishment of a stable equilibrium, the analysis of the adsorbed components leads to reproducible results. After 10 minutes, the adsorption of the analyt molecules on the SPME fibre is nearly finished; from that time onwards an optimal detector response is reached. Contrary to related investigations performed with *Allium* samples which have been processed by the simultaneous distillation/extraction (SDE) method, the SPME-GC results show little discrimination effects in the response of the individual sulphur substances. Only components with a higher amount of sulphur in the molecule, such as the dialk(en)yl trisulphides, have less tendency to adsorb on the fibre. Depending on the individual species, the investigated *Allium* accessions correspond more or less to a "garlic type" characterized by the main components diallyl disulphide and allyl methyl disulfide, or an "onion type" containing predominantly dipropyl disulphide and (*E*)-1-propenyl propyl disulphide in the aroma fraction (Figure 2). Therefore, it can be assumed that these two

FIGURE 2. SPME gas chromatograms of *A. lineare* (above) and *A. altaicum* (below).

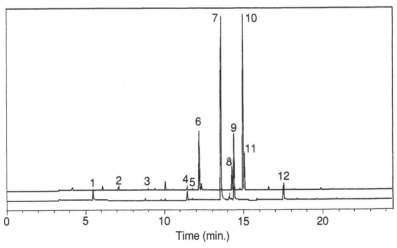

1: Propanethiol; 2: Allyl methyl sulphide; 3: Dimethyl disulphide; 4: Methyl propyl disulphide; 5: Dimethylthiophene; 6: Allyl methyl disulphide; 7: Dipropyl disulphide; 8: Allyl propyl disulphide; 9: (*E*)-1-Propenyl propyl disulphide; 10: Diallyl disulphide; 11: Allyl (*E*)-1-propenyl disulphide; 12: Dipropyl trisulphide.

chemotypes may also differ significantly with regard to the content of the four main important aroma precursors: allyl, 1-propenyl, propyl, and methyl cysteine sulfoxide. Similar results have been obtained recently from measurements of the species *A. obliquum*, *A. senescens*, *A. saxatile*, *A. globosum*, *A. chevsuricum* and *A. altyncolicum* (12).

Since the described method is rapid and reproducible and needs only small sample amounts (min. 1 g), it is an efficient tool to study the variability of the valuable dialk(en)yl(poly)sulphides in *Allium* gene bank evaluations.

REFERENCES

1. Zohri, A.N., K. Abdel-Gawad, and S. Saber. 1995. Antibacterial, antidermatophytic and antioxigenic activities of onion (*Allium cepa* L.) oil. *Microbiological Research* 150:167-172.

2. Elnima, E.I., S.A. Ahmed, A.G. Mekkawi, and J.S. Mossa. 1983. The antimicrobial activity of garlic and onion extracts. *Die Pharmazie* 38:747-748.

3. Dankert, J., T.F. Tromp, H. de Vries, and H.J. Klasen. 1979. Antimicrobial activity of crude juices of *Allium ascalonicum*, *Allium cepa*, and *Allium sativum*. *Zentralblatt für Bakteriologie, Parasitenkunde, Infektionskrankheiten und Hygiene* A 245: 229-239.

4. Yin, M.C. and S.M. Tsao. 1999. Inhibitory effect of seven *Allium* plants upon three *Aspergillus* species. *International Journal of Food Microbiology* 49:49-56.

5. Bordia, T., N. Mohammed, M. Thomson, and M. Ali. 1996. An evaluation of garlic and onion as antithrombotic agents. *Prostaglandins Leukotrienes and Essential Fatty Acids* 54:183-186.

6. Block, E. 1992. The organosulfur chemistry of the genus *Allium*–Implications for the organic chemistry of sulfur. *Angew. Chem. Int. Ed. Engl.* 31:1135-1178.

7. Koch, H.P. and L.D. Lawson. 1996. Garlic. The science and therapeutic applications of *Allium sativum* and related species. 2nd edn. Williams & Wilkins, Baltimore.

8. Kumari, K., B.C. Mathew, and K.T. Augusti. 1995. Antidiabetic and hypolipidemic effects of S-methyl cysteine sulfoxide isolated from *Allium cepa* L. *Indian Journal of Biochemistry & Biophysics* 32:49-54.

9. Makheja, A.N. and J.M. Bailey. 1990. Antiplatelet constituents of garlic and onion. *Agents and Actions* 29:360-363.

10. Ohsumi, C., T. Hayashi, K. Kubota, and A. Kobayashi. 1993. Volatile flavor compounds in an interspecific hybrid between onion and garlic. *J. Agric. Food Chem.* 41:1808-1810.

11. Schulz, H., H. Krüger, J. Liebmann, and H. Peterka. 1998. Distribution of volatile sulfur components in an interspecific hybrid between onion (*A. cepa* L.) and leek (*A. porrum* L.). *J. Agric. Food Chem.* 46:5220-5224.

12. Schulz, H., H. Krüger, N. Herchert, and E.R.J. Keller. 2000. Vorkommen flüchtiger Sekundärmetabolite in ausgewählten *Allium*-Wildtypen. *J. Appl. Bot.* 74(3/4):119-121.

The Effects of Casein and Its Constituents on the Development of Tissue Culture and Somatic Embryogenesis from *Malva silvestris* L.: A Preliminary Study

Spiridon Kintzios
Emmanouil Papagiannakis
Georgios Aivalakis
John Konstas
Dimitrios Bouranis
Lito Christodoulopoulou

SUMMARY. We investigated the kinetics of common mallow (*Malva silvestris* L.) petiole tissue growth and development *in vitro*, focusing on somatic embryo development at different concentrations of casein (0, 100, 200, 500 mg L^{-1}). The results indicated the key role of casein (100-200 mg L^{-1}) and potassium in the induction of direct somatic

Spiridon Kintzios, Emmanouil Papagiannakis, Georgios Aivalakis, John Konstas, Dimitrios Bouranis, and Lito Christodoulopoulou are affiliated with the Department of Agricultural Biotechnology, Laboratory of Plant Physiology, Agricultural University of Athens, Greece.

Address correspondence to: Spiridon Kintzios, Department of Agricultural Biotechnology, Laboratory of Plant Physiology, Agricultural University of Athens, Iera Odos 75, 11855 Athens, Greece.

[Haworth co-indexing entry note]: "The Effects of Casein and Its Constituents on the Development of Tissue Culture and Somatic Embryogenesis from *Malva silvestris* L.: A Preliminary Study." Kintzios, Spiridon et al. Co-published simultaneously in *Journal of Herbs, Spices & Medicinal Plants* (The Haworth Herbal Press, an imprint of The Haworth Press, Inc.) Vol. 9, No. 2/3, 2002, pp. 211-215; and: *Breeding Research on Aromatic and Medicinal Plants* (ed: Christopher B. Johnson, and Chlodwig Franz) The Haworth Herbal Press, an imprint of The Haworth Press, Inc., 2002, pp. 211-215. Single or multiple copies of this article are available for a fee from The Haworth Document Delivery Service [1-800-HAWORTH 9:00 a.m. - 5:00 p.m. (EST). E-mail address: getinfo@haworthpressinc.com].

embryogenesis combined with a culture duration greater than 10 days. They also provided important information concerning morphogenic, physiological and biochemical aspects in addition to structural and developmental patterns in somatic embryogenesis from this still under-exploited medicinal plant species. *[Article copies available for a fee from The Haworth Document Delivery Service: 1-800-HAWORTH. E-mail address: <getinfo@haworthpressinc.com> Website: <http://www.HaworthPress.com> © 2002 by The Haworth Press, Inc. All rights reserved.]*

KEYWORDS. Amino acids, common mallow, casein hydrolysate, somatic embryogenesis, *Malva silvestris*

INTRODUCTION

Malva silvestris L. (Malvaceae), the common mallow, is an annual to short-lived perennial plant, well known for its medicinal value, mostly in a traditional sense. The leaves and flowers are the main plant parts used for various purposes, including the treatment of coughs and throat infections and other bronchial problems, as well as stomach and intestinal irritations. Mallows are abundantly distributed in temperate and tropical regions and virtually worldwide (1). Several secondary metabolites belonging to various chemical groups (such as flavonoid glycosides and mucilage) (3,4) have been isolated from various parts of the plant. We have previously studied (1) the potential of *in vitro* production of secondary metabolites from common mallow. Numerous spherical somatic embryos could be directly induced within only three days of culture on a MS medium supplemented with 1.8-18 µM 1-naphthalenacetic acid (NAA) and 6-benzyladenine (BA), along with 0.5 g L^{-1} casein hydrolysate. Somatic embryogenesis was favoured by higher light intensities (250 vs. 50 µmol m^{-2} s^{-1}), and heart-shaped embryos (1-2 mm long) were observed two weeks after culture initiation. In addition, the presence of five furanocoumarin-like constituents was detected in tissues cultured *in vitro*, while only one constituent could be detected in stems received from plants growing *in vivo*. Differences were also observed between tissues cultured under different photosynthetic flux densities (PPFDs).

In the present study, we examined the effect of different casein concentrations (0, 100, 200 and 500 mg L^{-1}) on the kinetics of common mallow petiole tissue growth and development *in vitro*, focusing on somatic embryo development. The effects of the individual amino acid constituents of casein were studied as well. Various parameters such as tissue fresh and dry weight, chlorophyll content, number of globular embryos per explant, dehydrogenase activity and accumulation of K^+, Ca^{+2} and Mg^{+2} ions were assayed.

MATERIALS AND METHODS

Petiole explants were received from young mallow plants (bearing 6-8 leaves) collected in the area of Attiki, Greece. Explants were surface sterilized for 12 min in 1% (w/v) sodium hypochlorite solution containing 1-2% Tween-80, then rinsed 3 times in sterilized distilled water. One centimeter long explant pieces were excised and inoculated onto Murashige and Skoog (MS) basal medium solidified with 0.8% agar and supplemented with 3% sucrose, 9 µM NAA, 9 µM BA and casein hydrolysate at various concentrations (0, 100, 200, 500 mg L^{-1}). In a separate experiment, and in order to study the effect of the individual amino acid constituents of casein, explants were inoculated on solid MS media supplemented with 3% sucrose, 9 µM NAA and 9 µM BA in combination with one of the 17 amino acid constituents of casein at their percentage concentration at 500 mg L^{-1} casein. Media were adjusted to pH 5.8 using 1 N NaOH or 1 N HCl, autoclaved at 121°C for 20 min, and poured into polystyrene 100 × 20 mm petri dishes (30 ml of medium/dish, 4 explants/dish). Inoculated dishes were sealed with Parafilm™ and incubated under a PPFD of 250 µmol m^{-2} s^{-1} (16/8 h photoperiod, from cool white fluorescent lamps). Fifty explants were cultured on each casein concentration or amino acid treatment. Numbers of responding (embryogenic) explants and globular embryos per explant, as well as tissue growth (tissue fresh and dry weight) were recorded ten days after culture initiation. Total chlorophyll was extracted in 80% (v/v) acetone and spectrophotometrically determined at 664 and 647 nm. Triphenyl-tetrazolium-chloride (TTC) assay was used as a viability assay for the callus pieces. TTC assays were performed at 25°C in the dark for 24 h in 3 ml of 0.6% (w/v) TTC in 50 mM phosphate buffer (pH 7.5). TTC formazan was extracted from the callus tissues with 95% (v/v) ethanol and its concentration was determined spectrophotometrically at 530 nm. The callus content in K, Ca and Mg was determined by atomic absorption spectrometry using a Perkin-Elmer spectrophotometer. Results were assessed by a standard analysis of variance for a randomized complete block design, using MS-STATISTICA software.

RESULTS AND DISCUSSION

The addition of casein to the culture medium at different concentrations very significantly affected both explant growth (p < 0.01), explant metabolism and somatic embryo induction and proliferation (p < 0.0001). The results of these effects are summarized in Figure 1. The induction of direct somatic embryogenesis increased with casein concentration, while explant growth (expressed as explant fresh and dry weight) declined. Total chlorophyll and potassium accumulation also increased with casein concentration; however, maximum explant metabolism (dehydrogenase activity) was observed at 100

FIGURE 1. Morphogenic and biochemical responses of mallow (*Malva silvestris* L.) petiole cultures to different casein concentrations, ten days after culture initiation.

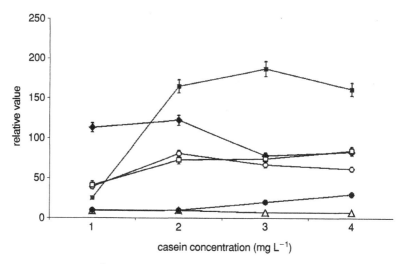

Key: (■) globular embryos mm^{-2} explant surface; (◆) explant fresh weight (g); (△) explant dry weight (g); (●) total chlorophyll accumulation (μM); (○) dehydrogenase activity (relative formazan absorbance); (□) K$^+$ accumulation (% dry weight).

mg L^{-1} casein. In particular, somatic embryo induction was highly positively correlated with K$^+$ accumulation ($r^2 = 0.930687$), whereas explant fresh and dry weight were negatively correlated with total chlorophyll accumulation ($r^2 = -0.95109$ and -0.96594, respectively). Potassium is the most abundant cation in the cytoplasm. It promotes the photosynthetic activity of plant cells by establishing the transmembrane pH gradient necessary for the synthesis of ATP in chloroplasts (2). It also facilitatses RuBP carboxylase activity, respiration and starch accumulation. Thus, both K$^+$ and chlorophyll accumulation at higher casein concentrations could have promoted somatic embryogenesis in mallow cultures through an increase of their metabolic and photosynthetic activity, as already reported for other species, such as melon, pepper and squash (3). The beneficial effect of casein on the induction of somatic embryogenesis from mallow petiole explants was verified by microscopical observations on explant sections derived ten days after culture initiation (analytical data not presented): abundant globular embryos were formed in cortex parenchyma. As far as the effects of individual amino acids are concerned, addition of tyrosine gave the best embryogenic response, followed by histidine and isoleucine. This response was lower than with casein addition; however, it indicated a possible implication of various biochemical pathways (such as the biosynthesis of auxins and phenylpropanoids, and the stimulation of photosynthesis) to the process of embryogenic induction. In conclusion, the results of the present study allow for further exploiting mallow petiole culture as a reliable, fast-responding model system for investigating the effect of various chemical factors on the induction of somatic embryogenesis from dicotyledonous plant species.

REFERENCES

1. Kintzios, S., E. Katsouri, D. Peppes and S. Koulocheri. 1998. Somatic embryogenesis and in vitro secondary metabolite production from common mallow (*Malva silvestris* L.) collected in Greece. *Acta Horticultura* 457:173-178.

2. Marschner, H. 1995. *Mineral Nutrition of Higher Plants*. 2nd edition. Academic Press, New York.

3. Kintzios, S., G. Hioureas, E. Shortsianitis, E. Sereti, P. Blouchos, C. Manos, O. Makri, N. Tarariva, J. Drossopoulos and C.D. Holevas. 1998. The effect of light on the induction development and maturation of somatic embryos from various horticultural and ornamental species. *Acta Horticultura* 461: 427-432.

Preliminary Evaluation of Somaclonal Variation for the *In Vitro* Production of New Toxic Proteins from *Viscum album* L.

Spiridon Kintzios
Maria Barberaki
Panagiotis Tourgielis
Georgios Aivalakis
An Volioti

SUMMARY. In an attempt at preliminary evaluation of the effect of somaclonal variation on the production of white berry mistletoe (*Viscum album* L.) proteins, one hundred individual callus tissues were assayed for their protein content by means of SDS-PAGE. Somaclonal variation in the qualitative aspect was relatively low (8%). However, variation in quantitative terms was high and in 10% of the calli a protein concentration six times higher than in stems was observed. These results indicate for the first time the possibility of applying tissue culture for the produc-

Spiridon Kintzios, Maria Barberaki, Panagiotis Tourgielis, Georgios Aivalakis, and An Volioti are affiliated with the Department of Agricultural Biotechnology, Laboratory of Plant Physiology, Agricultural University of Athens, Greece.

Address correspondence to: Spiridon Kintzios, Department of Agricultural Biotechnology, Laboratory of Plant Physiology, Agricultural University of Athens, Iera Odos 75, 11855 Athens, Greece.

[Haworth co-indexing entry note]: "Preliminary Evaluation of Somaclonal Variation for the *In Vitro* Production of New Toxic Proteins from *Viscum album* L." Kintzios, Spiridon et al. Co-published simultaneously in *Journal of Herbs, Spices & Medicinal Plants* (The Haworth Herbal Press, an imprint of The Haworth Press, Inc.) Vol. 9, No. 2/3, 2002, pp. 217-221; and: *Breeding Research on Aromatic and Medicinal Plants* (ed: Christopher B. Johnson, and Chlodwig Franz) The Haworth Herbal Press, an imprint of The Haworth Press, Inc., 2002, pp. 217-221. Single or multiple copies of this article are available for a fee from The Haworth Document Delivery Service [1-800-HAWORTH 9:00 a.m. - 5:00 p.m. (EST). E-mail address: getinfo@haworthpressinc.com].

217

tion of mistletoe protein extracts with increased antitumor/immuno-modulatory properties. *[Article copies available for a fee from The Haworth Document Delivery Service: 1-800-HAWORTH. E-mail address: <getinfo@ haworthpressinc.com> Website: <http://www.HaworthPress.com> © 2002 by The Haworth Press, Inc. All rights reserved.]*

KEYWORDS. Callus culture, mistletoe, lectin, somaclonal variation, viscotoxin, *Viscum album*

INTRODUCTION

Viscum album L. (Loranthaceae), mistletoe, was introduced in the treatment of cancer in 1917. Currently, there are a number of mistletoe preparations (e.g., Helixor, Iscador) used in many countries against different kinds of cancer (1). The anticancer properties of mistletoe have been attributed to both the cytotoxic proteins viscotoxins (a group of 5kDa basic polypeptides) and the three immunomodulatory mistletoe lectins (viscumins) ML I, ML II, and ML III, along with other, yet unidentified components of the crude preparation. Biotechnological applications could help standardize viscotoxin and lectin biosynthesis and increase the level of control over the quantity and the quality of the production (2). We have previously (3) succeeded in inducing callus and protoplast cultures from mistletoe leaves and stems in a large number of different growth regulator and media treatments. In the present study we report the results of a preliminary evaluation of the effect of somaclonal variation (i.e., *in vitro* genetic variability, which may be heritable or not) on the qualitative and quantitative production of proteins from mistletoe callus cultures. In this context, we also investigated the potential use of somaclonal variation for selecting and developing high-yielding cell lines as well as its effects on the standardization of *in vitro* synthesized mistletoe extracts.

MATERIALS AND METHODS

Flowering and non-flowering donor mistletoe (*Viscum album* L.) plants of the variety *abietis,* growing on fir trees on Mount Parnitha (Attiki, Greece) during the winter, were used in our study. Donor plants were kept at 4°C before the removal of explants. One centimeter long stem explants were surface disinfected in 1% (w/v) sodium hypochlorite for 15 min followed by 0.1% (w/v) mercuric chloride for 12 min. All solutions contained 2% Tween-80. Disinfected explants were finally rinsed four times in sterile distilled water. Then, 0.5-cm long stem explant segments were aseptically dissected and inoculated on a

Murashige and Skoog (MS) (4) basal medium supplemented with 3% (w/v) sucrose, 10 mg L^{-1} ascorbic acid, 4.95 µM NAA, and 2.82 µM BA, and solidified with 0.8% agar. Media were adjusted to pH 5.8 using 1 N NaOH or 1 N HCl, autoclaved at 121°C for 20 min and poured into polystyrene 100×20 mm petri dishes (30 ml of medium/dish, 4 explants/dish). A total of 200 stem explants were inoculated on each medium combination. Inoculated tubes and dishes were sealed with Nescofilm™ and incubated at 25°C under a PPFD of 50 µmol m^{-2} s^{-1} (16/8 h photoperiod, from cool white fluorescent lamps).

After six weeks of culture, callus tissues were subcultured on fresh induction medium for eight weeks before being assayed. Soluble mistletoe proteins were extracted from stems of donor plants and callus tissues (5 g) in 10 ml of Tris-EDTA buffer, pH 8.5 (50 mM Tris, 0.01 mM EDTA, 5% (v/v) glycerol, 20.5 mM NaCl, 100 mM (w/v) lactose and 2% (v/v) β-mercaptoethanol) and centrifuged twice at $15,000 \times g$, 4°C for 15 min. The lactose was included to avoid binding of viscumin to carbohydrate components in the homogenate. Proteins were then precipitated overnight at 4°C with 70% (w/v) ammonium sulfate then applied to a Sephadex 25 desalting column and eluted with buffer. Proteins were quantitated according to Bearden (1978) and their concentration (mg/L) was expressed relative to a bovine serum albumin standard solution. SDS-PAGE was performed according to Laemmli (1970) using a 10% polyacrylamide discontinuous gel. The gels were run at 200 V (30 mA) for 9 h. Protein markers (Pharmacia calibration kit) with a range 5-212 kDa were included. Proteins on gel were stained with silver staining. Callus fresh weight and protein concentration were recorded eight weeks after subculture. One hundred calli were analyzed. Results were assessed by a standard analysis of variance for a randomized complete block design using MS-STATISTICA software.

RESULTS AND DISCUSSION

Callus induction took place after two weeks and, as reported previously (3), somatic embryos at the globular stage were observed on pink-whitish callus tissues cultured for over four weeks on the induction medium. A significant variation in protein concentration ($p < 0.0001$, analytical data not shown) was observed between different callus tissues. Only 30% of the calli accumulated proteins at a concentration approximate to the average value of 2.6 µg/g fresh weight. On the contrary, in approximately 10% of the calli, a protein concentration of more than 9 µg/g was determined, a value six times higher than in donor stem tissues. Variation was also observed in culture growth; however, the majority (65%) of the calli had a fresh weight of approximately 1.6 g, which was the average value.

Electrophoretic protein patterns revealed the existence of six protein bands in mistletoe stems, found at the range of 53, 76, 116, 170 and 212 kDa (Figure 1), which possibly represented both dimeric and monomeric forms of viscumin (at approximately 114 and 57 kDa) (1) as well as other mistletoe proteins. In the majority of callus pieces investigated, these proteins were detected, except for in the 212 kDa band. However, in five callus lines, five additional protein bands were detected (three bands < 53 kDa, one band at approximately 100 kDa, and one at approximately 150 kDa). Somaclonal variation has been extensively studied in crop plants such as sugarcane, tobacco, rice, maize and barley (5). However, reports on the applications of somaclonal variation in medicinal plant species are extremely rare. Oksman-Caldentey et al. (6) observed considerable somaclonal variation in cell clones derived from mesophyll proto-

FIGURE 1. SDS-PAGE of mistletoe proteins from stems of donor plants and stem-derived callus tissues

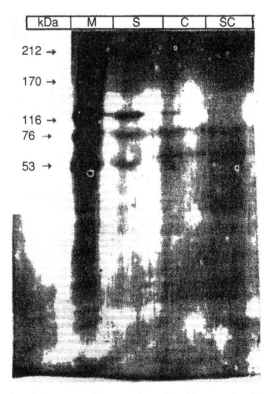

(M = marker proteins, S = mistletoe stem, C = non-variant callus, SC = somaclonally variant callus)

plasts of *Hyoscyamus muticus*. The clonal series contained a minor sub-population with relatively high (300×) scopolamine content. Zenk (7) also reported that 143 callus subclones of *Solanum laciniatum* possessed variation in their ability to produce alkaloids. The observed variation in *in vitro* mistletoe protein accumulation indicates for the first time the possibility of applying tissue culture for the production of mistletoe protein extracts with increased cytotoxic/immunomodulatory properties. Future investigations are planned in order to define the *in vitro* antitumor activity of novel variant mistletoe proteins.

REFERENCES

1. Olsnes, S., Stirpe, F., Sandvig, K. and Pihl, A. 1982. Isolation and characterization of viscumin, a toxic lectin from *Viscum album* L. (mistletoe). *J. Biol. Chem.* 257: 13263-13270.

2. Becker, H. and G. Schwarz. 1971. Callus cultures from *Viscum album*. A possible source of raw materials for gaining therapeutically interesting extracts. *Planta Medica* 20(4): 357-362.

3. Kintzios, S. and M. Barberaki. 2002. The biotechnology of *Viscum album*: Tissue culture, somatic embryogenesis and protoplast isolation. *In* A. Bussing, ed. *Medicinal and Aromatic Plants–Industrial Approaches: The Genus Viscum*. Harwood Publishers, Amsterdam (in press).

4. Murashige, T. and F. Skoog. 1962. A revised method for rapid growth and bioassays with tobacco tissue cultures. *Physiol. Plant.* 15: 472-497.

5. Larkin, P.J. and W.R. Scowcroft. 1981. Somaclonal variation–a novel source of variability from cell cultures for plant improvement. *Theor. Appl. Genet.* 60: 197-214.

6. Oksman-Caldentey, K.-M., H. Vuorela, A. Strauss and R. Hiltunen. 1987. Variation in the tropane alkaloid content of *Hyoscyamus muticus*. *Planta Med.* 52: 6-12.

7. Zenk, M.H. 1978. The impact of plant cell culture on industry. *In* T.A. Thorpe, ed. *Frontiers of Plant Tissue Culture*. Calgary: Association of Plant Tissue Culture. pp. 1-13.

The Effects of Light on Callus Growth and Somatic Embryogenesis from *Lavandula vera* and *Teucrium chamaedrys*: A Preliminary Study

Spiridon Kintzios
Iosif Papanastasiou
Panagiotis Tourgelis
Charalambos Papastellatos
Vlassios Georgopoulos
John Drossopoulos

SUMMARY. In order to investigate the effect of light on the tissue culture of lavender (*Lavandula vera* L.) and wall germander (*Teucrium chamaedrys* L.), leaf explants from greenhouse-grown plants were inoculated on a Murashige and Skoog (MS) medium (1) supplemented with 1 mg L^{-1} 2,4-D and 2 mg L^{-1} kinetin and incubated either (i) continu-

---

Spiridon Kintzios, Iosif Papanastasiou, Panagiotis Tourgelis, Charalambos Papastellatos, Vlassios Georgopoulos, and John Drossopoulos are affiliated with the Department of Agricultural Biotechnology, Laboratory of Plant Physiology, Agricultural University of Athens, Greece.

Address correspondence to: Spiridon Kintzios, Department of Agricultural Biotechnology, Laboratory of Plant Physiology, Agricultural University of Athens, Iera Odos 75, 11855 Athens, Greece.

[Haworth co-indexing entry note]: "The Effects of Light on Callus Growth and Somatic Embryogenesis from *Lavandula vera* and *Teucrium chamaedrys*: A Preliminary Study." Kintzios, Spiridon et al.. Co-published simultaneously in *Journal of Herbs, Spices & Medicinal Plants* (The Haworth Herbal Press, an imprint of The Haworth Press, Inc.) Vol. 9, No. 2/3, 2002, pp. 223-227; and: *Breeding Research on Aromatic and Medicinal Plants* (ed: Christopher B. Johnson, and Chlodwig Franz) The Haworth Herbal Press, an imprint of The Haworth Press, Inc., 2002, pp. 223-227. Single or multiple copies of this article are available for a fee from The Haworth Document Delivery Service [1-800-HAWORTH 9:00 a.m. - 5:00 p.m. (EST). E-mail address: getinfo@haworthpressinc.com].

ously in the darkness for 6 weeks or (ii) first in the darkness for 0-5 weeks and then under illumination (150 μmol m^{-2} s^{-1}, 16 hour light/8 hour dark photoperiod). Lavender callus growth was improved under increased incubation in darkness while somatic embryogenesis was remarkably reduced under the same conditions. Incubation in darkness did not affect the growth of wall germander cultures, but improved somatic embryogenesis. Moreover, a possible participation of boron in the effect of light on somatic embryogenesis for both species was indicated. *[Article copies available for a fee from The Haworth Document Delivery Service: 1-800-HAWORTH. E-mail address: <getinfo@haworthpressinc.com> Website: <http://www.HaworthPress.com> © 2002 by The Haworth Press, Inc. All rights reserved.]*

KEYWORDS. Boron, callus induction, light, mineral nutrition, somatic embryogenesis, *Lavandula vera, Teucrium chamaedrys*

INTRODUCTION

Somatic embryogenesis, the development of adventitious embryos from somatic (sporophytic) tissues, can be a valuable method for the micropropagation of various plant species. For medicinal plants, the availability of embryogenic cultures is particularly important, since they represent a high level of tissue differentiation, possibly linked to an increased rate of secondary metabolite accumulation. In the present report, we investigated the effect of light on the development of embryogenic cultures of two Lamiaceae species: lavender (*Lavandula vera* L.) and wall germander (*Teucrium chamaedrys* L.). We also investigated the relationship between somatic embryo induction and certain physiological parameters of callus growth, such as total chlorophyll, macronutrient and micronutrient accumulation.

MATERIALS AND METHODS

Embryogenic callus cultures of lavender (*Lavandula vera* L.) and wall germander (*Teucrium chamaedrys* L.) were derived from leaf explants (6th to 8th leaf) of greenhouse-grown plants. Prior to inoculation on the callus induction medium, explants were surface-sterilized in 0.1% (w/v) mercuric chloride, containing 2% Tween-80, for 12 min, rinsed three times in sterile distilled water and finally cut into 1 cm long segments. The basal medium of Murashige and Skoog (MS) (1) was used for the induction of callus and somatic embryos, solidified with 0.8% agar and supplemented with 3% (w/v) sucrose, 1 mg L^{-1} 2,4-dichlorophenoxyacetic acid (2,4-D) and 2 mg L^{-1} kinetin. Media were adjusted to pH 5.8 using 1 N NaOH or 1 N HCl, autoclaved at 121°C for 20 min and poured into polystyrene 100×20 mm^2 petri dishes (30 ml of medium/dish, 5 explants/plant/dish). A total of 105 explants from each species were inoculated on the medium. Inoculated dishes were sealed with Nescofilm™ and incubated at 25°C either (i) continuously under illumination or (ii) initially in darkness, and then progressively transferred under illumination (250 μmol m^{-2}s^{-1}, 16/8 h photoperiod, from cool white fluorescent lamps) in weekly intervals (each time three dishes from each species) during a total period of 6 weeks. Thus, cultures were exposed to dark incubation periods of different duration (0-6 weeks). After 6 weeks in culture, callus tissues with somatic embryos at the globular stage were aseptically removed from culture for fresh and dry weight determination. The numbers of globular embryos per cm^2 of callus surface were recorded. Total chlorophyll was extracted in 80% (v/v) acetone and spectrophotometrically determined at 664 and 647 nm. The callus content of K, Ca, Mg, Fe, Mn, Zn and Cu was determined by atomic absorption spectrometry using a Variant spectrophotometer (2). Boron was determined spectrophotometrically, after treatment with azomethine-H (3). Results were assessed by a standard analysis of variance for a randomized complete block design, using MS-STATISTICA software.

RESULTS AND DISCUSSION

The relative length of the incubation period under illumination significantly affected callus growth and somatic embryo induction and proliferation, although in a species-specific fashion (Figure 1). Lavender callus growth was improved under increased incubation in darkness while somatic embryogenesis was remarkably reduced under the same conditions. The duration of the dark period did not have any effect on the dry weight of the cultures; however, chlorophyll content was slightly affected, and a peak in its concentration was observed after four weeks in darkness. The relative length of the incubation in darkness did not significantly affect the growth (in terms of callus fresh weight) of wall germander cultures, although minimum growth was observed after four weeks in darkness. Both somatic embryogenesis and total chlorophyll accumulation were significantly ($p < 0.0001$) favoured by prolonged incubation in darkness, although culture dry weight was reduced under the same conditions. Moreover, the different treatments had a significant effect on the accumulation of the selected macro- and micronutrients in cultured tissues of both species. However, only boron concentration in callus tissues was closely related ($r^2 = 0.8772$-0.9425) to the *in vitro* embryogenic response of both species to the different length of incubation in darkness. Although the role of boron in plant nutrition is still not well understood, it could possibly affect

FIGURE 1. Relative somatic proembryo induction (proembryos cm^{-2} callus surface) from leaf explants of lavender (*Lavandula vera* L.) and wall germander (*Teucrium chamaedrys* L.) in response to the different length of culture incubation in darkness (white columns = lavender, black columns = wall germander).

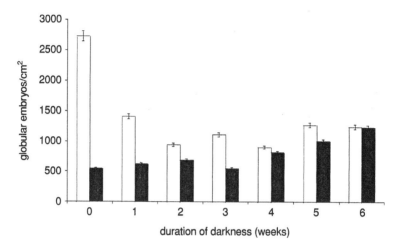

somatic embryogenesis through its complex interaction with phenol metabolism, indole acetic acid transport and tissue differentiation (4).

The results of the present study indicated the key role of light on the induction or suppression of somatic embryogenesis combined with a specific genotypic background. They also demonstrated a relationship between embryo induction and certain physiological parameters of callus growth. As previously reported, somatic embryogenesis from some species, such as cucumber (5), squash, melon and gardenia (6), is promoted under initial culture incubation in darkness. To our best knowledge this is the first report on somatic embryogenesis from these species, both of which are renown for their value as sources of essential oils for the perfume industry and traditional medicine (7). We feel that the information obtained in this study could contribute to the improvement of somatic embryogenesis from Lamiacae plant species and stimulate further progress in the application of this technique for the purposes of clonal propagation, *in vitro* germplasm conservation and secondary metabolite accumulation.

REFERENCES

1. Murashige, T. and F. Skoog. 1962. A revised method for rapid growth and bioassays with tobacco tissue cultures. *Physiol. Plant.* 15: 472-497.

2. Drossopoulos, J.B., D.L. Bouranis, S. Kintzios, G. Aivalakis, J. Karides, S.N. Chorianopoulou and C. Kitsaki. 1998. Effect of nitrogen fertilization on distribution

profiles of selected macro-nutrients in oriental field-grown tobacco plants. *J. Plant Nutrition* 22(3): 527-541.

3. Reinbott, T.M. and T.G. Blevins. 1995. Response of soyabean to foliar-applied boron and magnesium and soil-applied boron. *J. Plant Nutrition* 18: 179-200.

4. Marschner, H. 1995. *Mineral Nutrition of Higher Plants.* 2nd edition. Academic Press, New York.

5. Cade, R.M., T.C. Wehner and F.A. Blazich. 1988. Embryogenesis from cotyledon-derived callus of *Cucumis sativus* L. Cucurbit. *Genet. Coop. Rep* 11: 3-4.

6. Kintzios, S., G. Hioureas, E. Shortsianitis, E. Sereti, P. Blouchos, C. Manos, O. Makri, N. Tarariva, J. Drossopoulos and C.D. Holevas. 1998. The effect of light on the induction development and maturation of somatic embryos from various horticultural and ornamental species. *Acta Horticultura* 461: 427-432.

7. Podlech, D. 1996. *Herbs and Healing Plants of Britain and Europe.* HarperCollins, UK.

Studies on the Physiological Function of *In Vitro* Produced Antioxidants from Sage (*Salvia officinalis* L.): Effects on Cell Growth and Metabolism

Spiridon Kintzios
Maria Adamopoulou
Eleni Pistola
Katerina Delki
John Drossopoulos

SUMMARY. We investigated the effect of antioxidant phenolic compounds produced by sage (*Salvia officinalis*) callus cultures on some physiological parameters of the producing cells. Although cultures demonstrated a continuous growth during an incubation period of five weeks, the cell dehydrogenase activity, the cytochrome c oxidase activity and the respiration of isolated mitochondria declined. An analysis of methanolic extracts derived from the callus pieces indicated that the accumulation of phenolic compounds was correlated with mitochondrial activity, although the antioxidant activity (Fe^{+2} reduction) of the extracts was in-

Spiridon Kintzios, Maria Adamopoulou, Eleni Pistola, Katerina Delki, and John Drossopoulos are affiliated with the Department of Agricultural Biotechnology, Laboratory of Plant Physiology, Agricultural University of Athens, Greece.

Address correspondence to: Spiridon Kintzios, Department of Agricultural Biotechnology, Laboratory of Plant Physiology, Agricultural University of Athens, Iera Odos 75, 11855 Athens, Greece.

[Haworth co-indexing entry note]: "Studies on the Physiological Function of *In Vitro* Produced Antioxidants from Sage (*Salvia officinalis* L.): Effects on Cell Growth and Metabolism." Kintzios, Spiridon et al. Co-published simultaneously in *Journal of Herbs, Spices & Medicinal Plants* (The Haworth Herbal Press, an imprint of The Haworth Press, Inc.) Vol. 9, No. 2/3, 2002, pp. 229-233; and: *Breeding Research on Aromatic and Medicinal Plants* (ed: Christopher B. Johnson, and Chlodwig Franz) The Haworth Herbal Press, an imprint of The Haworth Press, Inc., 2002, pp. 229-233. Single or multiple copies of this article are available for a fee from The Haworth Document Delivery Service [1-800-HAWORTH 9:00 a.m. - 5:00 p.m. (EST). E-mail address: getinfo@haworthpressinc.com].

229

dependent from any other physiological parameter. These results might elucidate some aspects of the physiological function of *in vitro* produced phenolic antioxidants from sage. *[Article copies available for a fee from The Haworth Document Delivery Service: 1-800-HAWORTH. E-mail address: <getinfo@haworthpressinc.com> Website: <http://www.HaworthPress.com> © 2002 by The Haworth Press, Inc. All rights reserved.]*

KEYWORDS. Antioxidant, callus culture, cell metabolism, leaf explant, mitochondria, phenolic compound, *Salvia officinalis*

INTRODUCTION

Salvia officinalis L., the Dalmatian sage, has been identified as one of the most significant natural sources of (mainly phenolic) compounds with antioxidant properties, such as rosmarinic acid and carnosol (1). Previous experiments (2,3) have demonstrated the potential use of sage tissue culture as a cell factory for either an increased accumulation of rosmarinic acid or the production of novel antioxidants. Maximum rosmarinic acid accumulation in *S. officinalis* and *S. fruticosa* callus cultured on 4.5 µM 2,4-dichlorophenoxyacetic acid (2,4-D) and 4.5 µM kinetin (Kin) was 25.9 and 29.0 g/L, respectively. In the present study we investigated the effect of antioxidant phenolic compounds produced by leaf-derived callus cultures of sage on the growth and metabolism of the producing cells for an incubation period of five weeks.

MATERIALS AND METHODS

Sage (*S. officinalis*) plants grown in the greenhouse and bearing 6-8 leaves were used as the donor material. Callus cultures were established from one cm long leaf explants on a solid Murashige and Skoog (MS) (3,4) medium supplemented with 3% (w/v) sucrose, 10 mg L^{-1} ascorbic acid, 4.5 µM 2,4-D and 4.5 µM Kin. Media were adjusted to pH 5.8 using 1 N NaOH or 1 N HCl, autoclaved at 121°C for 20 min and poured into polystyrene 100 × 20 mm^2 petri dishes (30 ml of medium/dish, 4 explants/dish). Inoculated dishes were sealed with Parafilm™ and incubated for 5 weeks under a photosynthetic photon flux density of 150 µmol m^{-2} s^{-1} (16 hour light/8 hour dark photoperiod). Callus pieces were removed from the culture in weekly intervals and analyzed for fresh and dry weight. The metabolic state of culture cells was determined in two ways: First, the cell dehydrogenase activity (including $FADH_2$/NADH dehydrogenases) activity was assayed with triphenyl-tetrazolium-chloride (TTC): TTC assays were performed at 25°C in the dark for 24 h in 3 ml of 0.6% (w/v) TTC in 50 mM phosphate buffer (pH 7.5) (5). TTC formazan was extracted from the callus tissues with 95% (v/v) ethanol and its concentration was determined spectrophotometrically at 530 nm. Secondly, callus mitochondria were isolated by tissue homogenization in 2 mM HEPES buffer (pH 7.4, containing 70 mM sucrose, 220 mM mannitol and 0.5 g L^{-1} bovine serum albumin) and subsequent centrifugation (700 g × 10 min, then 11000 g × 30 min, 4°C). The cytochrome c oxidase activity of the supenatant was assayed spectrophotometrically (550 nm) in 0.1 M potassium buffer, pH 7.4, supplemented with. 6 µg cytochrome c (6), while their respiratory activity (oxygen uptake) was measured polarographically with a Clark-type electrode system at 25°C in 0.1 M potassium buffer, pH 7.4. Total phenolics were extracted from callus tissues in 80% MeOH and spectrophotometrically determined at 333 nm. The antioxidant capacity of the extracts was indirectly assayed by measuring the reduction of Fe^{+3} to Fe^{+2} and the subsequent formation of a red phenanthrolinium chloride-Fe^{+2} complex, which was spectrophotometrically determined at 510 nm. Results were assessed by a standard analysis of variance for a randomized complete block design, using MS-STATISTICA software. Five replications were used in each measurement.

RESULTS AND DISCUSSION

Although cultures demonstrated a continuous growth (expressed both as callus fresh and dry weight) during an incubation period of five weeks, the cell dehydrogenase activity (including $FADH_2$/NADH dehydrogenases) and the cytochrome c oxidase activity of isolated mitochondria declined, with a minimum value observed during the third week (Figure 1). During the same period mitochondrial respiration also declined rapidly. During each assay, oxygen consumption was increased after the addition of 150 µM ADP (analytical results not shown); however, the P/O ratios remained low (approx. 1.3), indicating that the isolated mitochondria did not retain a satisfactory respiratory control. Furthermore, the antioxidant activity *in vitro* was not correlated with the observed pattern of callus fresh and dry weight or any parameter of mitochondrial function. It was also independent from the accumulation of phenolic compounds, which increased during the first two weeks but declined rapidly thereafter. However, there existed a high correlation between the accumulation of total phenolics and the cell dehydrogenase activity, cytochrome c oxidase activity and mitochondrial respiration (r^2 = 0.83413, 0.91397 and 0.87403, respectively). On the contrary, callus dry weight was negatively correlated with oxygen uptake (r^2 = -0.9595).

In our laboratory, we have previously observed a time-dependent decline of dehydrogenase activity and ATP production in sychronized cell cultures of Jerusalem artichoke (*Ipomea* sp.) (unpublished results). Because of its central role in the production of ATP, decreases in mitochondrial function would lead

FIGURE 1. Patterns of physiological parameters of sage (*Salvia officinalis* L.) leaf callus cultures during an incubation period of five weeks, on MS + 10 mg L^{-1} ascorbic acid, 4.5 µM 2,4-D and 4.5 µM Kin.

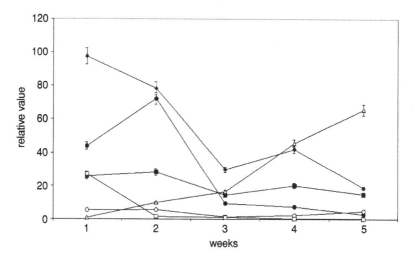

Key: (■) cytochrome c oxidase activity (relative absorbance × 100) (♦) dehydrogenase activity (relative formazan absorbance × 100) (△) callus fresh weight (× 0.01 g) (●) total phenolic accumulation (relative absorbance) (○) antioxidant capacity (relative absorbance) (□) oxygen consumption (µmol g^{-1} min^{-1})

to decreased energy production by the cell. Mitochondria also constitute the greatest source of reactive oxygen species (ROS), since their electron transport system consumes approximately 90% of the oxygen utilized by the cell and $O_2\cdot^-$ is produced by the electron transport system due to inefficiencies in oxidative phosphorylation (7). We currently analyze the accumulated phenolic substances by means of high-pressure liquid chromatography, in order to determine whether some compounds exert any particular, antioxidant-protective activity on sage mitochondria *in vitro*. Furthermore, we plan to investigate in detail changes in mitochondrial membrane lipid composition *in vitro*, as a function of time, along with *in vitro* ATP production. These studies, together with the present results, will hopefully elucidate some aspects of the physiological role of *in vitro* produced antioxidants from sage.

REFERENCES

1. Dorman, H.J.D, Deans, S.G., Noble, R.C. and Surai, P. (1995). Evaluation in vitro of plant essential oils as natural antioxidants. *J. Essential Oil Research.* 7, 645-651.

2. Hippolyte, I., Marin, B., Baccou, J.C. and Jonard, R. (1992). Growth and rosmarinic acid production in cell suspension cultures of *Salvia officinalis* L. *Plant Cell Reports.* 11, 109-112.

3. Kintzios, S., Nicolaou, A. and Skoula, M. (1998). Somatic embryogenesis and *in vitro* rosmarinic acid accumulation in *Salvia officinalis* and *S. fruticosa* leaf callus cultures. *Plant Cell Reports*. 18, 462-466.

4. Murashige, T. and Skoog, F. (1962). A revised method for rapid growth and bioassays with tobacco tissue cultures. *Physiol. Plant*. 15: 472-497.

5. Steponkus, P.L. and Lanphear, F.O. (1967). Refinement of the triphenyl tetrazolium chloride methods of determining cold injury. *Plant Physiol*. 42: 1423-1426.

6. Chirikjian, J.G. (1995). *Biotechnology. Theory and Techniques*. Vol. I. Jones and Bartlett Publishers. Sudbury, MA, USA. 204 pp.

7. Shigenaga, M.K. and Amess, B.N. (1994). Oxidants and mitochondrial decay in aging. *In* B. Frei, Ed. *Natural antioxidants in human health and disease*. Academic Press, NY, USA. pp. 63-106.

A Genetic Map of *Papaver somniferum* L. Based on Molecular and Morphological Markers

Petra Straka
Thomas Nothnagel

SUMMARY. An F_2 population obtained from a cross between a low-morphine poppy plant from our own selection line and a high-morphine plant from the Hungarian poppy variety 'Cosmos' was investigated. The segregation of six morphological traits was detected during the vegetative phase of plant development as well as on harvested capsules; 125 molecular markers were detected, 77 as AFLP and 48 as RAPD markers.

A total of 87 marker loci (66% of the total number) were placed in 16 linkage groups, which is five more than the haploid chromosome number of opium poppy (n = 11). Linkage analysis and map construction were performed by MAPMAKER 3.0. Eleven major linkage groups contained 4-12 loci whereas five minor groups comprised 2-3 loci. Only one morphological marker (*ns*) could be linked so far. *[Article copies available for a fee from The Haworth Document Delivery Service: 1-800-HAWORTH. E-mail address: <getinfo@haworthpressinc.com> Website: <http://www.HaworthPress. com> © 2002 by The Haworth Press, Inc. All rights reserved.]*

Petra Straka and Thomas Nothnagel are affiliated with the Federal Centre for Breeding Research on Cultivated Plants, Institute of Horticultural Crops, Neuer Weg 22-23, D-06484 Quedlinburg, Germany.

This project is supported by the Ministry of Cultural Affairs of the German State of Saxony-Anhalt.

[Haworth co-indexing entry note]: "A Genetic Map of *Papaver somniferum* L. Based on Molecular and Morphological Markers." Straka, Petra, and Thomas Nothnagel. Co-published simultaneously in *Journal of Herbs, Spices & Medicinal Plants* (The Haworth Herbal Press, an imprint of The Haworth Press, Inc.) Vol. 9, No. 2/3, 2002, pp. 235-241; and: *Breeding Research on Aromatic and Medicinal Plants* (ed: Christopher B. Johnson, and Chlodwig Franz) The Haworth Herbal Press, an imprint of The Haworth Press, Inc., 2002, pp. 235-241. Single or multiple copies of this article are available for a fee from The Haworth Document Delivery Service [1-800-HAWORTH 9:00 a.m. - 5:00 p.m. (EST). E-mail address: getinfo@haworthpressinc. com].

KEYWORDS. *Papaver somniferum*, genetic map, molecular markers, morphological markers

INTRODUCTION

In Germany the goal of poppy breeding and cultivation was, besides the extraction of alkaloids the use of poppy oil in food and in baker's ware. Poppy breeding was intensively accomplished till the 1960s. The variety 'Neuga' was the temporary final result of the German poppy breeding. The cultivation of poppy was continued till 1992. At present, on account of drug abuse, the cultivation of poppy is limited by restrictions of the German Federal Health Agency which fixed the maximal morphine content in *Papaver somniferum* L. to 0.01% in the dry capsule. For the reestablishment of the poppy cultivation in Germany, the development of low-morphine varieties was necessary. In the Federal Centre for Breeding Research on Cultivated Plants, the development of new poppy forms began in 1992. The development of low-morphine poppy based on the German ornamental variety 'Riesenmohn' was successful and resulted in several lines which have a mean content of morphine of M = 78 µg/g. These lines are a potential basis for breeding (4).

The successful cultivation of low-morphine poppy varieties, *P. somniferum* L., requires a practicable verification test for breeders. Linkage maps may be an important tool for marker assisted breeding. A high density of markers and the mapping of traits attractive for breeders are decisive prerequisites. The aim of the research project presented here is the development of a preliminary linkage map of poppy, *Papaver somniferum* L., based on molecular and morphological markers as the first step of mapping genetic loci corresponding with the morphine content.

MATERIALS AND METHODS

Plant material. An F_2 population obtained from a cross between a low-morphine poppy plant from our own selected line and a high-morphine plant from the Hungarian poppy variety 'Cosmos' was investigated. Plants were cultivated and evaluated in the field. The population comprised n = 59 plants.

RAPD and AFLP analysis. The DNA was isolated by modified CTAB extraction according to Porebski et al. (6). RAPD analysis of genomic DNA was performed using random decamer primer (Operon Techn. Inc., USA; Carl Roth GmbH, Karlsruhe, Germany). Amplification reactions were performed in volumes of 12.5 µl containing $10 \times NH_4$-reaction buffer (InViTek Berlin, Germany), 3.5 mM $MgCl_2$, 1 U Taq-Polymerase (InViTek, Berlin), 0.2 mM dNTP-Mix, 10 pmoles primer and 16 ng genomic DNA. The Multicycler of

MJ Research was used and programmed for 45 cycles of 1 min at 94°C, 1 min at 37°C and 2 min at 72°C for denaturing, annealing and primer extension, respectively. Amplification products were separated by horizontal polyacrylamide gel electrophoresis and stained by silver nitrate according to Haas et al. (1). The analysis of AFLP was performed using different primer pairs and the AFLP-Analysis System 1-Kit of GIBCO BRL Products. Goodness of fit with an expected segregation ratio of the RAPD and AFLP markers was estimated by χ^2-test. Markers which deviated from the expected ratios were also incorporated into the linkage analysis.

Genetic analysis of morphological traits. The segregation of six morphological traits was detected during the vegetative phase of plant development as well as on harvested capsules: plant height, number of stigmas (*ns*), petal colour, peduncle hairiness, capsule shape, and dehiscence of capsules (Figure 1).

Linkage analysis and map construction. For linkage analysis and map construction, the MAPMAKER 3.0 programme was used (3) with a minimum LOD score of 3.0 and a maximum recombination value of 0.50. Linked markers were identified using two-point analysis, followed by subsequent ordering of markers using threepoint and multipoint analyses. Recombination frequencies were transformed into centimorgans by the Kosambi function (2). For the map drawing the DrawMap1.1 programme was used (5).

RESULTS

A total of 7 primer pairs and 12 random primers was used for the analysis of AFLP and RAPD; 125 polymorphic fragments were detected, 77 as AFLP and 48 as RAPD markers. Six morphological markers were evaluated. Most of the traits could be assign to two phenotypic classes. The observed segregation ratios for morphological traits, AFLP and RAPD mainly corresponded to the expected monogenic inheritance (Tables 1 and 2).

There were 87 marker loci placed in 16 linkage groups. A minimum LOD score of 3.0 and a maximum recombination value of 0.50 were used. Eleven major linkage groups contained 4-12 loci whereas five minor groups comprised 2-3 loci. Only one morphological marker, the number of stigmas (*ns*), could be linked (Figure 2).

DISCUSSION

The segregation analysis of RAPD and AFLP and morphological polymorphic markers in the progeny of a cross between low-morphine and high-morphine

FIGURE 1. Some morphological traits of the F$_2$ population of poppy, *Papaver somniferum* L.: (1) number of stigmas, (2) dehiscence of capsules, (3) peduncle hairiness, (4) capsule shape.

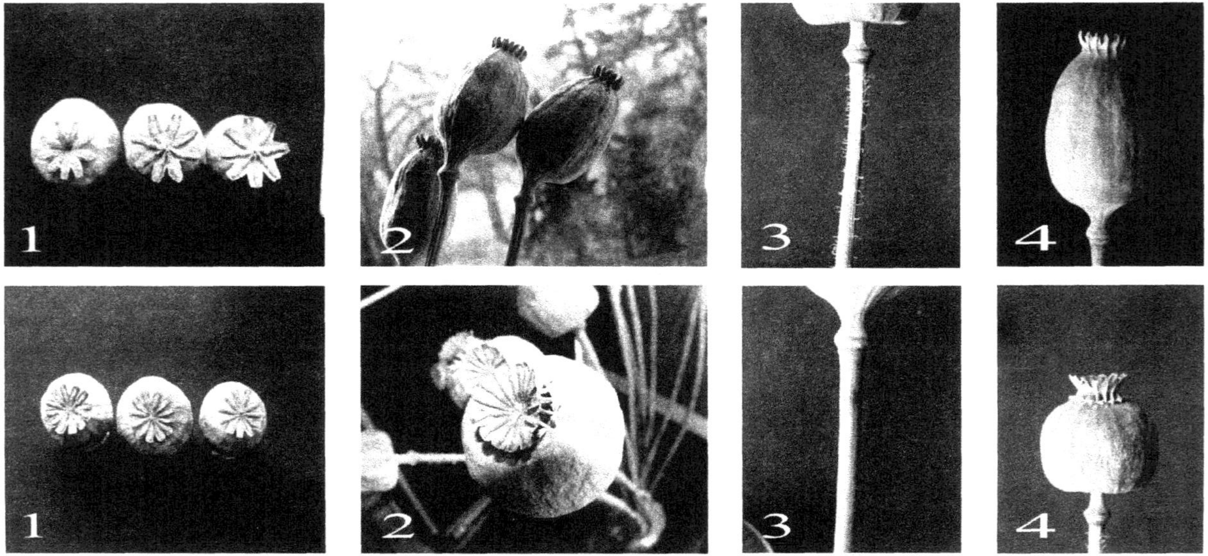

TABLE 1. F_2-segregation analysis of morphological traits.

Trait	Locus	P1-f	P2-m	Total	χ^2
Plant height	*ph*	32	21	53	6.04*
Number of stigmas	*ns*	41	8	49	1.97
Petal color	*pc*	15	38	53	0.31
Peduncle hairiness	h_p	14	40	54	0.02
Capsule shape	*cs*	6	43	49	4.25*
Dehiscence of capsules	*dc*	12	38	50	0.03

* Significant at $p > 0.05$.

TABLE 2. Observed markers.

Type	observed markers			deviated	mapped markers		
	total	maternal	paternal	segregation	total	maternal	paternal
Morphological	6			2	1		
AFLP	77	38	39	4	59	29	30
RAPD	48	26	22	8	27	16	11

genotypes allowed the construction of a linkage map of *Papaver somniferum* L. A total of 87 from 131 evaluated marker loci, i.e., 66%, could be integrated in 16 linkage groups with a total size of 779.2 cM. The average distance between markers is 8.9 cM. The number of markers per linkage group ranged from 2-12; two markers showed a cosegregation. The linkage map contained 16 groups, i.e., five linkage groups more than the haploid chromosome number of opium poppy (n = 11). Most of the morphological markers could not be mapped so far, except the *ns* marker locus. This linkage map is preliminary concerning the number of molecular marker loci and the number of morphological markers. On the basis of the extended F_2 population it will be developed further. The results of the morphine analysis carried out at this time will be integrated.

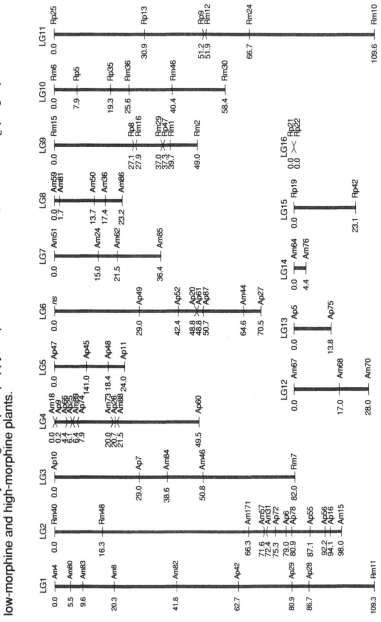

FIGURE 2. Preliminary linkage map of poppy, *Papaver somniferum* L., based on F$_2$ progeny of a cross of low-morphine and high-morphine plants.

REFERENCES

1. Haas, H., B. Budowle, and G. Weiler, 1994. Horizontal polyacrylamide gel electrophoresis for the separation of DNA-fragments. *Electrophoresis* 15(2):153-158.

2. Kosambi, D.D., 1944. The estimation of map distances from recombination values. *Ann. Eugen* 12:172-175.

3. Lander, E.S., P. Green, J. Abrahamson, A. Barlow, M.J. Daly, S.E. Lincoln, and L. Newburg, 1987. MAPMAKER: An interactive computer package for constructing primary genetic linkage maps of experimental and natural population. *Genomics* 1:174-181.

4. Nothnagel, Th., P. Straka, and W. Schütze, 1996. Selection of a low morphine poppy *Papaver somniferum* L. *Proceedings of International Symposium of Breeding Research on Medicinal and Aromatic Plants,* June 30-July 4, 1996, Quedlinburg, pp. 120-123.

5. Ooijen v., J.W., 1994. Drawmap: A computer program of drawing genetic linkage maps. *Journal of Heredity* 85(1):66.

6. Porebki, S., L.G. Bailey, and B.R. Braun, 1997. Modification of a CTAB DNA extraction protocol for plants containing high polysaccharide and polyphenol components. *Plant Molecular Biology Reporter* 15(1):8-15.

BIODIVERSITY AND CONSERVATION OF MEDICINAL AND AROMATIC PLANTS RESOURCES

Challenges and Opportunities in Enhancing the Conservation and Use of Medicinal and Aromatic Plants

S. Padulosi
D. Leaman
P. Quek

S. Padulosi is affiliated with the International Plant Genetic Resources Institute (IPGRI), Regional Office for Central & West Asia and North Africa, c/o ICARDA, P.O. Box 5466, Aleppo, Syria (E-mail: s.padulosi@cgiar.org).

D. Leaman is affiliated with the Medicinal Plant Specialist Group, IUCN–the World Conservation Union, Species Survival Commission, c/o Canadian Museum of Nature, P.O. Box 3443, Station D, Ottawa, Ontario K1P 6P4, Canada (E-mail: djl@green-world.org).

P. Quek is affiliated with the International Plant Genetic Resources Institute (IPGRI), Regional Office for Asia, Pacific and Oceania, P.O. Box 236, UPM Post Office, Serdang 43400, Selangor, Malaysia (E-mail: p.quek@cgiar.org).

[Haworth co-indexing entry note]: "Challenges and Opportunities in Enhancing the Conservation and Use of Medicinal and Aromatic Plants." Padulosi, S., D. Leaman, and P. Quek. Co-published simultaneously in *Journal of Herbs, Spices & Medicinal Plants* (The Haworth Herbal Press, an imprint of The Haworth Press, Inc.) Vol. 9, No. 4, 2002, pp. 243-267; and: *Breeding Research on Aromatic and Medicinal Plants* (ed: Christopher B. Johnson, and Chlodwig Franz) The Haworth Herbal Press, an imprint of The Haworth Press, Inc., 2002, pp. 243-267. Single or multiple copies of this article are available for a fee from The Haworth Document Delivery Service [1-800-HAWORTH 9:00 a.m. - 5:00 p.m. (EST). E-mail address: getinfo@haworthpressinc.com].

SUMMARY. Medicinal and aromatic plants represent a consistent part of the natural biodiversity endowment of many countries around the world. These species provide an important contribution to health, local economies, cultural integrity and ultimately the well-being of people, particularly the rural poor and fragile social group, e.g., the elderly, children and women. Their role has been increasingly acknowledged over the last decade which has brought about a new attitude towards these and other species, variously marginalized and neglected by research and conservation efforts. At the same time, threats to the genetic diversity and species survival have also increased due to unsustainable exploitation and loss of habitats. Progress in pharmacognosy has not been accompanied by equal advancements in knowledge of the distribution, genetic diversity, ecology and conservation of these species. Priority setting for species selection and understanding of resource management needs for these species is constrained by limited capacities and attention devoted to basic research. Current trends in support to environmental friendly agriculture and greater participation of local people in the sustainable conservation and use of natural resources are leading towards innovative approaches to enhance the use of medicinal and aromatic plants, more participatory in nature and hence more focused on local needs. The increasing number of organizations involved in medicinal and aromatic plants, the access to new tools for biodiversity prospecting, characterization and data analyses along with the change of traditional conservation systems towards more use-oriented initiatives should be seen as opportunities for revising research goals and partnership, create greater synergies at the national level and a more conducive policy environment. The future of medicinal and aromatic plants rests on today's ability to resolve the conflicts between conservation and use, and the shift towards more resource-based agriculture increasingly challenged by the globalization of economies. *[Article copies available for a fee from The Haworth Document Delivery Service: 1-800-HAWORTH. E-mail address: <getinfo@haworthpressinc. com> Website: <http://www.HaworthPress.com> © 2002 by The Haworth Press, Inc. All rights reserved.]*

KEYWORDS. Medicinal plants, aromatic plants, MAP, biodiversity, conservation, cultural heritage

INTRODUCTION

Medicinal and aromatic plants belong to a category of species characterized by multiple uses. Many species, used for food, ornamental, timber, etc., are in fact employed also for medicinal and aromatic purposes and this fact is often the cause of disagreements among workers over the classification of these species in a well defined use-oriented category of crops. Regardless the "exclusivity" of their use, the sheer numbers of MAP recorded from around the world (1) are a clear indication that these species indeed represent one of the largest endowments of useful biodiversity on earth.

Most MAP species can be defined as being underutilized because of the poor attention paid to them by research and conservation agencies (2). Such status of neglect greatly contrasts with their high popularity at the local level, which has been recorded since time immemorial. The first evidence of use of medicinal plants dates back 60,000 years, according to archeological evidence found in a Neanderthal cave in Iraq (3). The Greek Dioscorides (50 BC) described in detail some 500 medicinal species (almost 10% of the indigenous plant diversity known to occur in Greece today), and Galen (ca. 200 AD) compiled some 304 herbal remedies from known plants popular at his time (4).

According to the World Health Organization (WHO) some 35,000 to 70,000 species have been used so far as medicaments, a figure corresponding to 14-28% of the 250,000 plant species estimated to occur around the world (5), and equivalent to 35-70% of all species used world-wide (1). Within the global market, more than 50 major drugs have originated from tropical plans alone (3), confirming once again how precious for our lives are tropical areas around the world.

Over the last decade or so, among international agricultural organizations, the overarching goal of food security has been increasingly seen in the context of the wider scope of achieving better livelihood. The concept of *food security* has been evolving into a more *consistent* concept of *nutrition security*, which refers more explicitly to the quality of food, thus providing greater visibility to the importance of vitamins and micronutrients as necessary components of the human diet (6). We could therefore say that the role of medicinal plants in improving the livelihood of millions of peoples is thus seen with greater attention thanks also to the shift of the global research paradigm, calling today for better food for fighting hunger, malnutrition and poverty in order to achieve the ultimate goal of well being for all (7).

More than 80% of the world population currently relies on traditional remedies, including the use of local plants to treat various ailments and diseases (8). Medicinal plants as herbal remedies are an essential resource base particularly for the more fragile groups, the poor ones, for whom pharmaceutical drug preparations are not affordable.

MAP species are an important source of income: in the last 10 years, India's exports of medicinal plants have increased three-fold, the production during the 1992-95 period being 33,000 t (equivalent to US$ 46 million) (3,9). Greater economic benefits from MAP are gained by developed countries, for which these plants represent a multi-million dollar industry: in 1996, in the USA market alone, the MAP business was estimated to US$ 1.3 billion (10).

HOW WELL ARE MAP BEING CONSERVED?

The recognition of the important role of MAP in the livelihood and income generation in both developing and developed countries is not matched by a parallel increased attention towards their better conservation.

The globalization process–bringing closer once distant cultures and economies–is having dramatic impacts on the trade of plant species that are being marketed. Once harvested in the wild using sustainable community-driven systems, many medicinal and aromatic plants are now subject to increased demands from the market, for which harvests are now largely un-monitored, leading to dangerous levels of over-exploitation (11).

The slow harvest-recovery of many MAP species, coupled with often destructive harvesting systems, reduce dramatically the regeneration capacity of populations or single individuals. *Prunus africana* is the classical example to represent such a situation affecting hundreds of species around the world: highly sought for its medicinal properties to treat benignant prostate cancer, this tree species is being stripped of its bark to the extent that specimens eventually die (12). *Prunus africana* is included since 1975 in the Convention of International Trade in Endangered Species of Wild Fauna and Flora–CITES (information on conservation activities on this species can be found at the site of the WWF, UNESCO, Kew Gardens Initiative, People and Plants at <http://griffin.rbgkew.org.uk/peopleplants/activities/africa/index.html>). Similarly for the Himalayan yew (*Taxus baccata*), a coniferous species also heavily traded for its content in taxol, compound used to treat ovarian cancer, large quantities of this plant are being gathered from the wild and marketed in spite of the bans imposed by Asian governments to protect this endangered species (3). Indeed, the lack of a monitoring system on unregulated and unsustainable harvests of MAP often lead to a *fait accompli* of the genetic erosion caused to these species and, in the worse cases, to their extinction.

From the research stand point, basic taxonomic description and ethnobotanic studies on medicinal and aromatic plants are quite well represented in official or gray literature. Excellent floras specifically referring to medicinal plants are, for instance, available in many countries around the world. In Morocco more than 12 books specifically addressing the distribution, description and local uses of medicinal plants have been published so far along with more than 80 University Theses on MAP-related subjects prepared by Moroccan students over 20 years (13). Similar statements can be said for research investigations related to essential oil compounds in aromatic plants which account for a good amount of publications as well. The results of a survey on the publications published during the period 1972-1995, using the abstracts of CAB International (14), illustrate this aspect further:

- Published literature on MAP increased 5-fold in 20 years (from 1,000 papers in the 1970s to 4,000/year in the mid-1990s).
- 50,000 relevant papers appeared in 2,500 serials and in 45 languages. Half of these papers appeared in just 10 serials.
- Medicinal plants accounted for ca. 50%; herbs, spices and essential oil plants for 25%.
- More than 10,000 genera were covered in relevant papers published.

The field of research on MAP which is definitively the most neglected by scientists so far is that addressing the conservation of genetic resources. The roots of such a situation are to be found in the fact that major crops and commodities have "monopolized" the attention of gene bank managers and researchers for the last 20 to 30 years, leaving scarce means to be spent on all other species. Not only inadequate measures have been taken for the conservation of these resources, but very little research has been carried out to shed light on their geographical distribution, ecological requirements, reproductive biology (pollination and dispersal mechanisms), taxonomy and phylogenetic relationships within their gene pools, etc. All these factors play an essential role in influencing the survival of the species in natural conditions and are even more crucial in determining the consequences of man-made harvests (15). Furthermore, studies related to seed characteristics (e.g., to ascertain their recalcitrant or orthodox nature), their optimal field conservation requirements, their ability to reproduce *in vitro*, and other topics whose understanding is crucial for guiding *ex situ* management of MAP germplasm collections, are very poor (16). Given the scarcity of information in this field, worthwhile is the mentioning of the IPGRI publication *Seed Storage Behaviour: A Compendium* (17), which provides an introduction to seed storage physiology and a selective summary of the literature on seed survival in storage, for over 7,000 species from 251 families, including numerous MAP species.

Changes in germplasm collecting strategies have been taking place in the last 30 years as a result of better understanding of the role that biodiversity as a whole plays in improving our lives. For instance, the *species approach* that greatly characterized early germplasm expeditions had been replaced by the mid-1980s by the *gene pool approach*. Today, the call for a more holistic approach in biodiversity conservation (18) has contributed to bring about an additional new approach, the *ecosystem approach*, which is currently adopted by several international organizations along with and in complementation of the gene pool approach (19).

At the onset of a new century, an account of what has been done in the area of conservation of MAP species is very timely. To that regard, the study carried out by IPGRI (19) on the extent of conservation of so called minor crops (i.e., all those species not staples or main commodity–thus including MAP) in gene banks is particularly useful to describe the situation at the global level. The analysis was conducted in 1997 using the IPGRI conservation database that reported at that time a total of 5,013,258 accessions conserved in some 300 *ex situ* gene banks around the world. Out of this number, only 375,336 accessions (equivalent to 7% of the total holdings) were referred to as "minor crops." This figure was obtained as the difference between the total number of holdings and the number of accessions belonging to major crops, using the FAO classification of major crops provided in the State of the World Report (6). In addition, the study analyzed only data of those accessions for which complete information was made available by gene banks (which meant that some 3,081 entries–corresponding to 342,097 accessions–could not be considered because they were incomplete). With regard to the question of how well each of these species are represented in terms of their infra-specific diversity, the analysis focused on the number of accessions for each of the 6,639 species whose total holdings corresponded to the 375,336 accessions just referred above: more than 80% of minor species conserved around the world (ca. 5,351 species) are represented by just 1 to 10 accessions (Figure 1).

In addition, the little attention paid to MAP by germplasm collectors has also meant that the few accessions gathered for these species have been col-

FIGURE 1. Representation of minor crops in *ex situ* gene banks (see text for explanation).

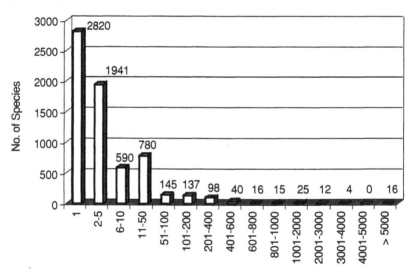

Classes (no. of accessions per species)

lected–in general–in an incidental manner, therefore using inadequate collecting methodology for sampling genetic diversity at the population or individual plant level.

An example to illustrate the point regarding the contradiction between the popularity of MAP species and the scarce conservation paid to them by *ex situ* gene banks is that of basil. Basil (*Ocimum basilicum*) is among the most widely used spices in Italy, where it is an essential ingredient of many food preparations, including pasta and pizza, symbols of the Italian culinary tradition. Although originating in Asia (20), all races described by Danert in 1959 can actually be seen also in cultivation in Italy (21,22). The conservation of such a popular spice is nevertheless ensured mainly through local cultivations and home gardens. The National Italian gene bank, the Germplasm Institute (National Research Council–CNR), currently holds in its collections just 41 accessions of basil of which 39 originated in Italy and 2 from foreign countries (Laghetti, pers. comm.). Given the increasing replacement of old landraces with new cultivars taking place in Italy (22), the strengthening of the *ex situ* collections to support farm conservation activities for basil in Italy should indeed be given a high priority.

The analysis just described represents an attempt to characterize the magnitude of a phenomenon that has been recognized in its gravity in many international fora where recommendations on most endangered species and issues regarding sustainable use of MAP have been extensively reiterated. In addition to the 1975 CITES Convention and the resolutions made by WHO (emphasizing the important role of traditional systems of medicine and calling for a comprehensive approach to medicinal plants), the meetings listed in Table 1 can be considered milestones of awareness raising and/or consolidation of action plans for the conservation and sustainable use of medicinal and aromatic plants within the last decade or so. For a detailed summary on the recommendations made in each of these meetings, the reader is referred to Leaman et al. (23).

According to the Directory for Medicinal Plants Conservation prepared by IUCN (24), some 138 organizations are involved in the conservation of medicinal plants. A subsequent study carried out by the Botanic Gardens Conservation International (BGCI) has led to the production of the Directory of Botanic Gardens Medicinal Plant Collections (25), which lists 460 botanic gardens involved at various levels in the conservation of both medicinal and aromatic plants.

In order to gather more detailed information on the status of conservation of MAP in gene banks, IPGRI, with the support of IDRC, has carried out a survey study contacting 21 selected organizations around the world (23). Nine replies were eventually received. Table 2 summarizes basic information about the number of species and accessions maintained by the respondents along with information on priority needs that have been indicated in their replies.

TABLE 1. Milestones in awareness raising/consolidation of MAP conservation work.

Conference	Venue	Year	Main issues
WHO-IUCN-WWF International Consultation on the Conservation of Medicinal Plants	Chiang Mai, Thailand	1988	• Conservation guidelines • Greater partnership • Role of IK
United Nations Conference on Environment and Development (UNCED)	Rio de Janeiro, Brazil	1992	• Conservation • Sustainable use • Sharing of benefits
I World Congress on Medicinal and Aromatic Plants for Human Welfare	Maastricht, The Netherlands	1992	• Documentation • Sustainable harvesting practices • Legal frameworks for use • Commercialization guidelines
IV International Technical Conference of FAO on Plant Genetic Resources	Leipzig, Germany	1996	• Promotion of underutilized species • Plan of Action for PGRFA
II World Congress on Medicinal and Aromatic Plants for Human Welfare	Buenos Aires, Argentina	1997	• IK documentation • Better use practices • Conservation & Partnership • Inventory of MAP
Medicinal Plants for Survival: International Conference on Medicinal Plants	Bangalore, India	1998	• Priority setting • Complementary conservation • Participatory approaches • Use promotion
International Conference of the Global Forum of Agricultural Research (GFAR)	Dresden, Germany	2000	• Role of underutilized species • Chain production system
International Conference on Medicinal Plants, Traditional Medicine & Local Communities in Africa: Challenges and Opportunities of the Millennium	Nairobi, Kenya	2000	• Inventory • Use-enhancing methods • Community based approaches • Documentation/dissemination of information • Networking
Conference of the Parties (COP) of the CBD	Nairobi, Kenya	2000	• Indigenous knowledge
SAT21–International Conference on Science and Technology for Managing Plant Genetic Diversity in the 21st Century (SAT21)	Kuala Lumpur, Malaysia	2000	• Complementary conservation • Methodologies and techniques • Priority setting

TABLE 2. Needs as indicated by MAP gene banks around the world.

Organization	Country	No. of species	No. of accessions
Gatersleben Gene Bank (IPK) Need: • More research on conservation biology of MAP • Evaluation/research for breeding purposes	Germany	931	17,757
Research Institute of Medicinal Plants Need: • Call for a greater international collaboration for *ex situ* conservation	Poland	7	33
Centro Nacional de Conservavion de recursos Fitogenetics Need: • Funds • More research on conservation biology of MAP	Venezuela	30	60
Biodiversity Institute Need: • Documentation of orally transmitted IK on MAP • Systematising information gathered so far	Ethiopia	35 (app.)	1282
Agricultural Research Council (ARC) Need: • Funds • Clarify access policy to MAP • More research on conservation biology of MAP • More research on complementary conservation approaches	Republic of South Africa	60	560 (app.)
Zhejing Research Institute of Chinese *Materia Medica* Need: • Greater international cooperation • More research on conservation methods	China	200	50,000
National Bureau of Plant Genetic Resources (NBPGR) Need: • More research on conservation biology of MAP • Research on *in vitro* conservation method • Surveys and collecting • Documentation • Research on agro-techniques for use promotion	India	600	849
Agricultural Research Organization, Newe-Ya'ar Research Center Need: • Funds • Greater research/conservation on endemic species	Israel	12 (app.)	350
Indian Institute of Horticultural Research (IIHR) Need: • Funds • Research on impact of *ex situ* conservation on MAP • Characterization of traits of economic interest	India	11	227

From analyses of the responses received by the surveyed organizations and on the basis of data from previous works, some general remarks can be made to re-affirm a few important points for furthering the conservation and sustainable use of MAP:

- *An holistic approach* in conservation (a combination of *ex situ* and *in situ*) for MAP species is needed in order to meet–*inter alia*–biological requirements (specific ecological needs that cannot be recreated in *ex situ* gene bank facilities) and users' needs (e.g., availability of germplasm for improvement and other research purposes),
- For both *ex situ* and *in situ* conservation methods, studies are needed to understand the *reproductive biological requirements* of many species along with setting up of protocols for *in vitro* multiplication (to address the poor multiplication ability of many MAP),
- Commercial promotion of MAP in most cases leads to overexploitation: systems for *better monitoring* the distribution of these species, and early-warning system assessments of dangerous levels of genetic erosion should be put into place,
- The number of accessions for each MAP species maintained in *ex situ* collections is still very low. The need for a *proper geographical representation* of species in gene banks across their area of distribution and proper representation of their genetic diversity within populations should be reiterated,
- In the medicinal gardens of Southeast Asia, it has been observed that the diversity of MAP species has been decreasing over the last 30 years (3). Such a fact should be seen as a further justification for *closer partnership* among all actors involved in conservation at the national level, so as to share the burden of conservation of MAP and ensure a more sustainable system for their safeguard.

The greatest challenges in promoting better conservation and use of MAP are faced by developing countries, where research infrastructure and human capacities of National Programmes are often poor. Challenges and opportunities in the promotion of medicinal plants in developing countries have been addressed recently at the *The International Conference on Medicinal Plants, Traditional Medicine and Local Communities in Africa: Challenges and Opportunities of the New Millennium.* This meeting was convened by the Environment Liaison Centre International (ELCI) and the Global Initiative for Traditional Systems (GIFTS) of Health of the University of Oxford, UK and the Commonwealth Working Group on Traditional & Complementary Health Systems. It was held in Nairobi, Kenya, from the 16th to the 19th of May 2000 and attended by more than 150 international delegates representing major stakeholders from 40 countries from all over the world. The recommendations of this Conference (Box 1) were submitted to the 5th Meeting of the Conference of the Parties (COP5) of the CBD which was held at the same time in Nairobi.

CONSERVATION OF MAP IN PERSPECTIVE

The 1996 IUCN and 1998 BGCI Directories, along with additional data-bases and institution profiles available on the Internet (relevant web pages can be found at the "Internet Directory for Botany: Conservation, Threatened Plants" at <http://www.helsinki.fi/kmus/botcoms.html>), are contributing substantially to fill the gaps in knowledge on who-is-doing-what in the area of conservation and sustainable use of MAP at the local, national, regional and international levels. Furthermore, these publications represent powerful tools to foster collaboration among a plethora of actors active in this field.

Greater efforts should be deployed, however, for the recognition and the proper documentation of the role played by the millions of rural people and forest dwellers around the world, custodians and stewards of MAP, who have been managing this diversity tirelessly for generations. Women and elderly people have been especially involved in safeguarding this genetic and cultural heritage, managing harvests from the wild, maintaining local cultivations and home gardens and developing appropriate ways to deploy MAP diversity to meet family needs.

As we proceed into a new high-technologically driven era, it may sounds strange to many people that *in situ*/on-farm conservation (home gardens, sacred gardens, natural reserve, community managed areas, etc.) will continue to be the most important method to conserve genetic resources of medicinal and aromatic plants. In fact, given the large number of MAP needing conservation on one end, and the limited financial resources for *ex situ* conservation on the other, decentralized conservation initiatives involving local communities are likely to play the most strategic role in the future within the context of a complementary conservation approach. Additional elements supporting this vision include the need for sustainability in conservation (support to local communities can greatly enhance such work), the need for conserving the culturally-rich and precious indigenous knowledge (IK) associated with MAP (on-farm conservation allows this) and the need for growing species in environments most suitable for the production of the parts used (e.g., natural reserve and agro-ecological niches).

During the IPGRI's Conference on Science and Technology for Managing Plant Genetic Diversity in the 21st century, held in Malaysia in June 2000, presentations given by gene bank managers around the world clearly indicated that, due to the current limited financial resources of most *ex situ* gene banks, it is un-

BOX 1. Furthering the use of medicinal plants in Africa in the new millennium.

Outcomes of the 2000 Nairobi Conference

Recommendations

Establish an International Working Group for promoting, monitoring and assessing the conservation, management and sustainable use of medicinal plants and traditional medicines used for human and livestock health care by providing appropriate financial mechanisms in order:

1) **To support** an African inventory and genetic assessment of medicinal plants; (2) Community-based in-situ conservation and management of ecosystems with high medicinal plant species diversity; (3) Development of sustainable harvesting guidelines of wild medicinal resources; (4) Identification and development of cultivation/propagation practices; (5) Information dissemination, networking, education and awareness raising programs; (6) Interdisciplinary research into the efficacy, safety, cultural importance and use of traditional medicine and adding value to the medicinal plants and phytomedicines; and (7) Co-ordinating and catalyzing the existing activities relating to medicinal plants and traditional health systems at international level.

2) **To establish** an appropriate mechanism for the effective involvement of indigenous and local communities in redefining and monitoring intellectual property rights in the context of traditional medicine.

3) **To establish** a mechanism to formalise, monitor and regulate the trade in medicinal plants and herbal products, in order to guarantee local communities fair and equitable access to benefits flowing from the trade; and to ensure that the trade is transparent and sustainable.

Follow-Up Actions

1. African governments and the Organisation of Africa Unity (OAU) should adopt a Decade devoted to the promotion and development of medicinal plants, traditional medicines and pharmacopoeia in Africa.
2. The Conference recommends to the Government of the Republic of Togo and His Excellency Gnassingbe Eyadema, President of the Republic and Head of State to kindly submit the Conference Declaration on the Decade on Medicinal Plants and Traditional Medicines in Africa to the attention of the forthcoming OAU Summit.
3. Governments, NGOs and International Agencies should take urgent steps to support activities and strategies for Research and Development on Medicinal Plants and Traditional Medicine relating to addressing priority diseases in Africa such as HIV/AIDS and Malaria.
4. To use existing electronic networking opportunities (e.g., "Phytomedica" worldwide emailing list) in order to promote constructive relationships and collaboration between diverse stakeholders and interested parties active in the field of medicinal plants, traditional medicine and pharmacopoeia.
5. African Governments in partnership with NGOs and International Agencies should find ways to support the implementation of the recommendations and the comprehensive Regional Action Plan outlined by the Regional Workshops on Medicinal Plants and Traditional Medicine in Africa jointly organised by the Environment Liaison Centre International (ELCI) and the International Development Research Centre (IDRC) held successively on 17-21 November 1997 in Conakry, Republic of Guinea for African French speaking countries and on 14-18 April 1998 in Cape Town, South Africa for African English speaking countries.
6. To establish and convene a two year regular conference to review progress in implementing the Programme of the Decade and to plan for future action, focusing on priority and specific issues and objectives.

likely that these facilities will be significantly expanding their conservation targets in terms of interspecific and intraspecific diversity of plant biodiversity (26).

Although reduced in breadth, new germplasm collections (targeting also MAP) will, however, continue to be mounted, but these missions are expected to be more rationalized, thanks to new tools such as GIS, to better guide collecting expeditions and gather fewer but representative samples (27).

In the light of what is said above, it is clear that use will continue to be the driving force behind the conservation of MAP in the future as it has been in the past. The sustainable conservation-through-use of MAP is, however, challenged by over-exploitation: although there are more than 500 medicinal and aromatic species currently being cultivated, the largest majority of them continue to be harvested directly in nature (10). The reasons behind this situation are varied, including the difficulties faced in the propagation of the material, the considerable efforts needed for their cultivation vs. the low price of wild harvested material, users' preference over wild harvested material, etc. (11). On the other hand, the benefits for cultivating MAP are not limited to a more sustainable and regular supply of material, as quality and post-harvest handling are also enhanced (42).

A thorough analyses of the on-farm and *in situ* conservation issues goes beyond the scope of this paper, but it is felt that the three following examples (Boxes 2, 3 and 4) of organized local communities projects from Africa (Uganda) and Asia (India and Sri Lanka) summarize some of the challenges and the opportunities faced by the rural communities to improve their income and contribute at the same time to the sustainable conservation of these important resources.

TRENDS IN THE MANAGEMENT OF INFORMATION ON MAP

Information plays a central role in the work on plant genetic resources. The availability of accurate information on germplasm holdings in both *ex situ* and *in situ* conservation is crucial to allow for proper use (28). Reiteration of this point has been made in many fora, including the latest EUROGARD Congress organized by the BGCI on April 2000 in Spain (for further information on the outcomes of this meeting see <www.bcgi.ord.uk>) which recommended *inter alia* the development of better information dissemination systems for botanic gardens around the world as a way to enhance the use of plant genetic resources, including MAP species.

Over the last 15 years, information technology has brought about dramatic changes in the way germplasm-related data are being gathered, maintained and disseminated. According to FAO (6), 55 countries have reported in 1996 the need for improvement of the documentation systems to better handle informa-

BOX 2. Community project in Uganda <http://www.idrc.ca.adventure/index.html>.

Medicinal Plant in Uganda: An IDRC Supported Project

The Problem

A number of rural areas in Uganda rely on traditional healers and traditional medicinal plants for their health care. The pressure on these medicinal plants has been such that many of them are now on the endangered species list.

The Proposal

Previous research has identified 15 of the most endangered species of trees and shrubs that are useful in treating some of the main causes of morbidity and mortality in Uganda, i.e., malaria, respiratory tract infection, diarrhea, measles, malnutrition, worm infestations, skin infections, meningitis and tetanus. This project aims to increase the population of these 15 species in four districts, so that they are available for sustainable use by the local population. Researchers will work closely with practicing traditional healers, local herbalists, community leaders and villagers on plant propagation and the non-destructive harvesting of plant material.

The Impact

The Medicinal Plants Garden at Entebbe Botanical Gardens will be expanded in both area and species content, with seeds and other planting materials from the 15 target species. These will be used for reference and educational purposes, as a gene bank, and as a source of seeds. The long-term expectation is to reduce national dependency on imported drugs for conditions that can be treated with home-grown remedies, thereby saving scarce foreign exchange for the treatment of conditions that are not amenable to traditional medicine, such as tuberculosis.

BOX 3. Community forestry project in India <http://www.worldbank.org/html/ extdr/offrep/sas/ ruralbrf/medplant.htm>.

The Kerala Forestry Project

The conservation of medicinal plants is a key objective of biodiversity conservation components in several forestry projects in India being assisted by the International Development Association (IDA), the World Bank's concessionary lending arm. The Kerala Forestry Project, recently approved by IDA's board, is supporting a pilot program that involves tribal and other forest-dependent communities in the inventory, conservation, and sustainable development of medicinal plants. The four-year project, expected to cost US$47.0 million, is being financed with a US$39.0 million IDA credit and contributions totaling US$8.0 million from the state of Kerala and project beneficiaries. Project activities related to medicinal plants will cost US$0.2 million, or 0.4 percent of the total.

The project supports technological improvements for artificial propagation of endangered plant species; research and training in better harvesting and processing techniques; community management of plant propagation, harvesting, and marketing; analysis of marketing policies; establishment of community-managed, forest-based enterprises for income generation; and monitoring and evaluation of the status of these natural resources with the assistance of local communities. This pilot program will be implemented initially in five villages that are economically highly dependent on medicinal plants. The Kerala Forest Department and the Tropical Botanical Garden and Research Institute have formed a partnership to design and implement the program. The institute will take the lead in providing the technical expertise for taking plant inventories, developing processing techniques, and designing marketing strategies, and the department will help form community groups. While relatively modest, this program holds real promise of enhancing the sustainable management of the medicinal plant resources of Kerala in a way that will enable local communities to reap the economic benefits of these resources without depleting the forests and endangered plant species.

BOX 4. Community project on medicinal plants in Sri Lanka <http://www. worldbank.org/html/extdr/offrep/sas/ruralbrf/medplant.htm>.

The Sri Lanka Medicinal Plants Project
This project is the first approved by the World Bank that is focused exclusively on the conservation and sustainable management of medicinal plants. To be implemented between 1998 and 2002, the project is estimated to cost US$5.07 million. It is being financed by a grant of US$4.57 million from the Global Environment Facility Trust Fund and a contribution of US$0.5 million from the government of Sri Lanka. The World Bank is the implementing agency for the fund. The objectives of the project are to conserve important medicinal plants, their habitats, and genetic stock while promoting sustainable use. These goals will be achieved through three initiatives to:

1) Establish five medicinal plant conservation areas where plant collection from the wild is particularly intensive and develop a conservation strategy for each; implement village action plans to reduce dependency on harvesting from the wild; collect basic socioeconomic and botanical data; and promote extension and education on medicinal properties of species within these conservation areas.
2) Increase nursery capacity to develop the cultivation potential of select species and support research on propagation and field planting techniques.
3) Collect and organize existing information on plant species and their use and promote an appropriate legal framework through production of draft regulations to ensure the protection of intellectual property rights.

This project is expected to yield important environmental and social benefits. It will help conserve more than 1,400 medicinal plant species used in Sri Lanka, of which 189 are found only there and at least 79 are threatened. It will spread knowledge about sustainable growth, crop yields, biological cycles, and the danger of depleting plant resources; maintain critical habitats for medicinal plants; and increase the diversity and quantity of threatened species. The project will also preserve indigenous knowledge about medicinal plants and their use, promote policy and legal reforms, involve tribal people and local communities in efforts to reduce dependency on wild resources, and generate alternative income opportunities for the rural population. From a national perspective, the project will increase supplies of raw materials for traditional medicines, improve the availability and management of information, and promote human resource development in medicinal plant-related fields.

---

tion in their gene banks. Such a situation is likely to change within a decade or so, thanks to the advent of new user friendly software for PGR and the information "revolution" brought about by the Internet. Indeed, easier and cheaper access to information will have a tremendous impact in the way PGR have been traditionally handled so far.

Management information systems for PGR are today being developed to link in a friendly environment all types of possible data related to PGR, from GIS-generated information to gene bank data (holdings and their status, regeneration, distribution, etc.), from molecular markers data to marketing opportunities, from agronomical requirements to traditional uses (including traditional recipes). The world of MAP users will be increasingly "smaller," linking up distant people, projects, organizations, etc., who will be able to share data and benefit from each other's experiences as never in the past. According to FAO

(40), more than 40 databases dealing with MAP are currently on the market (some accessible also through the Internet), the majority of them being, however, accessible only through payment of subscription fees. Less extensive databases, but freely accessible on the Internet, include the following: *the tropical plant database* at <http://www.rain-tree.com/plants.htm>, *the online directory for medicinal plant conservation* at <http://www.dainet.de/genres/mpc-dir/>, *the plants of the Machhiguenga: an ethnobotanical study of Eastern Peru* at <http://www.montana.com/manu/>, *Algy's herb page* at <http://www.algy.com/herb/medcat.html>, *Information system on medicinal plants* at <http://www.ciagri.usp.br/planmedi/planger.htm>, *Medicinal Plants of the Quijos–Quichua (Equador)* at <http://www.public.iastate.edu/~cbutter/ethnofra.htm>, *A Guide to Medicinal and Aromatic Plants* at <http://www.hort.purdue.edu/newcrop/med-aro/default.html>, and *Medicinal plants and their properties* at <http://world.std.com/~krahe/html1.html>, etc.

Among the most important initiatives for the provision of on-line information on biodiversity (including MAP) are UNESCO's Man and the Biosphere Species Databases (Box 5) and the Biodiversity Conservation Information System (BCSI), a joint initiative (launched in 1997) among 9 international organizations to provide users with better "data for better decisions" (Box 6).

However, as the globalization process proceeds, the number of partners and thus of information generated by them is also increasing dramatically. Additional challenges in information management for MAP include: (1) scarcity of data on sources of information, (2) access to data on an *ad hoc* basis, and (3) incompatibility of a great wealth of data that is available by hundred of organization, government agencies and individuals but in a dispersed fashion (29).

Among the contributions made by IPGRI in the area of documentation of MAP, mention should be made of the PlantGeneCD Abstracts published by CABI: during the period 1972 to 1997, PlantGeneCD published more than 8,500 entries on MAP alone (30).

The need to have some mechanisms for creating links among data sets and assisting MAP users in locating the needed information is becoming paramount and, to that regard, worthwhile recalling here is the initiative promoted

BOX 5. UNESCO's Man and the Biosphere Species Databases <http://ice.ucdavis.edu/mab/>.

The Information Center for the Environment, in association with the U.S. Man and the Biosphere (U.S. MAB) Program, is developing databases of vascular plant and vertebrate animal occurrences on the world's biosphere reserves and other protected areas. Currently, the MABFlora (for vascular plants) and MABFauna (for vertebrate animals) databases contain records from over 660 protected areas in 97 countries. The MABFlora and MABFauna databases are continually updated as additional data are received.

BOX 6. Partners of the Biodiversity Conservation Information System (BCSI)
<http://biodiversity.org/members.html>.

1. *Bird Life International*	Members from 80 countries, dealing with bird species conservation
2. *Botanic Garden Conservation International (BGCI)*	Network of 450 botanic gardens (http://www.rbgkew.org.uk/BGCI/)
3. *Conservation International*	Agency established in 1987 to address conservation of ecosystems and species (http://www.conservation.org)
4. *International Species Information System*	Network of 450 zoos and aquariums (http://www.worldzoo.org)
5. *IUCN–The World Conservation Union*	74 member countries, 1905 government agencies, 699 NGOs, 34 affiliated from 136 countries (http:/iucn.org/)
6. *The Nature Conservancy*	International Conservation organization working with a network of more than 85 countries (http://www.tnc.org)
7. *TRAFFIC*	Main source of information on trade in wildlife and its products. Joint IUCN and WWW programme (http://traffic.org)
8. *Wetland International*	48 countries membership–Network of specialists, maintains various databases including that of the RAMSAR Convention (three web sites including http://wetland.agro.nl)
9. *World Conservation Monitoring Center (UNEP-WCMC)*	On 3 July 2000 WCMC became the UNEP Global Biodiversity Information and Assessment Center. As announced in a recent press release, as an integral part of UNEP, WCMC will assess the health of species and ecosystems, and threats to their survival. The Centre will also help nations to create their own biodiversity information systems, enabling them to develop science-based policy and regulations for the environment (http://www.unep-wcmc.org/July3)

by IDRC, which in addition to its support to various development projects in volving MAP has recently launched a global information network as an implementation follow-up to the International Meeting on Medicinal Plants held in Bangalore, India in 1998. The terms of reference of this initiative include the following objectives (23):

- Improve communication between regions (institutions, NGOs and regional networks)–more opportunities to share experience and expertise,
- More collaboration on project design and fund raising–less duplication of efforts,

- Improve access to information (particularly that found in costly databases, such as NAPRALERT and the WCMC threatened list),
- A shared database of information on use, conservation, distribution, etc.,
- Shared information on model access and benefits regimes,
- Wide dissemination of models, guidelines and strategies for conservation of medicinal plants,
- Better linkages to important opportunities to influence policy (e.g., the CBD, WTO, etc.),
- Better linkages to donors.

With regard to the debate on how to make sure that the new technology is not going to exclude local communities from the rest of the other actors involved in MAP conservation and use, some excerpts from an e-mail conference held in June 2000 by CEDARE (the Centre for the Environment and Development for the Arab Region and Europe based in Cairo, Egypt) addressing the role of information and communication technology would be relevant. On the issue of *how to reach the un-reached*, raised by a participant to this e-conference, the following two suggestions were endorsed by many scientists as important and feasible solutions:

- Use of complementary communication means, such as radio and television, combined with indigenous means of knowledge dissemination (e.g., through community or religious leaders/chief, village educators, extension workers),
- Access to and training in information communication technology for community leaders.

Further insights over the debate on challenges and opportunities for rural communities brought about by modern information and communication technology can be found at <http://www.rimba.com/spc/penanhomepage.html>.

DOCUMENTING INDIGENOUS KNOWLEDGE FOR MAP: AN EXAMPLE FROM ASIA

Traditions attached to crops and their products are increasingly having a stronger appeal to consumers (31). Today the economic valorization of genetic diversity must pass through the recognition of the cultural identity attached to it, which is highly responsible of the diversification process and ultimately to the survival of such diversity (32). Reference is made by Art. 8 of the Convention on Biological Diversity which requests countries to "preserve and maintain knowledge, innovations and practices of indigenous and local communities embodying traditional lifestyles relevant for conservation and sustainable use of biological diversity" (18).

It is estimated that today more than 7000 compounds are used by western medicine that largely originate from traditional uses in the past (33).

Germplasm without accompanying knowledge is not sufficient to promote better conservation and sustainable use of PGR. This is particularly true for medicinal plants whose qualities for treating health disorders have made them popular ingredients in prescriptions made by local healers and other traditional practitioners. Although it may be argued that the claimed virtues of medicinal plants used in traditional pharmacopea have not been confirmed by scientific investigations through standardized and reproducible methods, the fact that these species have been used by generations of users is indeed a sign of their importance for the livelihood of millions of people around the world.

The current debate over access to plant genetic resources has directed a lot of attention to the issue of farmers' rights and in general to the recognition of the role played by rural communities for the stewardship of these resources and the maintenance of the traditional knowledge associated with them.

Most efforts spent today in the area of farmer rights are linked to the use of germplasm and comparatively little attention has been paid over the issue of access to the information that is maintained along with associated material.

In the 18 May (2000) issue of *Nature* magazine, it was reported that the Indian Government launched a project for the establishment of a digital database of traditional knowledge. Such an effort is being made to address the issue of benefit sharing of IK for PGR and, particularly, for MAP species whose commercial exploitation by pharmaceutical companies could generate large economic benefits. The Indian Government intends to include this database in the patent classification system of the World Intellectual Property Organization (WIPO). The database will be available to patent offices worldwide–especially in the United States and Europe–so that data on any Indian plant can be obtained before patents are issued. Some 90 indigenous plants with medicinal or industrial uses have already been entered into the database, which is expected to be completed within this year, whereas all major plants are expected to be covered in the next two years.

On the same line as this Indian initiative, as a way to contribute towards a fair and equitable sharing of benefits resulting from the use of local PGR, IPGRI has initiated, in collaboration with the National Programme of China in Yunnan, a project on documenting the Indigenous Knowledge (IK) on PGR using the "IK Journal Concept." Such an approach is based on the understanding that the concerns of the indigenous communities over the issue of sharing benefits of PGR use need to be urgently addressed. An important element in addressing these concerns is the documentation of IK, which is the basis both for sharing knowledge and for recognition and establishing any property rights on PGR. IK documentation can bridge the gap between formal systems of knowledge and intellectual property, and indigenous knowledge and traditional rights. Researchers recognize useful knowledge in IK and carry out various studies of the indigenous communities and the knowledge they hold, seen as important and complementary to scientific knowledge (34).

IK has been always shared openly within communities. On the other end, researchers and outside communities have tried to fit IK into the realm of protection and property rights. Thus, they have transformed a simple system of knowledge sharing into a complex system of commercial gains and intellectual property rights (IPR). The current process of recognition of IPR for indigenous communities tends to assume that their benefits should be quantified in ways not compatible with the customs and culture of the local communities. In many cases, monetary returns are seen as the only form of compensation, which may not be in accordance with the needs of the indigenous communities.

The process of IK documentation. Currently, the documentation of IK involves the interaction of a scientist and a farmer, resulting in the scientist producing a paper of the collected IK reflecting his scientific orientation and/or interpretations. The result of this current approach is that not all of the farmer's knowledge is documented, as scientists tend to retain only those items perceived as scientifically sound and discard all those that are not. The paper produced by the scientist is therefore the result of *an interpreted IK*, where no citation is available for the farmer's contribution as these ideas have been collected orally and only acknowledged in the text. Subsequent reuse of information from this paper will have only the scientist's name in the citation or references. Contribution of the farmer is not acknowledged after the information is re-used.

The idea of the IK journal. The problem of recognition and contribution by the farmer can be overcome through a farmer's journal (IK journal). In this new approach, a paper containing the full text of the farmer's story, and authored by the farmer, is prepared with acknowledgment of the scientist's assistance. The scientist is then free to develop his/her own scientific paper in which there will be a specific citation of the farmer's paper used as source of information. The ability to cite from an IK journal would mean that subsequent re-use of the IK will require the citation to be quoted as in a scientific journal. With the ability to cite, recognition can be given to the information provider. The practical aspects of the IK journal need to be considered to see if such a system is practical.

Some practical aspects of the farmer's paper and IPR. When the scientist interviews the farmer, there will be specific objectives and subjects discussed. The knowledge of the farmer related to the subject could then be taped if this consists of an oral transmission. The farmer's paper will therefore have a title and a short abstract with a clear reference to the tape and its current location (e.g., at the scientist's institute/library). The abstract can be published in the science journal, hence allowing citations to be made. Since the full text is in the farmer's language, a copy is kept by the community and becomes easy for the community to access the knowledge stored on the tapes. This aspect of re-use of the knowledge of the community will assist the community to maintain as well as develop its knowledge. Scientists will initially assist in the process, and when the community sees the benefits of the system, valuation of its knowledge and the ease with which that knowledge can be captured and recognized, the community will keep recording and documenting its knowledge. The use of a tape recorder and player will make IK accessible to the community.

PRIORITIES AND PARTICIPATION

Over the last 5 to 10 years, priority setting has emerged as one of the most critical steps in guiding PGR work. With the shrinking of funds devoted to conservation, the raising of awareness on the great wealth of underused valuable genetic resources and the advent of new technologies for more effective surveying (from GIS to molecular markers), a proper setting of priorities is fundamental to make the best use of limited resources (35). Priorities are needed in the area of species selection but also to choose the right activities so as to ensure that the achievement of the ultimate goal is pursued in a consistent and cost effective manner.

Lists of MAP priority species have been developed by many national programmes and international organizations (258 species prioritized by the CITES (36)) on the basis of genetic erosion threat and economic opportunities. Species selection is biased in most cases towards the views set by the stakeholder group that led the prioritization process. There is thus a need to reiterate that the sustainability of conservation activities for MAP (like for many other groups of species) rests on a bottom-up participatory approach in which farmers and other community members share their ideas and their needs with other stakeholder groups so as to ensure that the species and the activities that will be prioritized will ultimately meet the interest of the primary stakeholders (37).

The involvement of primary stakeholders (farmers, rural and forest communities) is thus the key for the sustainable maintenance of MAP species, but at the same time it is also the vehicle for facilitating bio-prospecting of existing resources for their better use and safeguard (15,38,39).

Participation should also be seen in the context of the marketing and promotion of the MAP species. Such an approach would have as its most strategic part the so-called "filiere." Filiere is a French word used to define the link of all stakeholders and activities starting from the collection, use enhancement, and policy definition to marketing and commercialization. Such a chain of actors, which is needed at local, regional, national and international levels, will allow the coverage of research aspects but also marketing and policy issues usually dealt with in isolated fashion. The filiere concept can be considered an evolution of the networking concept for plant genetic resources to meet the specific need of MAP species. The filiere would thus bring about greater participation of local actors that would ensure the proper addressing of local needs, and the wider representation of stakeholders to ensure the participation of the food processing and marketing sectors as well as policy makers who have traditionally been always left aside from PGR activities.

Although the filiere will be made particularly by local, regional and national players, the role of international organizations is seen, however, as important to ensure that lessons learned in one region can also benefit other regions. The strengthening of the links among international stakeholders involved in the promotion of MAP is indeed strategic, to allow best use of existing capacities and promote synergism across regions.

As previously mentioned, farmers and forest dwellers are the source of information for revealing the potentials of these species, their distribution and local use. Participatory research should therefore be actively pursued among stakeholders, particularly in the following areas (7,41):

- analyses of constraints and development of strategic work plans for enhancing seed/germplasm selection and supply, production, processing, commercialization, marketing (*greater cooperation between private sector and extension workers*),
- characterization and evaluation of work using descriptor lists and farmers' criteria (*closer cooperation between informal associations/NGO and international and national research organizations*),
- development/strengthening the seed supply systems (*closer participation of farmers in government-led efforts*),
- participatory plant breeding and selection activities (*bridging the gap between farmers' needs and breeders' objectives*),
- Strengthening in-country processing of MAP products (*greater cooperation between private sector and local communities*).

REFERENCES

1. Heywood V. 1991. Conservation of germplasm of wild plant species. In: Conservation of Biodiversity for Sustainable Development, Norwegian University Press and Cambridge University Press, UK.

2. Eyzaguirre P., S. Padulosi and T. Hodgkin.1999. IPGRI's strategy for neglected and underutilized species and the human dimension of agrobiodiversity. In: Padulosi S.

(ed.). Priority setting for underutilized and neglected plant species of the Mediterranean region. Report of the IPGRI Conference, 9-11 February 1998, ICARDA, Aleppo, Syria. International Plant Genetic Resources Institute, Rome, Italy.

3. de Padua, L.S., N. Bunyapraphatsara and R.H.M.J. Lemmens (eds.). 1999. Plant Resources of South East Asia. 21(1). PROSEA, Bogor, Indonesia.

4. Baumann H. 1993. The Greek plant world in myth, art and literature. Timber Press, Portland, Oregon.

5. Farnsworth N.R. and D.D. Soejarto. 1991. Global importance of medicinal plants. In: O. Akerele, V. Heywood and H. Synge (eds.). The conservation of medicinal plants. Proceedings of an International Consultation, 21-27 March 1988, Chiang Mai, Thailand. Cambridge, UK. Cambridge University Press, pp. 25-51.

6a. FAO. 1997. Human Nutrition in the developing World. FAO Food and Nutrition Series No. 29. Food and Agricultural Organization of the United Nations, Rome, Italy.

6b. FAO. 1996. Report on the State of the World's Plant Genetic Resources for Food and Agriculture, prepared for the International Technical Conference on Plant Genetic Resources, Leipzig, Germany, 17-23 June 1996. Food and Agriculture Organization of the United Nations, Rome, Italy.

7. Srivastava J., J. Lambert and N. Vietmeyer. 1995. Medicinal plants: a growing role in development. Washington, DC, USA; Agricultural and Natural Resources Department Division, The World Bank.

8. Bannerman, R.H. 1982. Traditional medicine in modern health care. World Health Forum 3(1):8-13.

9. Nicket W. and E. Sennhauer. 1999. Medicinal plants local heritage and global importance. <*http://www.worldbank.org/html/extdr/offrep/sas/ruralbrf/edplant.htm*>

10. Lange D. 1998. Europe's medicinal and aromatic plants: their use, trade and conservation. Cambridge, UK, TRAFFIC International.

11. Kuipers S.E. 1997. Trade in medicinal plants. In: Medicinal plants for forest conservation and health care. Non-Wood Forest Products Vol. 11. Food and Agricultural Organization of the United Nations (FAO), Rome, Italy.

12. Cunningham A.B and F.T. Mbenkum. 1993. Sustainability of harvesting *Prunus africana* bark in Cameron: a medicinal plant in international trade. People and Plants Working Paper No. 2. Paris, France, UNESCO.

13. Hmamouchi M. 1999. Les Plantes Medicinales et Aromatiques Marocaines. Imprimerie de Fedala, Mohammedia, Morocco.

14. Bhat, K.K.S. 1995. Literature published during the past two decades on medicinal, aromatic and other related groups of plants. *Acta Horticulturae* 390:11-17.

15. Cunningham A.B. 1990. People and medicines: the exploitation and conservation of traditional Zulu medicinal plants. Proceedings of the Twelfth Plenary Meeting of AETFAT, Hamburg, September 4-10, 1988. Mitteilungen aus dem Institut fur-Allgemeine Botanik, Hamburg, Germany.

16. Schipmann U. 1997. Medicinal Plant Conservation Bibliography. Vol. 1. Bonn. Medicinal Plant Specialist Group.

17. Hong T.D., S. Linington and R.H. Ellis. 1996. *Seed Storage Behaviour: A Compendium* (Handbook for Genebank No. 4). 656 pages. International Plant Genetic Resources Institute, Rome, Italy.

18. UNCED. 1992. Convention on Biological Diversity. United Nations Environment Programme.

19. Padulosi S., G. Ayad, and J. Wessels. 1999. The use of agro-biodiversity for desert development. Proceedings of the VI International Conference on the Development of Dry Lands, 22-27 August 1999, Cairo, Egypt.

20. Maass H.I. 1986. Labiatae. In: J. Schultze-Motel, Rudolf Mansfelds Verzeichnis landwirtschaftlicher und garrtnerischer Kultupflanzen (ohne Zierpflanzen)-Akademie-Verlag, Berlin, Germany.

21. Fiori A. 1923. Nuova Flora analitica d' Italia, Bologna.

22. Hammer K., H. Knupffer, G. Laghetti and P. Perrino. 1992. Seeds from the past. A catalogue of crop germplasm in South Italy and Sicily. CNR, Istituto del Germoplasma, Bari, Italy.

23. Leaman D.J., H. Fassil and I. Thormann. 1999. Conserving medicinal and aromatic plant species: identifying the contribution of the International Plant Genetic Resources Institute. Study commissioned by the International Development Research Centre (IDRC). IPGRI, Rome, Italy.

24. Kasparek M., A. Groger and U. Schippmann. 1996. Directory for Medicinal Plants Conservation: Networks, Organizations, Projects, Information sources. IUCN/SSC Medicinal Plant Specialist Group and German Federal Agency for Nature Conservation, Bonn, Germany (http://www.dainet.de/genres/mpc-dir/).

25. Dennis F. and P.S. Wyse Jackson. 1998. Directory of Botanic Gardens Medicinal Plant Collections. Botanic Gardens Conservation International (BGCI).

26. Van Hintum Th.J.L., N.R. Sackwille Hamilton, J.M.M. Engels and R. van Treuren. 2000. Accession management strategies: splitting and lumping accessions. Proceedings of the IPGRI's Conference on Science and Technology for Managing Plant Genetic Diversity in the 21st century, 12-16 June 2000, Kuala Lumpur, Malaysia (in press).

27. Guarino L. 2000. Geographic Information Systems and the conservation and use of plant genetic resources. Proceedings of the IPGRI's Conference on Science and Technology for Managing Plant Genetic Diversity in the 21st century, 12-16 June 2000, Kuala Lumpur, Malaysia (in press).

28. IPGRI. 1999. Diversity for Development. The new strategy of the International Plant Genetic Resources Institute. IPGRI, Rome, Italy.

29. BCIS. 1998. Biodiversity Conservation Information System. Botanic Gardens Conservation International (http://www.biodiversity.org).

30. Guarino L., S. Padulosi, A.S. Ouedraogo, and R. Arora. 1997. Relevance of IPGRI's work to the conservation of aromatic and medicinal plants, WOCMAP II Congress, Argentina, 1997.

31. Anonymous. 1993. Atlante dei prodotti tipici: le conserve (Atlas of typical food products: pickles). Ministero Agricultura e Foreste, Rome, Italy. Franco Angeli Publishers.

32. Chambers R. et al. 1986. Farmer First: Farmer Innovation and Agricultural Research. IT Publication, London.

33. Stiles D. 1994. Tribals and trade: a strategy for cultural and ecological survival. Ambio 23(2):106-111.

34. Eythorsson E. 1993. Sami Fjord Fisherman and State: Traditional Knowledge and Resource Management in Northern Norway. In: J.T. Inglis (ed.). Traditional Ecological Knowledge, Concepts and Cases.

35. Padulosi S. 1999. Criteria for priority setting in initiatives dealing with underutilized crops in Europe. In: Gass, T., Frese, F. Begemann and E. Lipman (compilers). "Implementation of the Global Plan of Action in Europe–Conservation and Sustainable Utilization of Plant Genetic Resources for Food and Agriculture." Proceedings of the European Symposium, 30 June-3 July 1998, Braunschweig, Germany. International Plant Genetic Resources Institute, Rome.

36. Schipmann U. 1999. Medicinal plants significant trade study (CITES Project S-109). Unpublished draft.

37. Williams J.T. and Z. Ahamad. 1997. Priorities for medicinal plants research and development in Pakistan. Draft report for the IDRC Medicinal Plants Network. New Delhi. IDRC, Canada.

38. Cunningham A.B. and B.J. Huntley. 1994. Combining skills: participatory approaches in biodiversity conservation. Botanical diversity in southern Africa. Proceedings of a conference on the Conservation and Utilization of Southern African Botanical Diversity, Cape Town, South Africa, September 1993. National Botanical Institute, Pretoria, South Africa.

39. Dutfield G. 1997. Between a rock and a hard place: indigenous peoples, nations states and the multinationals. In: Medicinal plants for forest conservation and health care. Non-Wood Forest Products Vol. 11. Food and Agricultural Organization of the United Nations (FAO), Rome, Italy.

40. Bhat K.K.S. 1997. Medicinal plant information database. In: Medicinal plants for forest conservation and health care. Non-Wood Forest Products Vol. 11. Food and Agricultural Organization of the United Nations (FAO), Rome, Italy.

41. Padulosi S., T. Hodgkin, J.T. Williams, and N. Haq. 2000. Underutilized crops: trends, challenges and opportunities in the 21st Century. Proceedings of the IPGRI's Conference on Science and Technology for Managing Plant Genetic Diversity in the 21st century, 12-16 June 2000, Kuala Lumpur, Malaysia (in press).

42. Palevitch D. 1991. Agronomy applied to medicinal plant conservation. In: O. Akerele, V. Heywood and H. Synge (eds.). The conservation of medicinal plants. Cambridge University Press, UK.

Genetic and Chemical Relations Among Selected Clones of *Salvia officinalis*

Elvira Bazina
Antonios Makris
Carla Vender
Melpomeni Skoula

SUMMARY. Previous study on *Salvia fruticosa* has shown that different essential oil profiles are maintained during cultivation and that differences were generally in agreement with the degree of genetic diversity of the clones as indicated by RAPD analysis. In this study, the genetic patterns, assessed by RAPD markers and volatile oil composition, are compared in a range of *Salvia officinalis* plants. Genetic distances generated by the RAPD analysis were not in agreement with the distances generated by the volatile oil composition. This is possibly because often quite different monoterpene products can be generated by small changes in terpene synthase sequences within a species. *[Article copies available for a fee from The Haworth Document Delivery Service: 1-800-HAWORTH. E-mail address: <getinfo@haworthpressinc.com> Website: <http://www. HaworthPress.com> © 2002 by The Haworth Press, Inc. All rights reserved.]*

Elvira Bazina, Antonios Makris, and Melpomeni Skoula are affiliated with the Department of Natural Products, Mediterranean Agronomic Institute of Chania, Greece.

Carla Vender is affiliated with the ISAFA-Forest and Range Management Research Institute, Piazza Nicolini n° 6, 38050 Villazzano-Trento, Italy.

Address correspondence to: Melpomeni Skoula, Department of Natural Products, Mediterranean Agronomic Institute of Chania, Alsyllion Agrokepion, P.O. Box 85, 73100 Chania, Greece.

The present work was financially supported through the EU funded research project FAIR3-CT96-1914.

[Haworth co-indexing entry note]: "Genetic and Chemical Relations Among Selected Clones of *Salvia officinalis*." Bazina, Elvira et al. Co-published simultaneously in *Journal of Herbs, Spices & Medicinal Plants* (The Haworth Herbal Press, an imprint of The Haworth Press, Inc.) Vol. 9, No. 4, 2002, pp. 269-273; and: *Breeding Research on Aromatic and Medicinal Plants* (ed: Christopher B. Johnson, and Chlodwig Franz) The Haworth Herbal Press, an imprint of The Haworth Press, Inc., 2002, pp. 269-273. Single or multiple copies of this article are available for a fee from The Haworth Document Delivery Service [1-800-HAWORTH 9:00 a.m. - 5:00 p.m. (EST). E-mail address: getinfo@haworthpressinc.com].

KEYWORDS. *Salvia officinalis*, Lamiaceae, RAPD, essential oils, volatile oils, hierarchical cluster analysis, dendrogram

INTRODUCTION

It has been shown previously that clones of *Salvia fruticosa* from populations collected in different parts of Crete exhibit different essential oil yields and profiles and maintain these profiles during cultivation (1,2). The differences were generally in agreement with the degree of genetic diversity of the clones as indicated by RAPD analysis (3). In this study, the genetic patterns, assessed by RAPD markers and volatile oil composition, are compared in a range of *Salvia officinalis* plants of different origin, but all grown in the ISAFA farm of Trento, Italy.

MATERIALS AND METHODS

Plant material was clonally-propagated *Salvia officinalis* plants of different origin, but all grown in the ISAFA farm of Trento. Nine clones were of Italian origin, the tenth came from Albania. Among the Italian clones, six were selected in Northern Italy (TN and Valda clones), the other two in Florence (FI). The Albanian landrace (Skodra) came from Northern Albania. A sample of *Salvia fruticosa* (VR09) from Crete was included as an outgroup. Isolation of genomic DNA extraction and PCR amplification was carried out as described in (1) using twenty random decamer oligonucleotide primers, purchased from Operon Technologies (Table 1). Amplified bands were scored as binary variables; [1] for presence, [0] for absence of a band. Only distinct well-resolved stable bands were considered for resolving (2). Hierarchical cluster analysis using the dice method was used for the amplified bands from PCR as described in Skoula et al. (1) and Weising et al. (2). Volatile oils were analyzed using headspace GC, and GC-MS as in Skoula et al. (3)

RESULTS AND DISCUSSION

The banding patterns obtained from the RAPD analysis were highly polymorphic, and hierarchical cluster analysis showed, as expected, that *S. fruticosa* was genetically most distinct from all the *S. officinalis* clones (Figure 1). Amongst the latter, the geographically distant Albanian landrace (from Skodra in N. Albania) was also genetically quite distant from all the Italian clones, which clustered together, although the Northern (Valda) clones were not as closely related as might have been expected.

TABLE 1. Sequences of the primers used for the RAPD analysis of *Salvia officinalis*.

Primer	Sequence	No. of bands generated
OPB-1	5'-GTTTCGCTCC-3'	60
OPB-2	5'-TGATCCCTGG-3'	36
OPB-3	5'-CATCCCCCTG-3'	32
OPB-4	5'-GGACTGGAGT-3'	85
OPB-5	5'-TGCGCCCTTC-3'	74
OPB-6	5'-TGCTCTGCCC-3'	42
OPB-7	5'-GGTGACGCAG-3'	29
OPB-8	5'-GTCCACACGG-3'	50
OPB-9	5'-TGGGGGACTC-3'	29
OPB-10	5'-CTGCTGGGAC-3'	79
OPB-11	5'-GTAGACCCGT-3'	58
OPB-12	5'-CCTTGACGCA-3'	49
OPB-13	5'-TTCCCCCGCT-3'	52
OPB-14	5'-TCCGCTCTGG-3'	29
OPB-15	5'-GGAGGGTGTT-3'	48
OPB-16	5'-TTTGCCCGGA-3'	40
OPB-17	5'-AGGGAACGAG-3'	89
OPB-18	5'-CCACAGCAGT-3'	65
OPB-19	5'-ACCCCCGAAG-3'	41
OPB-20	5'-GGACCCTTAC-3'	45

The volatile composition of each clone is shown in Table 2. Hierarchical cluster analysis of the quantitative variation of oil composition produced a somewhat different pattern from that obtained from the RAPD analysis (Figure 2). *Salvia fruticosa* was again distinct on account of its high 1,8-cineole and low thujone content. *S. officinalis* plants formed 3 clusters: the first was characterized by medium 1,8-cineole level (9.5-18.8%), high α-thujone content (38.3-47.2%) and by low camphor content (0.9-3.4%); the second cluster was characterized by low levels of 1,8-cineole (3.1-3.9%), medium levels of α-thujone (28.7-39.4%) and medium levels of camphor (5.6-5.0%); the third cluster was represented only by one plant, FI-12, that had a very high content of camphor (17.5%). Within *S. officinalis*, the Albanian clone was generally

FIGURE 1. Hierarchical cluster analysis dendrogram of the selected *S. officinalis* plants and one *S. fruticosa* (outgroup) based on the variation of RAPD patterns obtained with twenty oligonucleotide primers.

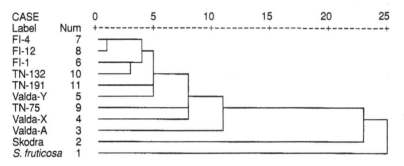

TABLE 2. Qualitative and quantitative composition of the volatile oils of selected clones of *Salvia officinalis*.

Compounds	FI-4	FI-12	TN-75	TN-132	TN-191	Valda-A	Valda-X	Valda-Y	Skodra	*S. fruticosa*
α-thujene	0.8	0.4	1	0.7	0.9	0.6	0.7	0.9	0.7	1.2
α-pinene	2.5	10.8	4.3	4.6	5.4	5.2	9	5.9	10.1	6.7
camphene	0.4	18.2	4.1	0.6	2.6	1.5	10	0.8	7.6	0.4
β-pinene	13.1	6	18.6	-	8	9.3	22	17.2	13.7	17
β-myrcene	1.6	1.5	1.4	1.7	2.2	2.3	0.9	1.5	1.6	19
α-terpinene	0.3	-	0.4	0.5	0.7	0.6	0.3	0.7	0.6	0.6
p-cymene	0.3	0.4	0.4	0.3	0.4	0.7	0.2	0.3	0.3	0.2
limonene	0.8	2.3	1.3	1.1	1.4	1.8	1.2	1.2	1.4	0.7
1.8-cineole	16	3.8	3.1	18.8	12.6	16.3	3.9	9.5	17	40.4
γ-terpinene	0.5	0.3	0.8	0.9	1.3	1	0.6	1.3	1.1	1.2
cis-sabinene-hydrate	0.3	0.2	0.3	0.3	0.3	0.2	0.2	0.2	0.2	0.4
α-terpinolene	-	-	0.2	-	0.3	0.2	0.1	0.2	0.3	0.2
trans-sabinene-hydrate	0.2	-	0.2	-	0.2	-	0.1	0.2	0.1	0.3
α-thujone	42.3	28.8	39.4	38.3	49.2	41.9	28.7	47.2	29.1	0.1
β-thujone	5	2.4	7.7	6.3	6.5	6.2	4.5	4.3	3.2	0.2
camphor	0.9	17.4	6	1.3	3.4	3.2	5.6	1.5	5.6	0.2
borneol	-	1.8	0.2	-	0.3	-	4.2	-	0.7	0.4
terpinene-4-ol	0.4	0.3	0.2	-	0.12	-	0.1	0.1	0.1	0.1
β-caryophyllene	5.1	1.4	3.4	4.8	0.7	3.9	0.8	2.3	2.2	0.6
α-humulene	6.4	1.5	4.8	5.3	1.7	4.4	4.7	3.4	3.2	1.2

FIGURE 2. Hierarchical cluster analysis dendrogram of the selected *S. officinalis* plants and one *S. fruticosa* (outgroup) based on the variation of the volatile compounds.

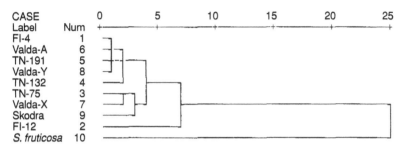

intermediate in percentage content of most of the main terpenes while the extremes in volatile composition were represented by geographically much closer plants (Table 2).

The differences between the two groups of results are more marked than in our previous studies with *S. fruticosa* (1), but are consistent with biochemical and genetical studies on terpene biosynthesis (4). These studies have shown that, sometimes, quite different monoterpene products and hence different chemical profiles can be generated by small changes in terpene synthase sequences within a species, whilst, on the contrary, synthases generating the same product in different taxa not infrequently show much larger sequence differences. Hence, a simple quantitative analysis of essential oil composition is not necessarily appropriate for estimating genetic proximity even in closely related taxa. More detailed knowledge of the enzymes involved in the synthesis of the *Salvia* terpenoids is needed.

REFERENCES

1. Skoula M, El-Hilali I, Makris A (1999). Evaluation of the genetic diversity of *Salvia fruticosa* Mill. clones using RAPD markers and comparison with the essential oil profiles. *Biochem. Syst. Ecol.* 27, 559-568.

2. Weising K, Nybom H, Wolff K, Meyer W (1995). DNA fingerprinting in plants and fungi. CRC Press, London.

3. Skoula M, Abbes JE, Johnson CB (2000). Genetic variation of volatiles and rosmarinic acid in populations of *Salvia fruticosa* Mill. growing in Crete. *Biochem Syst. Ecol.* 28, 551-561.

4. Bohlmann J, Meyer-Gauen G, Croteau R (1998). Plant terpenoid synthases: molecular biology and phylogenetic analysis. *Proc. Natl. Acad. Sci. USA* 95, 4126-4133.

The Essential Oil Composition
of Algerian Zaâtar:
Origanum spp. and *Thymus* spp.

Zahia Houmani
Samir Azzoudj
George Naxakis
Melpomeni Skoula

SUMMARY. *Origanum* spp. and *Thymus* spp. growing spontaneously in Algeria are collected from wild populations and are sold in the local markets under the same or similar vernacular name, *zaâtar* (*Origanum*) or *zhitra* (*Thymus*). *Thymus willdenowii* Boiss. and *Thymus algeriensis* Boiss. & Reuter are mostly used as condiments while *Origanum floribundun* Munby and *Origanum vulgare* L. ssp. *gladulosum* (Desf.) Ietswaart are used against diarrhoea and other digestive and respiratory system disorders, as well as additive to forrage as an appetite stimulant. All four species were quite rich in essential oils; the analyses of the oils showed that

Zahia Houmani and Samir Azzoudj are affiliated with the Laboratory of Medicinal and Aromatic Plants, Agronomic Institute, University of Blida, Algeria.

George Naxakis and Melpomeni Skoula are affiliated with the Department of Natural Products, Mediterranean Agronomic Institute of Chania, Greece.

Address correspondence to: Melpomeni Skoula, Department of Natural Products, Mediterranean Agronomic Institute of Chania, P.O. Box 85, 73100 Chania, Greece.

The authors thank CIHEAM and EU-DG 1 for partial financial support through the cooperative research network on Identification, Conservation, and Use of Wild Plants of the Mediterranean Region–The MEDUSA Network.

[Haworth co-indexing entry note]: "The Essential Oil Composition of Algerian Zaâtar: *Origanum* spp. and *Thymus* spp.." Houmani, Zahia et al. Co-published simultaneously in *Journal of Herbs, Spices & Medicinal Plants* (The Haworth Herbal Press, an imprint of The Haworth Press, Inc.) Vol. 9, No. 4, 2002, pp. 275-280; and: *Breeding Research on Aromatic and Medicinal Plants* (ed: Christopher B. Johnson, and Chlodwig Franz) The Haworth Herbal Press, an imprint of The Haworth Press, Inc., 2002, pp. 275-280. Single or multiple copies of this article are available for a fee from The Haworth Document Delivery Service [1-800-HAWORTH 9:00 a.m. - 5:00 p.m. (EST). E-mail address: getinfo@haworthpressinc.com].

275

all were rich in the compounds of the carvacrol pathway (p-cymene, γ-terpinene, carvacrol, thymol, and their methyl-ethers). Only minor qualitative, but considerable quantitative, variation was found within and between the species that comprise zaâtar: the major compounds of *O. floribundum* were p-cymene (31%), thymol (9.9%) and carvacrol (35.0%); the major compounds of *O. vulgare* ssp. *gladulosum* were γ-terpinene (13.6%), thymol-methylether (16.3%), carvacrol-methylether (11.4%) and thymol (26.1%); the major compounds of *T. willdenowii* were p-cymene (15.2%), thymol (15.1%) and carvacrol (51.3%); finally, from the two samples of *T. algeriensis* analyzed, one was rich in linalool (78.8%) and the other was rich in thymol (62.7%). *[Article copies available for a fee from The Haworth Document Delivery Service: 1-800-HAWORTH. E-mail address: <getinfo@haworthpressinc.com> Website: <http://www. HaworthPress.com> © 2002 by The Haworth Press, Inc. All rights reserved.]*

KEYWORDS. *Origanum floribundum, Origanum vulgare* ssp. *glandulosum, Thymus algeriensis, Thymus willdenowii*, Lamiaceae, essential oils

INTRODUCTION

Origanum spp. and *Thymus* spp. grow spontaneously in the humid and semi-arid zones of Algeria. They are collected from wild populations and are sold in the local markets under the same or similar vernacular name, *zaâtar* (*Origanum*) or *zhitra* (*Thymus*). This often creates confusion regarding the commercial product. *Thymus* is mostly used as a condiment while *Origanum* is used against diarrhoea and other digestive and respiratory system disorders, as well as additive to forrage as an apetite stimulant.

The present study concerns the following species that are restricted to the west side of the Mediterranean Region (1,2):

- *Origanum floribundun* Munby, endemic to Algeria,
- *Origanum vulgare* L. subsp. *gladulosum* (Desf.) Ietswaart (syn. *Origanum glandulosum* Desf.), endemic to Algeria and Tunisia,
- *Thymus willdenowii* Boiss. (syns. *Thymus hirtus* Willd., *Thymus cilliatus* ssp. *albiflorus* and *Thymus diffusus* Bentham.), endemic to the Iberian peninsula, Algeria and Morocco, and
- *Thymus algeriensis* Boiss. & Reuter (syns. *Thymus hirtus* ssp. *algeriensis* (Boiss. & Reuter) Murb. and *Thymus zattarelus* Pomel), endemic to Libya, Tunisia, Algeria and Morocco.

MATERIALS AND METHODS

Plant material of *Origanum floribundum* (4 samples), *Thymus willdenowii* (12 samples) and *T. algeriensis* (2 samples) were purchased from 10 herbalists at the market of Blida (45 km south of Alger). *O. vulgare* ssp. *glandulosum* (2 samples) was collected from the region of Medea (120 km south of Alger). The commercial material is sold in bags of 50 kg; in most of the cases it is a mixture of more than one species. Plant material was carefully selected to be uniform. Herbarium specimens are kept at the University of Blida, and taxa were identified according to Quezel and Santa (2).

Essential oils were obtained from dry leaves by hydrodistillation, dried over anhydrous Na_2SO_4, and analyzed by CG and GC-MS. Qualitative composition was carried out by a HP-GC 5890 II coupled with an MS VG TRIO 2000, and using a DB5 column, 30 m $\times$ 0.25 mm $\times$ 0.25 µm. Injector temperature was 230°C; oven temperature was programmed at an initial temperature of 60°C for 3 min and then increasing at a rate of 3°C/min up to 230°C and remaining there for 20 min. Five µl of essential oil in pentane (10% v/v) was injected, carrier gas was He with 1.6 ml/min flow rate; split ratio was 1:30. Mass spectra were taken at 70 eV, with 1scan/sec from 40 until 230 m/z. Components were identified by comparing retention indices and spectral data from electronic libraries (4,5). Quantitative composition was carried out by a HP-GC, 5890 II, equipped with a FID; column was the same as above; injector and detector were set at 230°C and 250°C, respectively; carrier gas and oven programme were the same as above.

RESULTS

The essential oil yield from leaves of *O. floribundum* ranged from 2.5 to 3.5% v/w; the yield of leaves of *O. vulgare* ssp. *glandulosum* was between 1.5 and 3.0% v/w; *T. willdenowii* yield ranged from 2.2 to 4.0% v/w and *T. algeriensis* yielded from 1.4 to 4.2% v/w.

The analyses of the essential oils of the above taxa has shown that they all are rich in the compounds of the carvacrol pathway (p-cymene, γ-terpinene, carvacrol, thymol, and their methyl-ethers) (Tables 1 and 2). Even though the essential oils of all taxa had only minor qualitative differences, considerable quantitative variation was found within and between the species that comprise zaâtar:

- the major compounds of *O. floribundum* were p-cymene (13.7%-48.7%), γ-terpinene (8.7-18.4%), thymol (5.5-14.2%) and carvacrol (28.3-41.8%);

- the major compounds of *O. vulgare* ssp. *gladulosum* were p-cymene (6.7-10.3%), γ-terpinene (8.7-18.4%), thymol-methylether (tr-32.6%), carvacrol-methylether (tr-22.9%), thymol (21.0-31.2%) and carvacrol (2.4-7.4%);
- the major compounds of *T. willdenowii* were p-cymene (9.5-22.5%), γ-terpinene (tr-7.8%), thymol (1.1-57.5%) and carvacrol (3.1-67.7%);
- finally, from the two samples of *T. algeriensis* analyzed, one was rich in linalool (78.8%) and thymol (4.7%) and the other was rich in p-cymene (8.8%), γ-terpinene (6%), thymol (62.7%) and carvacrol (3.6%).

TABLE 1. Essential oil composition of *Origanum*.

Compounds	*O. floribundum* mean	*O. floribundum* st. dev	*O. vulgare* ssp. *glandulosum* mean	*O. vulgare* ssp. *glandulosum* st. dev
α-thujene	0.5	0.1	0.6	0.2
α-pinene	0.6	0.0	0.4	0.1
camphene	0.1	0.0	0.0	0.0
sabinene	0.1	0.0	0.2	0.0
β-pinene	0.8	0.1	0.4	0.0
β-myrcene	0.8	0.5	1.0	0.3
α-terpinene	1.0	0.6	1.1	0.3
p-cymene	31.3	11.7	8.5	1.2
limonene	0.4	0.0	0.5	0.1
γ-terpinene	8.9	5.6	13.6	3.2
cis-sabinene-hydrate	0.4	0.1	0.1	0.0
trans-sabinene-hydrate	0.4	0.1	0.1	0.0
linalool	1.1	0.0	0.7	0.0
sabina ketone	0.0	0.0	4.1	0.2
borneol	0.2	0.0	2.1	1.3
terpinen-4-ol	0.5	0.1	0.6	0.0
p-cymen-8-ol	0.6	0.4	0.0	0.0
thymol-methyl-ether	0.2	0.1	16.3	10.9
carvacrol-methyl-ether	1.4	0.6	11.4	7.6
trans-sabinene hydrate acetate	0.2	0.1	0.1	0.0
p-cymen-7-ol	0.3	0.1	0.1	0.0
thymol	9.9	2.9	26.1	3.4
carvacrol	35.0	4.5	4.9	1.7
β-caryophyllene	0.6	0.4	0.7	0.1
β-humulene	0.3	0.1	0.1	0.0
germacrene-D	0.0	0.0	1.1	0.0
bicyclogermacrene	0.0	0.0	0.3	0.0
β-bisabolene	0.3	0.1	0.5	0.1

TABLE 2. Essential oil composition of *Thymus*.

Compounds	T. willdenowii		T. algeriensis	
	mean	st. dev	mean	st. dev
α-thujene	0.3	0.2	0.5	0.3
α-pinene	1.4	0.4	0.4	0.1
camphene	0.2	0.0	0.3	0.0
sabinene	0.2	0.0	0.1	0.1
β-pinene	0.4	0.2	1.2	0.3
β-myrcene	0.6	0.3	0.7	0.3
α-terpinene	0.7	0.2	0.5	0.3
p-cymene	15.2	3.1	5.5	2.2
limonene	0.5	0.1	0.2	0.1
β-phellandrene	0.6	0.5	0.5	0.3
γ-terpinene	2.6	2.4	3.2	1.9
cis-sabinene-hydrate	0.3	0.3	0.9	0.5
trans-sabinene-hydrate	0.0	0.0	0.1	0.1
linalool	2.2	0.8	40.2	25.8
borneol	0.4	0.1	0.4	0.1
terpinen-4-ol	0.4	0.0	0.2	0.1
carvacrol-methyl-ether	0.5	0.3	0.3	0.1
geraniol	0.5	0.5	1.0	0.7
thymol	15.1	14.2	33.7	19.4
carvacrol	51.3	16.1	2.2	0.9
β-caryophyllene	1.3	1.3	2.7	0.1
trans-bergamotene	0.2	0.1	0.0	0.0
β-humulene	0.2	0.1	0.1	0.0
D-germacrene-4-ol	0.5	0.4	0.0	0.0
caryophyllene-oxide	0.4	0.1	0.3	0.0

DISCUSSION

In this study, the analysis of essential oil from *Origanum floribundum* is reported for the first time; previously analyzed *O. vulgare* ssp. *glandulosum* from Italy was found to be carvacrol-rich (5). Both *Origanum* species analyzed here belong to the infrageneric group C; *O. floribundum* belongs to the section *Elongataspica* Ietswaart and *O. vulgare* ssp. *glandulosum* belongs to the section *Origanum* (6); their essential oil profiles are in agreement with their taxonomic grouping (7). Algerian samples of *Thymus willdenowii* have similar profiles to the Moroccan samples of the same species (8); the linalool-rich type of *T. algreriensis* is here reported for the first time while Moroccan samples of this species were rich in thymol or carvacrol (8); however, linalool chemotypes are not uncommon in *Thymus* (9).

The above findings show that, apart from the morphological differences among *zaâtar* or *zhitra* plants, there is high variability in the chemical composition that possibly implies variable effectiveness of their therapeutical properties.

REFERENCES

1. Greuter W., Burdet H.M. and Long G. 1986. Med-Checklist. In Editions de Conservatoire de jardin Botanique de la Ville de Geneve, 3.

2. Quezel P. and Santa C. 1963. Nouvelle flore de l'Algérie et des régions désertiques méridionales, 2, CNRS Edit., Paris.

3. Adams R. 1995. Identification of essential oil components by gas chromatography/mass spectrometry. Allured Publishing Corporation. Carol Stream, Illinois.

4. Wiley Library. 1989. John Wiley and Sons, Inc.

5. Melegari M., Severi F., Bertoldi M., Benvenuti S., Circetta G., Morone Fortunato I., Bianchi A., Leto C. and Carrubba A. 1995. Chemical characterization of essential oils of some *Origanum vulgare* L. subspecies of various origin. Rivista Italiana Eppos, 16, 21-28.

6. Ietswaart J.H. 1980. A taxonomic revision of the genus *Origanum* (Labiatae). Leiden Botanical Series 4. Leiden University Press, The Hague.

7. Skoula M., Gotsiou P., Naxakis G. and Johnson C.B. 1999. A chemosystematic investigation on the mono- and sesquiterpenoids in the genus *Origanum* (Labiatae). *Phytochemistry* 52 (3), 1-9.

8. Benjilali B., Hammumi M. and Richard H. 1987. Polymorphisme chimique des huiles essentielles de thym du Maroc. 1. Characterization de composantes. Science des Alimentes, 7, 77-91.

9. Franz C. 1993. Genetics. In: Hay and Waterman (eds) Volatile oil crops: their biology, biochemistry and production, 63-96.

Evaluation of Different Madder Genotypes (*Rubia tinctorum* L.) for Dyestuff Production

Simone Siebenborn
R. Marquard
I. Turgut
S. Yüce

SUMMARY. *Rubia* accessions, derived from different locations in the wild flora of Turkey and from Western Europe, have been described by different morphological characteristics, root yield and dye content. Plants from the Turkish location of Cesme have already been selected by means of root yield. The database was computed by cluster- and factor-analysis. Compared to the genotypes from Cesme and Western Europe, the plants from the Turkish location of Usak are low growing with small leaves and numerous shoots. They show good values for root yield and dye content, although no selection steps have been done before. *[Article copies available for a fee from The Haworth Document Delivery Service: 1-800-HAWORTH. E-mail address: <getinfo@haworthpressinc.com> Website: <http://www. HaworthPress.com> © 2002 by The Haworth Press, Inc. All rights reserved.]*

Simone Siebenborn and R. Marquard are affiliated with the Institut for Agronomy and Plant-Breeding I, Justus-Liebig University, 35390 Giessen, Germany.

I. Turgut is affiliated with Menderes University, Tarla Bitkileri Bölümü, Aydin, Turkey.

S. Yüce is affiliated with Ege University, Tarla Bitkileri Bölümü, 35100 Izmir-Bornova, Turkey.

[Haworth co-indexing entry note]: "Evaluation of Different Madder Genotypes (*Rubia tinctorum* L.) for Dyestuff Production." Siebenborn, Simone et al. Co-published simultaneously in *Journal of Herbs, Spices & Medicinal Plants* (The Haworth Herbal Press, an imprint of The Haworth Press, Inc.) Vol. 9, No. 4, 2002, pp. 281-287; and: *Breeding Research on Aromatic and Medicinal Plants* (ed: Christopher B. Johnson, and Chlodwig Franz) The Haworth Herbal Press, an imprint of The Haworth Press, Inc., 2002, pp. 281-287. Single or multiple copies of this article are available for a fee from The Haworth Document Delivery Service [1-800-HAWORTH 9:00 a.m. - 5:00 p.m. (EST). E-mail address: getinfo@haworthpressinc.com].

KEYWORDS. *Rubia tinctorum*, dyestuff production, morphological characters, cluster-analysis

INTRODUCTION

Common madder (*Rubia tinctorum* L.), the roots of which were used for the famous "Turkish-Red" dyeing process, is one of the most important dyeplants. It is assumed to have originated from the southeastern regions of the Mediterranean. From this area it spread to central Europe, where extensive areas were placed under cultivation for the dying industry up to the end of the 19th century. From the beginning of the 20th century onwards, the discovery of synthetic dyes completely ousted the use of natural ones. In recent years, in response to growing demands for natural fibres dyed with the traditional natural colouring agents, an increasing market for vegetable dyes was established. In Germany, several projects on the cultivation of dye plants have been started (3). In Turkey, madder is collected from local wild habitats in eastern Mediterranean regions, and even imported from Asia. Thus, various problems occur, especially concerning the quality of dyestuff. The dyeing agent in madder root consists of a mixture of various hydroxyanthraquinones, mainly Alizarin and Purpurin. The main problem in processing is constituted by the vast range of variation in dye content and composition found in the wild species, which result in uncontrollable colour variation in the final product. Development of plants with a homogeneously high yield and uniform dye content is vital for commercial madder cultivation (1).

MATERIALS AND METHODS

Seeds collected from different locations in the wild flora of Western Turkey were sown at the experimental station in Antalya (Figure 1). While the seeds from Usak and Isparta did not germinate, 13 Cesme accessions with high root yields could be selected out of 96 single plant descendants after 3 years of cultivation. At a later time, we received new seeds collected from the wild flora of Usak and also from Western Europe–mainly from various botanical gardens in Germany.

During the period from 1998 to 1999, these genotypes were grown in pots at the experimental station at Rauischholzhausen (Germany) over 2 years, as were the seed descendants of the 13 selected genotypes from Cesme. During that time, various morphological characteristics (flowering time, plant height, leaf size, etc.) were described for 103 single plants. Root yield and dye content were determined after harvesting in the second year. The method for determination of dye content by means of photometric measurement was developed at our institute (2). The database was computed by cluster and factor analysis, implementing the statistics program SPSS.

RESULTS AND DISCUSSION

Figures 2 and 3 show the variation of root yield and dye content of the 96 single plant descendants from Cesme.

FIGURE 1. Collection-areas and places of cultivation.

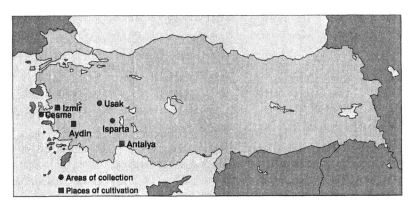

FIGURE 2. Variation of root yield among *Rubia* single-plant descendants from Cesme.

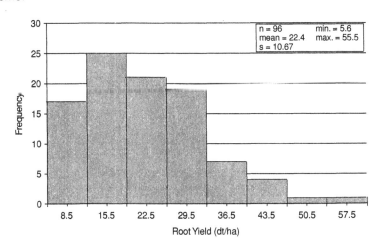

FIGURE 3. Variation of dye content among *Rubia* single-plant descendants from Cesme.

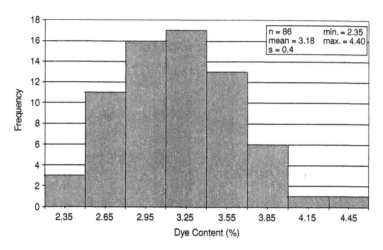

Out of these, 13 single plant descendants were propagated by root cuttings. They were selected according to the criterion of root yield. However, having determined the dye content, it appears that not only the average root yield (41.8 dt/ha) but also the average dye content (3.48%) of the 13 selections is clearly higher than the corresponding mean values of the entire Cesme population (22.4 dt/ha root yield and 3.18% dye content). As the genotypes were culti-vated in one replication, these differences were not tested for significance. More than half of the 103 *Rubia* plants grown in Rauischholzhausen in 1998 died over winter, thus only 42 genotypes could be harvested in 1999. Accord-ing to the dendrogram of the cluster analysis, the *Rubia* accessions could be di-vided in 6 groups (Figure 4).

In this division, the accessions from botanical gardens (cluster IV) are sepa-rated from Turkish cultivars (Cesme and Usak) mainly by variables describing the herbal parts of the plant. They show a tendency towards higher growth and larger leaves, especially if compared to the small-sized Usak genotypes (clus-ter II and III). The Usak accessions in cluster III show good values for root yield with sufficient dye content.

Nine factors were extracted by means of factor analysis; 3 of them show Eigenvalues > 1 and explain 73% of the total variance (Table 1). In Table 2 the factor loadings for all variables are described. Only factor loadings > 0.5 are listed because variables with values < 0.5 are not really correlated to a specific factor. Factor 1 represents the shape of herbal plant parts, showing negative correlation to number of shoots and positive correlation to stem diameter, plant and leaf size. Factor 2 is correlated to dye content and root morphology

FIGURE 4. Dendrogram of different *Rubia* accessions.

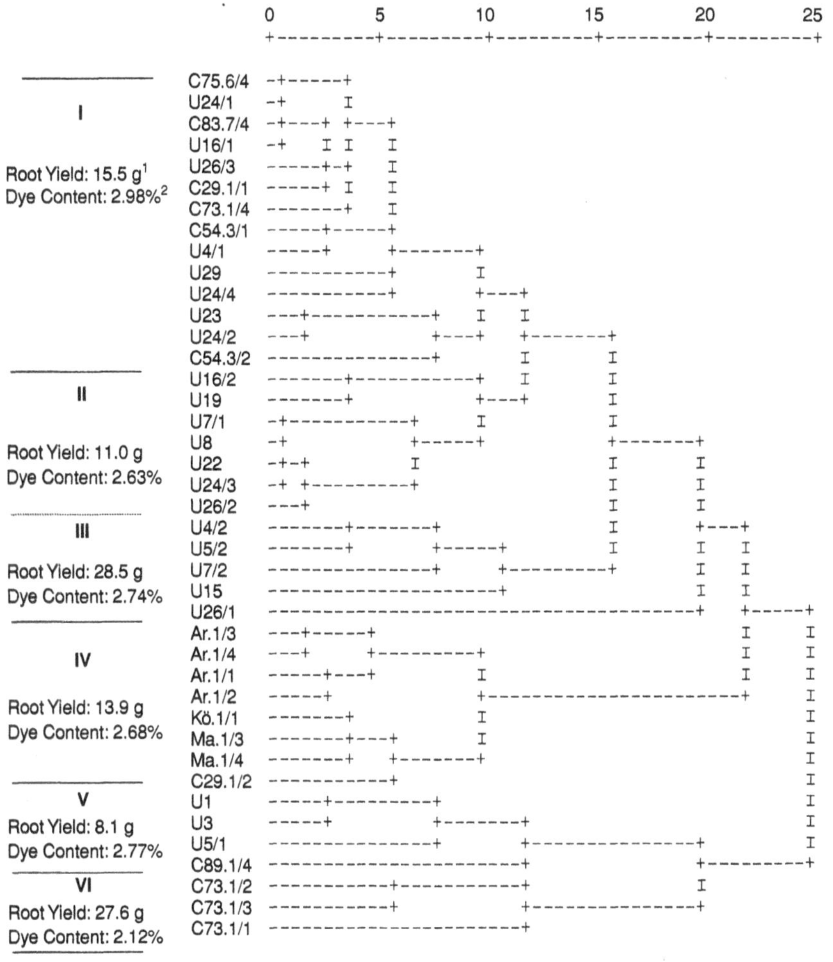

```
                        0         5        10        15        20        25
                        +---------+---------+---------+---------+---------+

 _____    C75.6/4 -+-----+
      I         U24/1    -+      I
                C83.7/4 -+---+ +---+
                U16/1    -+   I I   I
Root Yield: 15.5 g¹     U26/3    -----+-+   I
Dye Content: 2.98%²     C29.1/1  -----+ I   I
                C73.1/4  -------+   I
                C54.3/1  -----+-----+
                U4/1     -----+       +-------+
                U29      -----------+       I
                U24/4    -----------+     +---+
                U23      ---+-----------+   I   I
                U24/2    ---+          +---+  +-------+
                C54.3/2  -----------------+   I       I
 _____    U16/2    -------+-----------+   I       I
      II        U19      -------+       +---+       I
                U7/1     -+-----------+   I       I
Root Yield: 11.0 g      U8       -+          +-----+       +-------+
Dye Content: 2.63%      U22      -+-+       I       I       I
                U24/3    -+ +---------+       I       I
                U26/2    ---+               I       I
 _____    U4/2     -------+-------+       I       +---+
      III       U5/2     -------+       +-----+       I       I   I
                U7/2     -----------------+       +---------+   I   I
Root Yield: 28.5 g      U15      ---------------------+       I   I
Dye Content: 2.74%      U26/1    ---------------------------------+       +-----+
 _____    Ar.1/3   ---+-----+               I       I
      IV        Ar.1/4   ---+       +---------+               I       I
                Ar.1/1   -----+---+   I               I       I
Root Yield: 13.9 g      Ar.1/2   -----+       +---------------------+       I
Dye Content: 2.68%      Kö.1/1   -------+       I               I
                Ma.1/3   -------+---+       I               I
                Ma.1/4   -------+   +-------+               I
 _____    C29.1/2  -------------+               I
      V         U1       ----+---------+               I
                U3       -----+       +-------+       I
Root Yield: 8.1 g       U5/1     ----------------+       +---------------+       I
Dye Content: 2.77%      C89.1/4  ----------------------------+       +---------+
 _____    C73.1/2  ----------+-----------+       I
      VI        C73.1/3  -----------+       +---------------+
Root Yield: 27.6 g      C73.1/1  -----------------------+
Dye Content: 2.12%
```

1. average root yield per cluster (g/single plant)
2. average dye content per cluster
Ar, Kö, Ma: *Rubia* accessions from botanical gardens; C: *Rubia* accessions from Cesme/Turkey; U: *Rubia* accessions from Usak/Turkey

and factor 3 to plant size, beginning of flowering period and root yield. Single plants with high values (values > 1) for factor 1 show a specific growth habit in the herbal plant parts: The plants are very high and form one dominant main shoot. The leaves are comparatively large. These are exactly the *Rubia* plants which are integrated in Cluster IV after cluster analysis.

TABLE 1. Results of the principle components analysis for factor extraction in the experiment with different *Rubia* genotypes.

Factor	Eigenvalue		
	Total	% of variance	cum. %
1	3.510	39.00	39.00
2	1.777	19.75	58.75
3	1.289	14.32	73.06
4	0.906	10.06	83.13
5	0.464	5.16	88.28
6	0.384	4.26	82.55
7	0.285	3.17	95.72
8	0.204	2.27	97.99
9	0.181	2.01	100.00

TABLE 2. Factor loadings of the variables used to characterize the *Rubia* genotypes. Method for factor extraction: principle components analysis with varimax rotation.

	Factor loadings		
	fac 1	fac 2	fac 3
Number of shoots	−0.852		
Stem diameter	0.847		
Leaf length	0.811		
Leaf width	0.788		
Plant height	0.624		0.563
Dye content		−0.831	
Root morphology		0.761	
Start of flowering			−0.838
Root yield			0.536

Most plants with low values for factor 1 (values < −1) are clustered in Cluster III and could be described as low-growing, with small leaves and numerous shoots. This applies only to *Rubia* genotypes from Usak. Compared to the plants in cluster IV, which are mainly accessions from Western Europe, they are superior in root yield and dye content. The single plants of the Cesme population are mainly collected in Cluster I and VI. Cluster I consists of Usak and Cesme genotypes and do not show high values for any of the three factors, but the dye content is comparatively high. The accessions in Cluster VI are characterized by a specific root morphology: they develop many hairy roots, resulting in high root weight but low dye content. These plants show high factor values for factor 2 which is related to dye content and root morphology.

REFERENCES

1. Marquard R. & S. Siebenborn 1999. Das Rot das aus der Wurzel kam, DFG-Forschung 2/99, S. 16-18.

2. Siebenborn S., R. Marquard, S. Yüce & I. Turgut 1998. Untersuchungen zur Inkulturnahme von Färberkrapp (*Rubia tinctorum* L.). 5. Symposium deutsch-türkischer Agrar- und Naturwissenschaftler in Antalya 1997; Verlag Ulrich. E. Grauer Stuttgart, 193-198.

3. Vetter A., G. Würl, A. Biertümpfel 1997. Auswahl geeigneter Färberpflanzen für einen Anbau in Mitteleuropa. Zeitschrift für Arznei- und Gewürzpflanzen 4/97, 186-192.

Variation of the Anthocyanin Content in *Sambucus nigra* L. Populations Growing in Portugal

F. G. Braga
L. M. Carvalho
M. J. Carvalho
H. Guedes-Pinto
J. M. Torres-Pereira
M. F. Neto
A. Monteiro

SUMMARY. The analytical results concerning the identifications of four different chromatographic profiles of *Sambucus nigra* based on HPLC analysis of the anthocyanins present in the berries are described. That information can be used to select the best plants for anthocyanin pigments production and also to determine the best harvest time. *[Article copies available for a fee from The Haworth Document Delivery Service: 1-800-HAWORTH. E-mail address: <getinfo@haworthpressinc.com> Website: <http://www.HaworthPress.com> © 2002 by The Haworth Press, Inc. All rights reserved.]*

F. G. Braga, L. M. Carvalho, and M. J. Carvalho are affiliated with the Dep. de Química; H. Guedes-Pinto is affiliated with the Dep. de Genética e Biotecnologia; J. M. Torres-Pereira is affiliated with the Dep. Eng. Biológica e Ambiental; Universidade de Trás-os-Montes e Alto Douro, Vila Real, Portugal.

M. F. Neto and A. Monteiro are affiliated with the Direcção Regional de Agricultura de Trás-os-Montes, Quinta do Valongo, Mirandela, Portugal.

Address correspondence to: F. G. Braga, Dep. de Química, Universidade de Trás-os-Montes e Alto Douro, Apartado 202, 5001-911 Vila Real, Portugal.

The authors would like to thank the PAMAF-IED 8113 for funding this work.

[Haworth co-indexing entry note]: "Variation of the Anthocyanin Content in *Sambucus nigra* L. Populations Growing in Portugal." Braga, F. G. et al. Co-published simultaneously in *Journal of Herbs, Spices & Medicinal Plants* (The Haworth Herbal Press, an imprint of The Haworth Press, Inc.) Vol. 9, No. 4, 2002, pp. 289-295; and: *Breeding Research on Aromatic and Medicinal Plants* (ed: Christopher B. Johnson, and Chlodwig Franz) The Haworth Herbal Press, an imprint of The Haworth Press, Inc., 2002, pp. 289-295. Single or multiple copies of this article are available for a fee from The Haworth Document Delivery Service [1-800-HAWORTH 9:00 a.m. - 5:00 p.m. (EST). E-mail address: getinfo@haworthpressinc.com].

KEYWORDS. *Sambucus nigra* L., anthocyanins, HPLC, principal component analysis

INTRODUCTION

Sambucus nigra L. (Caprifoliaceae), often referred to as elderberry, is a small tree, endemic in central and northern Europe where it is still a common sight in country gardens. The fruit of the tree is a bright black berry (Figure 1), known for many centuries to possess valuable medicinal properties in addition to its common uses in preparing jellies, elderberry wine and imparting a brilliant red colour to grape wines (4).

A widely used Gypsy remedy for coughs and colds (perhaps because the berries are a good source of vitamins and bioflavonoids), historically it has also been used as a treatment for skin ailments and for the relief of burns, eczema and rashes. Even though the medicinal use of *S. nigra* dates to the fifth century BC and is found in the writings of Hippocrates, Dioscorides and Pliny,

FIGURE 1. Some details of the flowers and berries of *S. nigra*.

the most exciting use of the fruit is in its application against influenza. Recent research conducted in Israel has indicated potent antiviral properties of some species (1).

In northern Portugal (Varosa Valley), the elder tree takes profit from the excellent edaphoclimatic conditions and its intensive culture is increasing. The objective of this investigation is to characterize which local cultivar is the best for commercial anthocyanin production by HPLC analysis of the pigments present in the berries.

MATERIALS AND METHODS

Sample Selection. During the autumn of 1998, 15 elders were selected and marked, belonging to five different populations of the Varosa Valley (Salzedas (S), Ucanha (U), Dalvares (D), Cimbres (C) and Ferreirim (F)) according to the farmers description. From each tree about 200 ripe fruits (3 cymes) were sampled in the first week of September 1999. The berries were immediately frozen with liquid nitrogen. Later, the frozen fruits were separated manually from the stems, weighted, lyophilized and triturated. The resultant powder was used for the anthocyanin extraction.

Extraction of Elderberry Anthocyanins. About 5 grams of each powder sample was individually extracted with 25 mL of 0.1% MeOH-HCl at room temperature. After four repeated extractions, the combined extracts were filtered under reduced pressure through a 0.45-μm membrane filter (Ekicrodisc 13, Gelman Science, Germany) and 20 μL of the filtrate was injected into a Gilson HPLC system (306 Pump, 151 Single Wavelength UV/Vis Detector-Gilson, Inc., USA).

HPLC Analysis. The column (4.6 × 150 mm) used was a 5-μm Nucleosil C18 (Macherey-Nagel, Germany); 0.5% (v/v) phosphoric acid in water (A) and 0.5% phosphoric acid in 60% tetrahydrofuran (B) were used as solvents. A linear gradient between 10% B and 100% B over a period of 25 minutes at a flow of 0.9 mL/min was established. Detection was made at 520 nm.

Statistical Analysis. PCA is one of the most used chemometric methods for data reduction and exploratory analysis on high dimensionality data sets. The main goal of PCA is to obtain a small set of principal components (latent variables) that contain most of the variability on these data sets. The new sub-space defined by these principal components leads to a model that is easier to interpret than the original data set. From these results, it should be possible to highlight several characteristics and correlate them to the physicochemical properties of the samples (3). The PCA data matrix was composed of 15 objects (trees from different location and within the same location with different characteristics) and 7 variables (peaks obtained from the chromatograms). These main peaks were area normalized to 100%. The original variables (peaks) were then normalized to zero mean and unit standard deviation, in order to give the same weight to all variables in the analysis. The cluster analysis of the PC1 and PC2 scores was done using the complete linkage algorithm with euclidian distances. The software used for the statistical analysis was Statistica®.

RESULTS AND DISCUSSION

In Figure 2 can be seen a bidimensional representation of PC1 and PC2 scores for the samples studied. As can be observed, there was four different groups related to the anthocyanin profile confirmed by the cluster analysis (Figure 3). In order to determine which chemical variables are responsible for such discrimination, the PC1 and PC2 loadings can be observed in Figure 4. It can be seen that samples D1 and F2, have a mean content of the pigment with retention time of 18.4 minutes higher than all other samples (this is the principal source of discrimination of this two samples, according to PC1). The discrimination of the other three groups is due to PC2. PC2 is highly positively correlated with the cyanidin 3-sambubioside-5-glucoside quantity and negatively correlated with the cyanidin 3-glucoside. It can be seen that samples F3, U2, S2, C2 and C1 have a mean content of cyanidin 3-glucoside higher than the other samples, but have a lower content of cyanidin 3-sambubioside-5-glucoside (see Table 1).

Although the nature of the anthocyanins found in each berry studied did not change, the proportion of the anthocyanins is very different, and in the samples

FIGURE 2. Graphic representation of PC1 and PC2 scores for the samples studied.

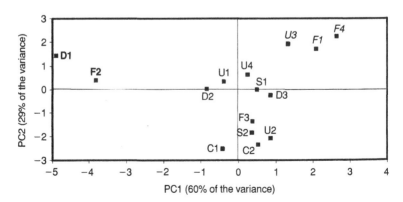

FIGURE 3. Cluster analysis of the PC1 and PC2 scores for the samples studied.

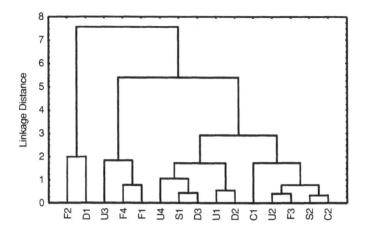

FIGURE 4. Graphic display of the PC1 and PC2 loadings (retention times), where it is possible to see in which way the pigments account for profile discrimination.

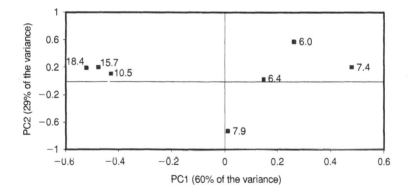

studied four distinct situations could be found (Figure 5). This difference in the overall quality of the anthocyanins can be very important in industrial processing (drying) and applications, as it is known that the chemical and light stability of the different anthocyanins can be very distinct (5). This aspect should be further exploited.

TABLE 1. Correspondence between the chemical structure, retention time, PC1 and PC2 for each anthocyanin (Figure 4).

Retention time (min.)	PC1	PC2	Anthocyanin (4)
6.0	0.26	0.58	Cyanidin 3-sambubioside-5-glucoside
6.4	0.15	0.03	Cyanidin 3,5-diglucoside
7.4	0.48	0.21	Cyanidin 3-sambubioside
7.9	0.01	−0.73	Cyanidin 3-glucoside
10.5	−0.43	0.10	Cyanidin 3-O-(6-O-Z-p-coumaroyl-2-O-β-D-xylopyranosyl)-β-D-glucopyranoside-5-O-β-D-glucopyranoside
15.7	−0.48	0.21	Cyanidin 3-O-(6-O-E-p-coumaroyl-2-O-β-D-xylopyranosyl)-β-D-glucopyranoside-5-O-β-D-glucopyranoside
18.4	−0.52	0.20	Cyanidin 3-O-(6-O-E-p-coumaroyl-2-O-β-D-xylopyranosyl)-β-D-glucopyranoside-β-D-glucopyranoside

FIGURE 5. Four different profiles of anthocyanins were identified in 15 trees of *S. nigra*.

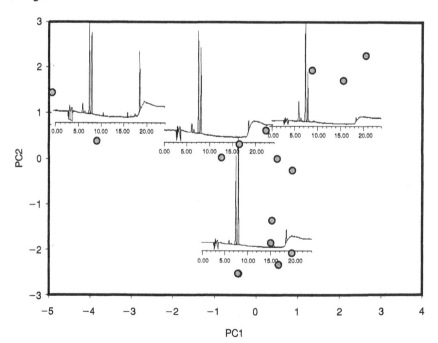

REFERENCES

1. 6th International Congress for Infectious Disease, April 26-30, 1994, Prague, Czech Republic.

2. Inani, O., Tamura, I., Kikuzaki, H. and Nakatan, N. 1996. Stability of anthocyanins of *Sambucus canadensis* and *Sambucus nigra*. *J. Agric. Food Chem.*, 44: 3090-3096.

3. Jolliffe, I. T. 1986. *Principal Component Analysis*. Springer-Verlag, New York, USA.

4. Mindell, E. 1992. *The Herb Bible*. Simon and Schuster, New York, USA.

5. Shahidi, F. and Naczk, M. 1995. *Food Phenolics: Sources, Chemistry, Effects and Applications*. Technomic Publishing Company, Inc., Basel, Switzerland.

Variability in the Content of Gamma-Linolenic Acid and Other Fatty Acids of the Seed Oil of Germplasm of Wild and Cultivated Borage (*Borago officinalis* L.)

Antonio De Haro
Vicente Domínguez
Mercedes del Río

SUMMARY. Two hundred and six *Borago officinalis* L. (Boraginaceae) accessions of cultivated and wild germplasm collections from different origins were evaluated for gamma-linolenic acid (GLA) and other fatty acids and seeds characters. GLA showed an important range of variation, from 8.7% to 28.6% of the seed oil. Oleic, linoleic and erucic acid also showed wide ranges of variation. White flowered cultivated genotypes had higher contents of GLA than blue flowered wild material collected in Spain. Blue flowered germplasm from Northern Europe showed higher values of erucic acid. No correlation was found between oil content and

Antonio De Haro and Mercedes del Río are affiliated with the Instituto de Agricultura Sostenible, CSIC, Córdoba, Spain.

Vicente Domínguez is affiliated with the Centro de Investigación Agraria, Logroño, Spain.

Address correspondence to: Antonio De Haro, Instituto de Agricultura Sostenible, CSIC, Apartado 14080, Córdoba, Spain.

[Haworth co-indexing entry note]: "Variability in the Content of Gamma-Linolenic Acid and Other Fatty Acids of the Seed Oil of Germplasm of Wild and Cultivated Borage (*Borago officinalis* L.)." De Haro, Antonio, Vicente Domínguez, and Mercedes del Río. Co-published simultaneously in *Journal of Herbs, Spices & Medicinal Plants* (The Haworth Herbal Press, an imprint of The Haworth Press, Inc.) Vol. 9, No. 4, 2002, pp. 297-304; and: *Breeding Research on Aromatic and Medicinal Plants* (ed: Christopher B. Johnson, and Chlodwig Franz) The Haworth Herbal Press, an imprint of The Haworth Press, Inc., 2002, pp. 297-304. Single or multiple copies of this article are available for a fee from The Haworth Document Delivery Service [1-800-HAWORTH 9:00 a.m. - 5:00 p.m. (EST). E-mail address: getinfo@haworthpressinc.com].

297

GLA. In spite of the positive correlation between GLA and erucic acid, genotypes with high oil and GLA content and low levels of erucic acid content could be identified for breeding purposes. *[Article copies available for a fee from The Haworth Document Delivery Service: 1-800-HAWORTH. E-mail address: <getinfo@haworthpressinc.com> Website: <http://www. HaworthPress.com> © 2002 by The Haworth Press, Inc. All rights reserved.]*

KEYWORDS. *Borago officinalis* L., borage, germplasm collection, fatty acid composition, gamma-linolenic acid, seed characters

INTRODUCTION

Borage (*Borago officinalis* L.) is an annual plant that has been used from ancient times for culinary and medicinal purposes (4). Recently, interest in borage has been renewed because its seeds contain a high percentage of gamma-linolenic acid (GLA, 18:3 Δ6,9,12), an essential fatty acid in increasing demand for its clinical and pharmaceutical applications. This fatty acid is an intermediate of indispensable compounds in the body, such as prostaglandin E1 and its derivatives (6).

Although seeds of many plants contain GLA, the most common commercial sources of this fatty acid are evening primrose (*Oenothera biennis* L.) and borage (5,7). The advantages of borage are its annual life cycle, higher oil content in the seed (24-34% as compared to 14-25% in *Oenothera*), as well as higher GLA proportion in the seed oil (23% in borage and 10.5% in *Oenothera*) (8).

The different parts of borage plants have been used for centuries in Spain as a vegetable. The leaves and petioles were cooked with legumes, the flowers were used in salads and the seed mixed with wine. Nowadays, white flowered borage is currently used as a vegetable in the north of Spain. However, no attention has been paid until now in Spain to the study of borage as a source of GLA. Moreover, the only germplasm evaluated for other traits has been the white flowered *material* used as a vegetable, neglecting the wild blue flowered material which could offer potential advantages of adaptability to Mediterranean conditions.

The objective of this work was to evaluate the existing variability for oil, GLA content and other seed characteristics of different sources of borage germplasm, including white flowered borage cultivated in Spain, blue flowered borage collected from roadside populations in Spain, and introductions from several countries of Europe.

MATERIALS AND METHODS

The material evaluated in this study consisted of borage seed from 206 accessions originating from three different sources, one from introductions from Europe and two from Spain. Twenty-one accessions were received from European botanical gardens and breeders: 5 from the UK and 4 each from Austria, Germany, Sweden and Netherlands. The Spanish stocks were from two well differentiated origins. The first was formed by 130 cultivars and populations of white flowered material cultivated as a vegetable, and provided by the Center of Agricultural Research of Rioja for this study. This material was originally developed by smallholdings and seed companies; it was further evaluated and selected for morphological and agronomical traits, and classified in six groups (Table 1). The Spanish second group consisted of 55 wild entries of spontaneous blue-flowered roadside populations collected in different localities of Andalusia (southern Spain).

Seed weight was determined by weighing 1000 seeds. Oil content was determined by nuclear magnetic resonance (NMR) of seed previously dried at 65°C for 60 hours. The NMR analyzer was standardized by using a clean high quality sample of borage oil extracted by the Soxhlet method. The fatty acid composition of the oil was determined on a bulk of 10 seeds/entry by simultaneous extraction and methylation (3), followed by gas-liquid chromatography (GLC) on a Perkin Elmer Autosystem (Perkin Elmer, Norwalk, CT) equipped with a flame ionization detector (FID) and a 2 m column packed with 3% SP-2310/ 2% SP-2300 on Chromosorb WAW. Fatty acids were identified by comparing the retention times of the borage methyl esters with those of known mixtures of methyl esters run on the same column under the same conditions. The GLA standard was purchased from Sigma Chemical Co. (L2378).

TABLE 1. Characteristics of cultivated white flowered entries.

Group	Plant Habit	Seed Production	No. of Entries
I	erect	very low (< 100 g/plant)	23
II	erect	low (100-150 g/plant)	26
III	prostrate	intermediate (150-200 g/plant)	28
IV	semierect	intermediate (150-200 g/plant)	22
V	erect	high (200-225 g/plant)	17
VI	semierect	very high (225-250 g/plant)	14

RESULTS AND DISCUSSION

Table 2 gives the seed weight, oil content and fatty acid composition of the different populations of borage evaluated in this work. Considering all the entries, there was an important range of variation for most of the characters studied. Thousand-seed weight showed an overall range of variation from 9.3 to 26.3 g. Oil content varied from 26.7 to 38.0%, and the fatty acid composition of the seed oil showed the following overall ranges: palmitic 6.5 to 18.9%, stearic 2.3 to 9.2%, oleic 9.9 to 49.1%, linoleic 21.3 to 48.8%, gamma-linolenic 8.7 to 28.6%, eicosenoic 1.8 to 9.0% and erucic 0.6 to 13.8%. These ranges are wider than those published in other studies (2,7,8). Blue flowered material showed lower values for oil and GLA and seed weight than the cultivated white flowered (WF) population.

Some entries from wild blue flowered populations collected in Spain (BFS) showed the highest mean and maximum values of oleic acid, 27.6 and 49.1%, respectively. All the other groups of entries exhibited similar mean values for this acid except for group V of white flowered material, with the lowest mean and minimum values of 15.5% and 9.9%, respectively. This group also showed the highest content of GLA, with a mean value of 25.2% and a maximum value reaching 28.6%. An additional advantage of group V is its high seed production (more than 200 g per plant) and high level of oil content (maximum value 38%). Linoleic acid did not present large differences between groups although the wild material collected in Spain (BFS) showed the lowest mean and minimum values. The high temperatures during seed formation in southern Spain, where these samples were collected, could be responsible for these "high oleic-low linoleic" ratios, as has been pointed out in other oil crops (1). Saturated fatty acids, palmitic and stearic, showed similar contents in all populations and they were similar to the values published in other studies (8). However, erucic and eicosenoic acids showed important differences between groups, specially erucic acid. All the white flowered groups showed similar values for this acid except group V, with the maximum mean and individual values of 2.6% and 3.3%, respectively. The BF material behaved differently depending on the population considered. The material collected in Spain showed a low mean content for erucic acid, 1.9%, and one entry with a minimum value of only 0.6% was found. The other group, blue flowered from Northern Europe (BFNE), showed a higher mean, 3.8%, and a very wide range for this acid, 1.9 to 13.8%. The highest content of erucic acid corresponded to material from the Netherlands and the lowest from Germany. The differences observed are important since this acid is undesirable for human consumption.

The correlations between the content of oil and the different fatty acids are listed in Table 3, separately for each population. Oil content did not show any

TABLE 2. Mean and range of 1000-seed weight, oil content,[a] and fatty acid composition[b] of white and blue flowered borage populations.

Borage Populations	n	1000-Seed Weight (g)	Oil Content (% d.m.)	Palmitic Acid (16:0)	Stearic Acid (18:0)	Oleic Acid (18:1)	Linoleic Acid (18:2)	γ-Linolenic Acid (18:3)	Eicosenoic Acid (20:1)	Erucic Acid (22:1)
WF I	23	15.6	34.3	11.1	4.0	19.5	37.2	22.2	4.1	1.8
		12.9-20.1	32.1-36.4	8.5-15.3	3.1-5.7	14.0-25.6	32.6-41.5	14.1-26.5	3.3-5.3	1.3-2.3
WF II	26	15.2	34.2	10.8	4.3	19.6	38.3	22.1	3.6	1.6
		12.7-19.2	31.8-35.8	8.8-18.7	2.4-9.2	14.9-30.1	26.6-44.5	8.7-27.1	1.8-5.8	1.0-2.0
WF III	28	15.1	34.5	10.6	4.1	19.1	38.3	22.3	3.9	1.7
		10.8-20.1	32.4-37.9	8.9-11.6	3.3-5.1	15.3-24.6	33.6-41.8	19.6-25.7	1.9-4.7	1.4-2.1
WF IV	22	14.8	34.2	11.1	3.6	18.6	38.6	23.6	3.8	1.6
		12.4-17.8	32.1-37.2	9.0-14.2	2.6-5.3	15.5-25.1	35.6-41.2	15.9-27.2	3.2-4.7	1.3-1.9
WF V	17	15.1	35.8	10.9	3.6	15.5	37.5	25.2	4.6	2.6
		13.8-17.7	33.2-38.0	10.1-12.1	2.3-4.8	9.9-18.9	35.2-41.1	21.9-28.6	3.5-5.4	1.3-3.3
WF VI	14	15.3	34.4	10.8	3.4	18.4	38.8	23.9	3.9	1.8
		12.8-19.1	32.5-37.2	9.8-12.3	2.5-4.6	16.1-21.4	35.1-41.9	21.3-26.8	3.3-5.2	1.2-2.7
BFS	55	14.1	30.7	10.1	4.3	27.6	32.7	18.9	4.3	1.8
		9.3-18.2	26.7-35.4	6.5-12.8	2.8-6.3	16.7-49.1	21.3-39.3	12.6-26.6	2.9-6.1	0.6-3.0
BFNE	21	23.1	29.6	10.82	4.64	18.6	36.15	20.4	5.6	3.7
		19.2-26.3	27.6-32.5	8.3-18.9	2.3-6.8	13.4-24.1	28.1-48.8	13.1-27.2	3.2-9.0	1.9-13.8

n = number of entries, WF = white flowered, BFS = blue flowered from Spain, BFNE = blue flowered from North Europe
[a] Oil content expressed as percentage in the seed
[b] Fatty acids are expressed as percentage of the total fatty acids (wt%) in the oil

301

TABLE 3. Correlation coefficients among oil and fatty acid content obtained from data of borage populations.

WF (I to VI) (n = 130)

	Oil	Palmitic	Stearic	Oleic	Linoleic	γ-Linolenic	Eicosenoic
Palmitic	0.22 **						
Stearic	0.15 NS	0.36 **					
Oleic	0.02 NS	0.23 **	0.59 **				
Linoleic	0.04 NS	−0.46 **	−0.38 **	−0.54 **			
γ-Linolenic	−0.16 NS	−0.44 **	−0.80 **	−0.79 **	0.35 **		
Eicosenoic	0.11 NS	0.35 **	0.11 NS	−0.15 NS	−0.36 **	−0.04 NS	
Erucic	−0.08 NS	0.21 *	−0.28 **	−0.54 **	−0.17 *	0.41 **	0.62 **

BFS (n = 55)

	Oil	Palmitic	Stearic	Oleic	Linoleic	γ-Linolenic	Eicosenoic
Palmitic	−0.19 NS						
Stearic	−0.02 NS	0.10 NS					
Oleic	−0.05 NS	−0.70 **	0.08 NS				
Linoleic	0.16 NS	0.60 **	0.13 NS	−0.93 **			
γ-Linolenic	−0.09 NS	−0.44 **	−0.12 NS	−0.89 **	0.73 **		
Eicosenoic	−0.09 NS	0.56 **	−0.12 NS	−0.72 **	0.56 **	0.64 **	
Erucic	−0.18 NS	0.66 **	−0.19 NS	−0.60 **	−0.49 **	0.45 **	0.67 **

BFNE (n = 21)

	Oil	Palmitic	Stearic	Oleic	Linoleic	γ-Linolenic	Eicosenoic
Palmitic	0.15 NS						
Stearic	0.12 NS	0.11 NS					
Oleic	0.05 NS	0.05 NS	0.58 **				
Linoleic	0.10 NS	−0.51 **	−0.22 NS	−0.15 NS			
γ-Linolenic	−0.12 NS	−0.17 NS	−0.68 **	−0.76 **	0.05 NS		
Eicosenoic	0.08 NS	−0.30 NS	0.11 NS	−0.07 NS	−0.57 **	−0.16 NS	
Erucic	−0.13 NS	0.01 NS	−0.06 NS	−0.27 NS	−0.51 **	−0.04 NS	0.57 **

* $P < 0.05$
** $P < 0.01$

significant correlation with the content of any fatty acid although it was positively correlated with seed weight (0.77 in WF, 0.64 in BFS, and 0.60 in BFNE, data not shown in Table 3). The absence of significant correlation between oil content and undesirable fatty acids, such as erucic acid, in all the groups studied, is favourable because it indicates that it is possible to select material with high oil and low erucic acid content. Similarly, the lack of correlation between oil and GLA content also indicates that it is possible to select simultaneously for high values of both traits.

The correlations between the different fatty acids showed some variations depending on the population considered. For instance, erucic acid and GLA were positively correlated in WF and BFS populations but not in the BFNE population. This positive correlation, also reported in a previous work (2), could be considered as unfavourable, since both high levels of GLA and low levels of erucic acid are important aims of selection in the breeding of borage. However, the contents of erucic acid in WF and BFS populations are relatively low compared with other reports. Erucic acid and oleic acid were also correlated (negatively) in WF and BFS but not in the BFNE population. Moreover, erucic acid and eicosenoic acid were always positively correlated, and oleic acid and GLA were negatively correlated in all the borage populations.

Entries with about 27% of GLA, oil content ranging from 32 to 37% and erucic acid values lower than 2.5% have been identified in the three populations. On the other hand, entries with low levels of GLA (< 15%), and entries with high erucic acid (> 13%) have been also identified (Table 4). These selections are being multiplied for further use in agronomic studies and breeding programs.

TABLE 4. Characteristics of selected entries.

Accession	Oil Content[a]	γ-Linolenic Acid[b]	Erucic Acid
WF-V-8	36,4	28.6	0.1
WF-V-13	37.6	27.2	1.8
BFS-47	34.0	26.6	0.9
BFNE-3	32.1	27.5	2.4
WF-II-18	34.7	8.7	1.5
BFNE-9	31.2	14.1	13.7

WF = white flowered, BFS = blue flowered from Spain, BFNE = blue flowered from North Europe
[a] Oil content expressed as percentage in the seed
[b] Fatty acids are expresed as percentage of the total fatty acids (wt%) in the oil

REFERENCES

1. Canvin, D.T. 1965. The effect of temperature on the oil content and fatty acid composition of the oils from several oil seed crops. *Can. J. Bot.* 43: 63-69.

2. Galwey, N.W., and A.J. Shirlin. 1990. Selection of borage (*Borago officinalis*) as a seed crop for pharmaceutical uses. *Heredity* 65: 249-257.

3. Garcés, R., and M. Mancha. 1993. One-step lipid extraction and fatty acid methyl esters preparation from fresh plant tissues. *Anal. Biochem.* 211: 139-143.

4. Gerard, J. 1597. *The History of Plants*, edited by Marcus Woodward, 1927, and published by Senate, Studio Editions Ltd., London, 1994, pp. 185-186.

5. Gunstone, F.D. 1992. Gamma-linolenic acid: Occurrence and physical and chemical properties. *Prog. Lipid Res.* 31: 145-161.

6. Horrobin, D.F. 1992. Nutritional and medical importance of gamma-linolenic acid. *Prog. Lipid Res.* 31: 163-194.

7. Janick, J., J.E. Simon, J. Quinn, and N. Beaubaire. 1989. Borage: A source of gamma-linolenic acid. *In* L.E. Cracker and J.E. Simon, eds. *Herbs, Spices and Medicinal Plants*, Vol. 4, Oryx Press, Phoenix, pp.145-168.

8. Muuse, B.G., M.L. Essers, and L.J.M Van Soest. 1988. *Oenothera* species and *Borago officinalis*: Sources of gamma-linolenic acid. *Neth. J. Agr. Sci.* 36: 357-363.

Geographic and Environmental Influences on the Variation of Essential Oil and Coumarins in *Crithmum maritimum* L.

Jan Burczyk
Krystyna Wierzchowska-Renke
Kazimierz Głowniak
Paweł Głowniak
Dariusz Marek

SUMMARY. Different plant materials (fruits and herbs of *Crithmum maritimum* L.) collected in five geographic regions (in Greece, France and cultivated in Poland) were examined for content and composition of essential oil and coumarins. Total content of essential oil in the powdered plant material was determined by steam water distillation and the quantity of main essential oil components was analyzed by gas chromatography. Influence of fertilization with Tytanit fertilizer containing titanium ions on the content of essential oil and concentration of four main

Jan Burczyk, and Paweł Głowniak are affiliated with the Department of Pharmacognosy and Phytochemistry, Silesian Medical University, 4 Jagiellońska Street, 41-200 Sosnowiec, Poland.

Krystyna Wierzchowska-Renke, and Dariusz Marek are affiliated with the Department of Biology and Pharmaceutical Biology, Medical University, 107 Gen. J., Hallera Avenue, 80-460 Gdańsk, Poland.

Kazimierz Głowniak is affiliated with the Department of Pharmacognosy, Medical University, 12 Peowiaków Street, 20-007 Lublin, Poland.

[Haworth co-indexing entry note]: "Geographic and Environmental Influences on the Variation of Essential Oil and Coumarins in *Crithmum maritimum* L." Burczyk, Jan et al. Co-published simultaneously in *Journal of Herbs, Spices & Medicinal Plants* (The Haworth Herbal Press, an imprint of The Haworth Press, Inc.) Vol. 9, No. 4, 2002, pp. 305-311; and: *Breeding Research on Aromatic and Medicinal Plants* (ed: Christopher B. Johnson, and Chlodwig Franz) The Haworth Herbal Press, an imprint of The Haworth Press, Inc., 2002, pp. 305-311. Single or multiple copies of this article are available for a fee from The Haworth Document Delivery Service [1-800-HAWORTH 9:00 a.m. - 5:00 p.m. (EST). E-mail address: getinfo@haworthpressinc.com].

compounds (α-pinene, p-cymene, limonene, β-pinene) was examined for variation of the content of essential oil in different parts of plant material collected from five natural sites.

Coumarins were analyzed after exhaustive extraction with petroleum ether and methanol. The obtained extracts were purified from polar components by liquid-liquid extraction and separated by column chromatography on silica gel 60 (230-400 mesh) stationary phases and then by preparative TLC on silica gel 60 plates, 0.5 mm thickness.

Several coumarin fractions have been obtained in this way. As homogeneous fractions, scopoletin and scoparone have been isolated. They have been confirmed by comparison of their chromatographic retention data with standards in TLC and RP-HPLC analysis; similar composition of coumarins, mainly in etheral and methanolic extracts from plant material collected in different natural sites (Crete, Malta, Cyprus, Tenerife, France) and cultivated in Poland, was established. *[Article copies available for a fee from The Haworth Document Delivery Service: 1-800-HAWORTH. E-mail address: <getinfo@haworthpressinc.com> Website: <http://www. HaworthPress.com> © 2002 by The Haworth Press, Inc. All rights reserved.]*

KEYWORDS. *Crithmum maritimum* L., coumarins, essential oil, chromatographic separation, isolation, geographical variation, fertilization

INTRODUCTION

Investigations of natural coumarins, including both simple coumarins and their glycosides, as well as furano- and pyranocoumarins, provide new information concerning an even wider spectrum of their biological activity, in addition to their well-known antibacterial, antifungal, anti-virus, anti-coagulant and photosensibilizing properties utilized in the therapy of vitiligo and psoriasis (1).

In recent years, reports have been published on their spasmolytic action, as blockers of calcium channels in the heart muscle, as inhibitors of the reverse transcriptase of HIV-1 as well as cytostatic activity relative to L-1210 and anticarcinogenic activity (2,3). Therefore, the search for new plant sources containing furano- and pyranocoumarins is an actual problem. The search is mainly directed to the Apiaceae family, the representatives of which are characterized by the highest content of these compounds.

Coumarins occur most abundantly in members of the Apiaceae family, among others, in *Crithmum maritimum* L. which grows in rocks and sands of sea shores. It occurs in Greece, Crete, also on the coasts of the Atlantic Ocean and the Mediterranean and Black Seas (4). It is mostly investigated for the content of volatile oils (5,6); the remaining phytochemical investigations are of the screening type (7).

The earliest component of *C. maritimum* L. investigated in detail was the essential oil; the first reports concerned the content of essential oils in the fruits in amounts of 0.7-0.8%, mainly dillapiole (9.4%) and D-terpene (12%). Some compounds of hydrocarbon type were also found. Detailed investigations of the essential oil indicated that its content in the material amounts to 1.5% and that it contains 57 substances of which 44 were identified (5), the main components being monoterpenes: limonene (22.3%), γ-terpinene (22.9%) and m-thymol (25.5%). In addition, a small amount of sesquiterpenes (2%) were found (β-bisalbolene, cuparene and limonene, triopsene and hydrocarbons C_{21}-C_{33}), totalling 4% of the oil.

Investigations carried out in 1994 gave a somewhat different chemical composition of the oil (6): 58 compounds were detected, of which 50 were identified, and the main components were: γ-terpinene (37%), m-thymol (29%), p-cymene (10%), β-pinene (10%). The discrepancies of the chemical compositions reported in these papers suggest that different chemotypes are present within the species *C. maritimum* L.

The investigations of Consulo and Ruberto (8) carried out in the years 1992-1999 reported also on the compounds obtained from the lipid extract: the first was polyacetylene C-17-fulcarinol (panaxynol); the second was fulcarindiol, earlier isolated from *Panax ginseng* and *Fulcaria vulgaris*, also present in many species of the Apiaceae and Araliaceae families.

Recently (1999), another compound from the C-17 polyacetylene group, optically active oily substance of general formula $C_{17}H_{26}O_2$ named Crithumdiol, was reported (9). Furthermore, in phytochemical screening investigations, the presence of flavonoids, coumarins and gums in *C. maritimum* L. was found (7). The material is also rich in microelements and vitamins, especially vitamin C. The antimicrobial activity of dillapiole and extracts from *C. maritimum* L. was examined (10).

MATERIALS AND METHODS

Plant material. The plant material investigated were the fruits and aerial parts of *C. maritimum* L. collected in five natural sites and, since 1999, in the Botanical Garden of the Medical University of Gdańsk, Northern Poland. Details of the times of gathering of the plant material used is given in Table 1. The plants investigated were grown in the Garden and in a greenhouse. Plants were in the first year of their vegetation which had been earlier, in the spring, sprayed three times (in two-week periods) with a 0.04% solution of Tytanit. The leaves of control plants were not sprayed.

TABLE 1. Content of essential oil in various organs of *Crithmum maritimum* L. collected from five natural sites.

No	Site	Date	Plant material	Content of oil % v/v
1.	Crete	30.IX.1998	fruits	1.8
2.	Cyprus	23.X.1999	fruits	1.4
3.	Malta	1.X.1999	fruits leaves	2.7 1.8
4.	Tenerife	25.V.1999	aerial parts	1.0
5.	France	23.VII.1999	aerial parts	3.2
6.	Poland	20.IX.1999	aerial parts	1.1

Gas chromatography. The plant material was dried in air at 18-21°C. In all samples the content of essential oil was determined according to Polish Pharmacopoeia FP V by steam distillation (indirect method using m-xylene), then the samples of oils were analyzed by capillary gas chromatography. Four monoterpenes–α-pinene, p-cymene, limonene and γ-terpinene–were also identified. Capillary gas chromatography of the oils was carried out using a Hewlett-Packard gas chromatograph equipped with a flame ionization detector under the following conditions: capillary 0.25 mm I.D., 30 m long, HP-Innowax; carrier gas: argon; column temperature range: 50-200°C, 20°C/min; carrier gas flow rate: 1ml/min; standards: γ-terpinene, α-pinene, p-cymene, limonene.

Identification of coumarins. The compounds isolated by LC and preparative TLC were identified on the basis of V_R determined in HPLC on Waters 996 PDA analysis on Knauer Lichrospher 100 RP 250 $\times$ 4 mm^2 column (50% and 60% aq. MeOH, flow rate 50 µl/min, PDA detection at 254 nm) by comparison with standards.

Densitometric TLC determinations. Plant extracts from *C. maritimum* L. herbs and fruits were applied on precoated 10 $\times$ 20 cm^2 silica Si60 or Kieselguhr F$_{254}$ plates (E. Merck, Darmstadt, Germany) with Desaga AS 30 applicator as streaks 5 $\times$ 1 mm^2 and eluted on the distance of 20 cm with the mobile phases: n-heptane-ethyl acetate 4:1, cyclohexane-ethyl acetate 87:13. The densitometry was carried out using a computer-controlled Desaga CD 60 densitometer. The scanning was carried out at predetermined λ_{max} = 254 nm at slit dimensions 0.1 $\times$ 0.4 mm^2. Meander scanning (6 mm, 2 $\times$ 3 mm^2) at the same slit dimensions was also applied. Some coumarin fractions were analyzed with LSIMS technique after dissolving in NBA or glycerol.

RESULTS

The petroleum ether and methanolic extract from the fruits and herbs of *C. maritimum* L. was found to be abundant in UV absorbing coumarins after TLC separation on silica and Kieselguhr plates. Extensive chromatography of coumarin fraction by column chromatography and preparative TLC analysis yielded two coumarins with blue fluorescence in UV light 254 nm and R_f values similar to scopoletin (R_f 0.31) and scoparone (R_f 0.64) standards. Identification of these two compounds isolated by LC and preparative TLC was confirmed by HPLC, densitometric analysis and mass spectrometry.

Among less polar compounds, furanocoumarins with yellow and brown fluorescence after TLC separation yielded a brown-yellow spot (R_f 0.82), probably indicating imperatorin. Small amounts were not sufficient for spectral analysis and structure elucidation.

The level of essential oil in examined plant material is closely related to the geographic zone of the growth of the collected plants (Table 1). The richest amount of essential oil was contained in fruits collected from Malta (2.8%) and Crete (1.8%) and in aerial parts growing on natural sites in France (3.2%).

The investigations on the effect of spraying of the leaves of *C. maritimum* L. with the fertilizer Tytanit have been carried out since 1999; the results relate to the aerial parts of the plant, in the first year of vegetation (Table 2).

DISCUSSION

Similar quantitative composition of coumarin compounds was found in each extract from plant material (herbs, fruits) collected in various geographical regions. Coumarins were present in both etheral and methanolic extracts in richest amounts in the herbs examined. Isolated and identified coumarins: scopoletin and scoparone ($[M + H]^+$ ions 193 and 207, respectively) were never described to be present in *C. maritimum* L. Scoparone displays interesting pharmacological immunosuppressive activity and *C. maritimum* can be a

TABLE 2. Percent content of the terpenes investigated in the oil from the leaves of *Crithmum maritimum* L.–samples from the garden.

Experiment	Compounds investigated				Content of oil % v/v
	α-pinene	p-cymene	limonene	γ-terpinene	
Plant sprayed with Tytanit	9.16	1.12	5.59	2.22	1.1
Control	0.78	0.72	4.76	1.30	0.6

source for the isolation of this compound on a large scale. It was found that plants fertilized with Tytanit (an ecological complex of titanium which activates the metabolic processes in plants, increasing the yield of fruits and seeds) better resisted low temperatures during winter (which indicates better resistance to frosts) and grew better in comparison to control plants. Thirty percent of the controls (not sprayed) perished during the winter, whereas all of the sprayed plants survived. In the spring, the spraying of *C. maritimum* L. was repeated as in the preceding year.

It follows from literature data (11) that titanium, the active element in Tytanit, accelerates growth of the seedlings and germination of the fruits. It encouraged us to carry out trial spraying with this fertilizer, since the seeds of plants of the Apiaceae family, to which *C. maritimum* L. belongs, germinate with difficulty which is a serious obstacle to their cultivation.

The experimental results indicate that the spraying of the leaves of *C. maritimum* L. with Tytanit had a profitable effect on the content of essential oil (the leaves of the sprayed plants contained considerably higher amounts of oil that the controls). The analysis of the oil obtained indicates that the spraying had an effect on the percentage content of the four terpenes investigated.

REFERENCES

1. Murray R., Mendez J., Brown S. 1982. The Natural Coumarins, Wiley, Chichester.

2. Harmalava P., Vuorela H., Tornquist K., Hiltunen R. 1992. Choice of solvent in the extraction of *Angelica archangelica* roots with reference to calcium blocking activity, *Planta Medica*, 58, 176-183.

3. Vlietinck A.J., DeBruyne T., Apers S., Pieters L.A. 1998. Plant-derived leading compounds for chemotherapy of human immunodeficiency virus (HIV) infection, *Planta Medica*, 64, 97-109.

4. Hedge I.C., Lamond J.M. 1972. *Crithmum* L. in flora of Turkey and the east aegean islands (P.H. Davis, ed.), Edinburgh University Press, *Edinburgh*, 4, 367.

5. Ruberto G. 1991. Composition of the volatile oil of *Crithmum maritimum* L., *Flavour and Fragrance Journal*, 6, 121-123.

6. Senatore F., De Feo V. 1994. Essential oil of a possible new chemotype of *Crithmum maritimum* L. growing in Campania (South Italy), *Flavour Fragrance Journal*, 9(6), 305-307.

7. Coiffard L., De-Roeck-Holtzhauer Y. 1994. *Crithmum maritimum* L. (Apiaceae): Hydromineral composition and phytochemical screening, *Ann. Pharm. Fr.*, 52(3), 153-159.

8. Consulo F., Ruberto G. 1993. Bioactive metabolites from Sicilian marine fennel *Crithmum maritimum, Journal of Natural Products*, 56(9), 1598-1600.

9. Ruberto G., Amico V. 1999. Crithmumdiol: A new C_{17}–acetylene derivative from *Crithmum maritimum* L., *Planta Medica* 65, 681-682.

10. Katsouri E., Demetzos C., Perdetzoglou D., Gazouli M., Tzouvelekis L.S. 1999. An interpopulation study of the essential oil of various parts of *Crithmum maritimum* L. growing in Amorgos island (Greece), *Book of Abstracts 2000 Years of Natural Products Research*, 416.

11. Borkowski J., Dyka B. 1998. Effect of titanium on plants (in Poland), *Ogrodnictwo (Warsaw)*, 5-6, 18.

Study of Autochthon
Camelina sativa (L.) Crantz in Slovenia

Janko Rode

SUMMARY. *Camelina sativa* (L.) Crantz (Cruciferae) or false flax, an old oil-seed crop, is gaining interest because of its low environmental impact and wide possibilities of use. The tradition of growing false flax is still present in Slovenia. In folk medicine, the oil is considered a good remedy for stomach ulcers, the treatment of burns, wounds, eye inflammations and as a tonic. A widespread investigation of false flax growing and technologies used was conducted. A pilot cultivation experiment with seeds of local population was done in the experimental field of the institute. Analysis of oil content and composition was performed.

The traditional way of false flax growing is very extensive. The field is prepared by ploughing the meadow to avoid problems with weeds. Sowing takes place at the end of March. The only fertilizer is manure. The plots are relatively small. The harvest takes place in July. Harvested plants are air dried and threshed. Seeds are kept in sacks and processed a short time before the oil is sold. The oil content of the seeds from the experimental field was around 33%. The content of fatty acids was typical, with high polyunsaturated fatty acid content. The crop showed some problems of a nonstabilized cultivar. The tradition of cultivating and using false flax could be a good basis for its introduction in sustainable crop rotation. The important role in knowing and preserving this crop in

Janko Rode is affiliated with the Institute of Hop Research and Brewing, SI-3310 Žalec, Slovenia.

[Haworth co-indexing entry note]: "Study of Autochthon *Camelina sativa* (L.) Crantz in Slovenia." Rode, Janko. Co-published simultaneously in *Journal of Herbs, Spices & Medicinal Plants* (The Haworth Herbal Press, an imprint of The Haworth Press, Inc.) Vol. 9, No. 4, 2002, pp. 313-318; and: *Breeding Research on Aromatic and Medicinal Plants* (ed: Christopher B. Johnson, and Chlodwig Franz) The Haworth Herbal Press, an imprint of The Haworth Press, Inc., 2002, pp. 313-318. Single or multiple copies of this article are available for a fee from The Haworth Document Delivery Service [1-800-HAWORTH 9:00 a.m. - 5:00 p.m. (EST). E-mail address: getinfo@haworthpressinc.com].

313

Slovenia is played by farmers. It is an ideal model for the introduction of on-farm conservation and management practices. *[Article copies available for a fee from The Haworth Document Delivery Service: 1-800-HAWORTH. E-mail address: <getinfo@haworthpressinc.com> Website: <http://www. HaworthPress.com> © 2002 by The Haworth Press, Inc. All rights reserved.]*

KEYWORDS. *Camelina sativa*, field production, false flax oil, medicinal use, Slovenia

INTRODUCTION

Camelina sativa (L.) Crantz, or false flax, is an old oil-seed crop. Its cultivation is evident from the neolithic through to Roman times (6). Despite the decline of cultivation in the Middle Ages, it remained as a weed in flax stands (11). In some countries of Europe the cultivation on smaller stands in specific circumstances persisted and competed with modern high yielding crops. The crop is gaining interest because of its low environmental impact and wide possibilities of use. There are some recent studies of field production (7,9,13,17), gene resources and breeding (14,16), and methods of molecular genetics (15) of *C. sativa*. The importance of the crop was confirmed by a European commission which funded a project with *C. sativa* through the Agro-Industrial Research program (3,17). The lack of clear utilization patterns of the crop was identified as one of the limiting factors in extending the production of *C. sativa* (11). Possibilities of use include the utilization of seed, oil, meal and fiber (1,11,12,13,17). The most promising is the use of the oil, with unique fatty acid content characterized by unsaturated fatty acids, as a health promoting food additive (17) and nutraceutical (1). Some traditional or folk uses of the plant and oil confirm that the false flax can also be considered a medicinal plant.

The field production technologies and uses are in many cases in the stage of investigation. But a lot can be learned by investigating the traditional way of *C. sativa* field production and uses. The tradition of false flax growing is still present in Slovenia.

MATERIALS AND METHODS

The widespread investigation of *C. sativa* growing and technologies used was conducted with the help of local extension service offices in the Koroška region of northern Slovenia. Since it is known that environmental conditions also affect the yield and quality, a pilot cultivation experiment with seeds of the local population from the southern part of the cultivation area was done in the experimental field of the institute in 1998 and 1999. In addition, the cultivar Ivan, preserved on one of the farms, was analyzed. Analysis of oil content and composition in the seeds was performed by gas chromatography according to standardized methods ISO 5508 and ISO 5509 by the laboratory of the Gea Oil Factory in Slovenska Bistrica.

RESULTS

There are still approximately 20 farmers maintaining *C. sativa* production. The number varies from year to year. Interest is growing because of introduction of sustainable or biological agricultural methods.

Field production of *C. sativa* is conducted on smaller plots (1500-3000 m^2). The extent of modern technology used differs from farmer to farmer. But there are still some farmers who are using the traditional way of cultivation which is very extensive. The field is prepared by ploughing the meadow or cleared areas in the woods. Choosing proper field locations in places with moderate air circulation helps to avoid possible problems with fungal diseases. The only fertilizer used is manure which is applied in the stage of soil preparation. Fertilizing with manure helps in ameliorating the soil properties and maintaining the optimal level of nutrients which should not be too high. The plots are relatively small. Early sowing in March or April with high density or broadcasting the seeds mixed with semolina helps in annual weed control. The seeds germinate earlier than annual weedy species. In the traditional practice there is no special treatment of the stands until harvest. The harvest takes place in July when the plants form typical heads with tiny seeds. The harvest begins when around 80% of the seed heads are yellow to light brown. The plants are harvested by cutting or pulling out of the ground. Harvested plants are left under a roof to dry. After that the plants are threshed, and the seeds are cleaned, dried and stored. The yields vary from 400 to 800 kg/ha. Seeds are kept in sacks and processed when needed. The traditional way of growing *C. sativa* has incorporated some practices to avoid problems with the crop and they contribute to the applicability for sustainable agriculture.

The main way of processing the seeds is the pressing of seeds for their oil. The seeds are milled in a coarse mass which is mixed with an equal volume of water, giving it a pasty appearance. The paste is roasted at 60-90°C, becoming "sandy." Then it is placed into a press and the oil is expressed. Oil is left to clear and then is filtrated through gauze. The seeds are processed shortly before selling. This traditional practice solves the problem of the oil's relatively short shelf life because of a high content of unsaturated fatty acids.

The oil produced from the seeds is partly used as an edible oil but most of it is used as a traditional home remedy. In folk medicine the oil is considered a good remedy for stomach and duodenal ulcers, the treatment of burns, wounds and eye inflammations. The oil is used in some cases of cancer and is considered a tonic. The demand is high and it is sold in local pharmacies. Some studies to standardize the procedures of identification and quality control were done (10). The content of the oil and its composition for two consecutive seasons from populations experimentally grown in the field of the institute are summarized in Table 1.

DISCUSSION

The use of *C. sativa* for self-medication is scarcely documented. *Hagers Handbook of Pharmaceutical Praxis* indicates some uses of the fresh herb or seeds in cataplasms and the herb against eye inflammations and cancer (8). Mentioning the traditional use of the oil as a remedy is also justified from an analytical point of view. A majority of the authors reviewed stress the unique composition regarding the high percentage of polyunsaturated fatty acids, especially the high content of α-linolenic acid (also known as Omega 3 acid). These properties of the oil make it suitable for direct self-medication use, con-

TABLE 1. Oil content and composition of *Camelina sativa* experimentally grown at the Institute of Hop Research and Brewing, 1998 and 1999. Results in relative area %.

	IHP 1998	IHP 1999	cv. Ivan
Oil	33.02%	31.83%	33.79%
C16:0	5.76	6.55	6.17
C18:0	2.59	2.74	2.73
C18:1	16.40	16.18	16.58
C18:2	18.73	19.18	16.93
C18:3	32.85	33.05	36.63
C20:1	15.83	15.50	14.32
C22:0	1.53	1.87	1.80
C22:1	3.02	1.58	1.85
C24:0	0.54	3.31	2.96

sumption and for health promoting food products (11,17,1). Also, the use of the oil in cosmetic products is indicated because of its healing properties.

The content of fatty acids is comparable to results from America (2) and results reviewed by Kuhlmann (7,9), but did not reach some parameters of new selected lines (16) and cultivars (1). The crop showed some of the problems of a nonstabilized cultivar. Uneven ripening and shedding of seeds were evident. Further study of local populations is required to identify the lines with favorable traits. Since *C. sativa* is a promising oil and medicinally interesting crop, a collection and testing of local seeds is planned.

Properties of the false flax oil from Koroška were investigated (10). The review and comparison with properties of the oil from French cultivars which were approved by their authorities for food use in France (1) is given in Table 2.

The properties of the Slovene false flax oil from local varieties are comparable to French oil but show variability and moderate differences in iodine, peroxide and saponification index. The content of tocopherols is higher in Slovene oil.

The tradition of cultivating and using *C. sativa* in Slovenia could be a good basis for reintroduction of practices of sustainable agriculture on the basis of farmers knowledge, new developments and research.

The important role in knowing and preserving this crop in Slovenia is played by farmers. On-farm conservation is a well established way of *in situ* conservation and management of crop biodiversity in Asia, Africa and Latin America (5). The need for *in situ* conservation of plants for food and agriculture in Europe was recognized and initiated at the Braunchweig conference in 1999 (4). The taskforce of IPGRI/GR elaborated the possibilities and recommendations for on-farm conservation in Europe at Isola Polvese in May 2000. *C. sativa* is an ideal model for introduction of on-farm conservation and management practices in Slovenia.

TABLE 2. Comparison of unrefined *Camelina sativa* oil characteristics.

Characteristics	Oil of Slovene origin	Oil of French origin
Acid index	1.7-3.9	1.8
Iodine index	148-150	142.0
Peroxide index	2.4-6.9	1.8
Saponification index	187-188	185.0
Total tocopherols (µg/g)	1151.82	900

REFERENCES

1. Bonjean A., F. Le Goffic, 1999. La cameline–*Camelina sativa* (L.) Crantz: Une oppotunité pour l'agriculture et l'industrie européennes. *OCL* 6(1), 28-33.

2. Budin J.T, M.V. Vreene, D.H. Putnam, 1995. Some compositional properties of *Camelina* (*Camelina sativa* (L.) Crantz) seeds and oils. *JAOCS* 72 (3), 309-315.

3. European Commission, 1995. Alternative oilseed crop–*Camelina sativa*. *In:* C. Mangan, B. Kreckow, M. Falnagan (eds.), *AIR Non-Food Projects*, 152-153.

4. Gass T., L. Frese, F. Begemann, E. Lipman (compilers), 1999. Implementation of Global Plan of Action in Europe–Conservation and Sustainable Utilization of Plant Genetic Resources for Food and Agriculture. Proceedings of the European Symposium, 30 June-3 July 1998, Braunschweig, Germany, International Plant Genetic Resources Institute, Rome, Italy.

5. Jarvis D., B. Sthapit, L. Sears (editors), 2000. *Conserving agricultural biodiversity in situ: A scientific basis for sustainable agriculture.* International Plant Genetic Resources Institute, Rome, Italy.

6. Knorzer K.H., 1978. Evolution and spread of Gold of Pleasure (*Camelina sativa* S.L.). *Berichte der Deutschen Botanischen Geselschaft* 91, 187-195.

7. Kuhlmann H., 1985. Untersuchungen über ertradgsleistung und Samenqualität von Leindotter (*Camelina sativa* (L.) Crantz). Dilomarbeit, Justus–Liebig Universität, Giessen.

8. List P.H., L. Hörhammer, 1972. *Hagers handbuch der pharmaceutischen praxis* vol. 3, 625-626, Springer Verlag, Berlin.

9. Marquard R., H. Kuhlmann, 1986. Investigations of productive capacity and seed quality of linseed dodder (*Camelina sativa* Crtz.). *Fette, Seifen Anstrichmittel* (Germany) 88, 245-248.

10. Pratnekar S., 1996. Fizikalne in kemijske lastnosti olja semen navadnega rička (*Camelina sativa* (L.) Crantz) = Physical and chemical properties of oil from seeds of *Camelina sativa* L. Crantz. B.Sc. Thesis, Pharmaceutical Faculty, University of Ljubljana, Slovenia.

11. Putnam D.H., J.T. Budin, L.A. Feild, W.M. Breene, 1993. Camelina: A promising low-input oilseed. *In:* J. Janock, J.E. Simon (eds.), *New Crops*, 314-322. Wiley, New York

12. Robinson R.G., W.W. Nelson, 1975. Vegetable oil replacement for petroleum oil adjuvants in herbicide sprays. *Econ. Bot.* 29, 146-151.

13. Robinson R.G., 1987. Camelina: A useful research crop and potential oilseed crop. *Minnesota Agr. Expt. Sta. Bul.*, 579.

14. Seehuber R., J. Vollmann, M. Dambroth, 1987. Application of single-seed-descent method in false flax to increase the yield level. *Landbauforschung Völkenrode* (Germany) 37, 132-136.

15. Tattersall A., S. Millam, 1999. Establishment and in vitro regeneration of the potential oil crop species *Camelina sativa*. *Plant, Tissue and Organ Culture* 55, 147-149.

16. Vollman J., A. Damboeck, A. Eckl, H. Schrems, P. Ruckenbauer, 1996. Improvement of *Camelina sativa*, an underexploited oilseed. *In*: J. Janick (ed.), *Progress in new crops*. ASHS Press, Alexandria, VA, USA.

17. Zubr J., 1996. New oilseed crop *Camelina sativa*. Book of Abstracts Third European Symposium on Industrial Crops and Products, 22-24, April 1996, Remis, France.

Biodiversity and Uses
of White Mustard (*Sinapis alba* L.),
Native to Israel,
as a Plant with Economic Potential

Zohara Yaniv
David Granot
Efraim Lev
Dan Schafferman

SUMMARY. Much interest has been expressed in the last several years in the possibility of developing new oilseed, medicinal and garden crops from wild Cruciferae, for human consumption and industrial uses. One of the promising species is *Sinapis alba* (white mustard) which is very common throughout Israel. *S. alba* is very well known in folk traditions in the Mediterranean area as a medicinal and spice plant. As a part of a general survey of Crucifer species in Israel, 280 seed accessions of *S. alba* were collected. In order to assess biodiversity within the species, several parameters were recorded, and it was found that oil quality–as expressed by fatty-acid profile–and other factors of agronomical impor-

Zohara Yaniv, David Granot, and Dan Schafferman are affiliated with the Department of Genetic Resources and Seed Research, Institute of Field and Garden Crops, ARO, The Volcani Center, P.O. Box 6, Bet Dagan 50250, Israel.

Efraim Lev is affiliated with the Department of Land of Israel Studies, Faculty of Jewish Studies, Bar-Ilan University, Ramat Gan, Israel.

[Haworth co-indexing entry note]: "Biodiversity and Uses of White Mustard (*Sinapis alba* L.), Native to Israel, as a Plant with Economic Potential." Yaniv, Zohara et al. Co-published simultaneously in *Journal of Herbs, Spices & Medicinal Plants* (The Haworth Herbal Press, an imprint of The Haworth Press, Inc.) Vol. 9, No. 4, 2002, pp. 319-327; and: *Breeding Research on Aromatic and Medicinal Plants* (ed: Christopher B. Johnson, and Chlodwig Franz) The Haworth Herbal Press, an imprint of The Haworth Press, Inc., 2002, pp. 319-327. Single or multiple copies of this article are available for a fee from The Haworth Document Delivery Service [1-800-HAWORTH 9:00 a.m. - 5:00 p.m. (EST). E-mail address: getinfo@haworthpressinc. com].

319

tance were affected by the site of origin. Cultivation of selected accessions under controlled conditions demonstrated the maintenance of biodiversity.

The level of genetic variability among eight accessions collected from two geographical locations was analyzed by RAPD (Random Amplified Polymorphic DNA), and a genetic distance between accessions from the two locations was found. Seeds from each accession were subsequently cultivated in three different climatic regions. After one year of cultivation, a diverging effect on the genetic polymorphism was observed.

The results indicate the importance of biodiversity conservation in the process of development and evaluation of germplasm of *S. alba* for the use of man. *[Article copies available for a fee from The Haworth Document Delivery Service: 1-800-HAWORTH. E-mail address: <getinfo@haworthpressinc. com> Website: <http://www.HaworthPress.com>* © 2002 by The Haworth Press, Inc. All rights reserved.]

KEYWORDS. Biodiversity, seed oils, Cruciferae, ethnobotany, RAPD

INTRODUCTION

Sinapis alba (*Brassica hirta, Brassica alba*, white mustard), a rape-seed relative currently grown mainly for condiment use, is very common throughout Israel. Seed accessions (280) were collected from the native flora in Israel and germplasm was evaluated (17). *S. alba* is known in both past and present folk traditions in the Mediterranean area as a medicinal and spice plant (14). In order to evaluate its ethnobotanical uses, a literary survey into old and present sources was conducted and is presented here. Biodiversity is a very important attribute in identifying traits that are useful and desirable for crop development.

The aim of the present study was to estimate the diversity of the genepool obtained by collecting accessions from the wild, and cultivating them. RAPD analysis was used for this purpose.

MATERIALS AND METHODS

Collection of specimens from the native flora. More than 70 different sites of *S. alba* habitats, throughout Israel, were identified and marked. Two hundred and eighty accessions were collected at full maturity. In most cases four different seed samples were taken from each site. Seeds were analyzed for fatty acid composition and accessions were selected for field observation.

Field observation trial. Seeds were sown at the Bet Dagan experimental station in small plots of 1.2 m², consisting of four rows with 30 cm between rows. Basic fertilization was conducted at the time of soil preparation, at rates of 100/100/50 N/P/K. Trifluralin was applied as a herbicide (2.5 kg/ha). During the experiment, seeds at full maturity were harvested manually, and the following information was recorded: days to flower (DTF), 1000-seed weight (SWT), seed yield/ha, oil content and fatty acid composition.

Field experiment at three geographical sites. Seeds of two *S. alba* lines (Bet Dagan and Nahariya) were sown at three sites: Bet Dagan, Jerusalem and the Golan Heights. These sites represent different geographical regions: Bet Dagan–Mediterranean climate, Jerusalem–higher elevation, Golan Heights–the highest and coldest site. Four replicates were tested for each line. Plot size and soil preparation were the same as previously described. Harvested seeds were oven dried and analyzed for oil content and fatty acid composition as described previously (17). Seeds from one collection from each region were individually grown and subjected to RAPD analysis.

RAPD analysis–Plant material. Several seed samples had previously been collected from two natural populations of *S. alba* located in two regions in Israel, Bet Dagan and Nahariya, which are approximately 120 km apart. Four accessions from each location were used for the current analysis (Figure 1).

DNA extraction. Seeds from each accession were grown and genomic DNA was isolated from each seedling separately, and prepared according to Doyle and Doyle (5).

DNA amplification. RAPD was carried out according to Yu and Pauls (18). The primers were random decamer oligonucleotides from Operon Technologies Inc. Nineteen Operon set-A and set-B primers were tested. The following 12 primers gave reproducible and informative marker patterns and were selected for the study: OPA-04, OPA-06, OPA-07, OPA-08, OPA-09, OPA-10, OPA-15, OPB-03, OPB-05, OPB-08, OPB-12, OPB-15. A 15-µl sample of each PCR reaction was run on a 1.4% agarose gel and visualized by ethidium bromide staining.

Data analysis. To increase the fidelity of the homology interpretation, bands on RAPD gels were scored as present or absent only within the same 36-lane gel. Nahariya and Bet Dagan populations were each represented by 13 plants (three accessions) in the comparison between the populations and by 20 plants (four accessions) in the cultivation analysis. The number of plants within each accession that shared a scored band varied between 0 and 5. These values were used for cluster analysis by Systat® software (12).

RESULTS AND DISCUSSION

Survey of literature regarding the ethnobotanical uses of S. alba. The survey was divided into two parts. In part one (Table 1), historical sources from

FIGURE 1. Accessions of *S. alba* collected in Israel.

Accession No.	Collection Site	% Erucic Acid
33		52.9
34	Bet Dagan	48.8
35		52.6
36		52.2
300		52.8
306	Nahariya	54.9
308		49.5
309		51.6

Relative genetic distance between *S. alba* accessions from two locations in Israel: Bet Dagan and Nahariya

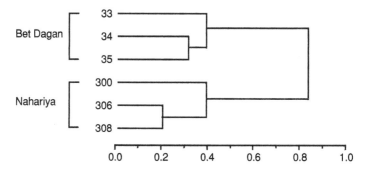

ancient times (centuries 9-7 BC) to the Middle ages were reviewed. References are cited in Table 1. In parallel, an ethnobotanical survey (Table 2) was conducted of the present-day uses of mustard. Relevant references are cited in Table 2.

Biodiversity among native S. alba accessions. Table 3 presents the natural variation in fatty-acid composition of 280 accessions of *S. alba* seed oils collected in Israel.

Agronomical and chemical variability of selected cultivated accessions. Selected accessions, chosen according to geographical distribution, were cultivated at the Bet Dagan experimental farm. Agronomical and chemical parameters were recorded (Table 4). There was great variability in agronomic characters. Days to flowering ranged from 82 to 114, a range of 32 days. Seed weights ranged from accessions with small seeds (2.1 g/1000 seeds) to those with large seeds (6.1 g/1000 seeds). Yield estimates ranged from 120 to 3600 kg/ha. However, chemical characteristics such as oil percentage (16.7-23.7%) and fatty-acid composition were much more uniform.

TABLE 1. Ancient medicinal uses of mustard.

Geographical location	Estimated periods (centuries)	Medicinal indications and uses
Assyria	9th-7th B.C.	swellings, jaundice, cough (13)
Babylon	7th-5th B.C.	swellings, jaundice, venereal diseases, lung problems (11)
Greece	4th-1st B.C.	digestive disorders, irritations (6)
Rome	1st-3rd A.D.	swellings of the tonsils, sciatica, leprosy (7)
Egypt	4th-13th A.D.	headache, flatulence, dyspepsia, hearing problems (13)
Iraq	10th-12th A.D.	flatulence, dyspepsia, leprosy, itching (11)
Levant	13th-16th A.D.	cough, jaundice, lung problems, pains, influenza, inflammations, rheumatism, skin diseases, fever (9)
Turkey	16th A.D.	pains, influenza, inflammations, rheumatism, skin diseases (1)

TABLE 2. Present day ethnobotanical uses of mustard.

Geographical location	Medicinal uses
Yemen	used as diuretic, analgesic, for arthritis, to treat disorder of menstruation and skin diseases (15)
Europe	to treat colds, tonsillitis, and as a sedative (4)
North Africa	used as a laxative, emetic, rubefacient and antirheumatic (2)
Iraq and Iran	used as an emetic and rubefacient (8)
Middle East	to treat various aches and inflammations (10)

The effect of geographical location on yield performance. Two selected lines of *S. alba* (Nahariya and Bet Dagan) were cultivated at three geographical locations: Bet Dagan, Jerusalem and the Golan Heights. Seed yield parameters were evaluated and the results are summarized in Table 5. The growing period was shortest (168-176 days) in Bet Dagan, and longer at cooler loca-

TABLE 3. Fatty acid composition of 280 accessions of *Sinapis alba* seed oils collected in Israel.

Fatty acid tested	Percentage of total	
	Mean	Variation[a]
Oleic acid (C18:1)	17.2	10.6-24.0
Linoleic acid (C18:2)	10.2	6.2-14.2
Linolenic acid (C18:3)	11.5	7.2-15.8
Eicosenoic acid (C20:1)	5.8	2.3-9.4
Erucic acid (C22:1)	47.7	36.6-58.8

[a] Variation is demonstrated with the lowest and the highest values observed in the population.

TABLE 4. Agronomical and chemical variability in selected accessions of *Sinapis alba* grown at Bet Dagan.

Parameter tested	Mean	Variation[a]
Days to flowering	98.4	82-114
Days to harvest	174.0	159-188
SWT/1000 (g)	4.2	2.1-6.1
Yield (kg/ha)	1,450	120-3,640
Oil content (%)	19.5	16.7-23.7
Fatty acids (% of total)		
Palmitic (C16:0)	3.0	2.6-3.4
Oleic (C18:1)	15.8	12.2-19.5
Linoleic (C18:2)	9.0	7.2-10.9
Linolenic (C18:3)	8.6	8.0-9.2
Eicosenoic (C20:1)	5.8	5.3-6.3
Erucic (C22:1)	50.8	46.2-55.4

[a] Variation is demonstrated with the lowest and the highest values observed in the population.

tions. In the Golan Heights, where the lowest growing season temperatures were registered, the growing period was 35 days longer.

Seed weight, too, is affected by the geographical location: there was an increase of 23-26% in the size of seeds maturing under the Golan Heights conditions. The highest yields were obtained at Bet Dagan. The line Bet Dagan produced the best yield: 5,330 kg/ha. The lowest yields of both lines tested

TABLE 5. Agronomic evaluation of two selected accessions of *Sinapis alba* in field trials at three sites (data are averages of four replicates).

Accession	Trial site[a]	Days to harvest	SWT/1000 (g)	Seed yield (kg/ha)	Oil content (%)	Erucic acid (%)
Nahariya	BD	176	3.9	3,690 a	24.0 b	52.2 a
	Jm	188	4.4	1,720 bc	20.0 c	50.6 ab
	GH	203	4.8	2,350 b	28.1 a	52.6 a
Bet Dagan	BD	170	4.4	5,330 a	24.4 ab	51.6 ab
	Jm	188	4.4	2,060 bc	19.8 b	49.3 b
	GH	203	5.4	2,670 b	26.6 a	52.9 a

[a] BD = Bet Dagan; Jm = Jerusalem; GH = Golan Heights.
Figures followed by a common letter do not differ significantly at $P < 0.005$ (Duncan's Multiple Range Test).

were produced at the Jerusalem site, a fact which cannot be explained only on the basis of differences in temperatures during seed-filling time; other factors, such as soil type, might be involved too. The highest oil percentages in the seeds (25-28% for both lines) were obtained in the Golan Heights, where the temperatures prevailing during seed maturation were coolest. The effect of lower temperatures on the oil content of seeds has been reported previously (3,16). The line Nahariya had the highest oil content. No significant differences were observed among the relative concentrations of erucic acid in both lines tested at the three geographical sites. During the critical stage of oil formation, temperatures at all three sites were high, so that a uniform level of 50-53% was obtained.

Diversity evaluation by RAPD analysis–comparisons between accessions from two locations in Israel, Bet Dagan and Nahariya. The comparison between the two locations was done with three accessions from each location. DNA was individually prepared from five seedlings of each accession. Each of the 30 DNA preparations was subjected to random DNA amplification (RAPD) with two different primers, OPA-04 and OPA-08. Twenty-two RAPD bands were obtained with the two primers, none of which was specific to any accession of the two collection sites. However, following statistical analysis, the accessions of Bet Dagan and Nahariya were found to cluster into two different groups (Figure 1), indicating a genetic distance between the two locations.

Effect of cultivation on the genetic polymorphism. The effect of cultivation on the genetic polymorphism was tested, following cultivation in the three climatic regions. Three samples, one from each region, were compared by RAPD with four samples from the original Bet Dagan collections. Five seedlings of each sample were tested. Forty-six RAPD bands were obtained with four different primers: OPA-04, OPB-03, OPB-05 and OPB-08. None of the bands was specific to any one of the samples. As shown in Figure 2, the cultivation in the two relatively high and cool climatic regions, Jerusalem and the Golan Heights, had a diverging effect.

A similar approach was taken with a Nahariya accession (Table 5), which was cultivated in the three climatic regions, Bet Dagan, Jerusalem and the Golan Heights, for one season. Three samples, one from each region, were compared with four samples from the original Nahariya accessions. Five seedlings of each sample were tested. Forty-one RAPD bands were obtained with four different primers: OPA-04, OPA-06, OPB-12 and OPB-15. None of the bands was specific to any one of the samples. As shown in Figure 2, the cultivated samples remained genetically closer to their parent. Nevertheless, cultivation in Jerusalem had a diverging effect.

In conclusion, use of the RAPD technique enabled us to show genetic distance between accessions of *S. alba* from two locations. In addition, cultivation in different climatic regions has a diverging effect which can be detected by RAPD.

FIGURE 2. Effect of cultivation of Bet Dagan and Nahariya accessions of *S. alba* on the genetic polymorphism.

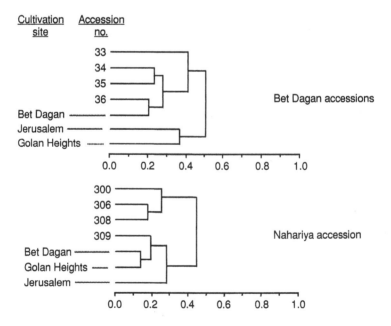

REFERENCES

1. al-Antaki, Daud. 1935. *Tadhkirat Ula li-'Lbab wa 'l-Jami al-Ujab*. (in Arabic) (The remembrance of the best of gates and the compilation of the great wonders). Bulaq, Cairo.

2. Boulos, L. 1983. *Medicinal Plants of North Africa*. Reference Publication Inc., Michigan.

3. Canvin, D.T. 1965. The effect of temperature on the oil content and fatty acid composition of the oils from several oilseed crops. *Can. J. Bot*. 43:63-69.

4. Dafni, A. 1984. *Eatable wild plants of Israel*. Society for Protection of Nature, Tel-Aviv.

5. Doyle, J.J. and Doyle, J.L. 1990. Isolation of plant DNA from fresh tissue. *Focus*. 12:13-15.

6. Grieve, M. 1994. *A Modern Herbal*. Tiger Books International, London.

7. Gunther, R.T. 1959. *The Greek Herbal of Dioscorides*. Hafner Publication Co., New York.

8. Hooper, D. 1937. *Useful plants and drugs of Iran and Iraq*. Field Museum of National History, Publ. 387, Chicago.

9. Ibn al-Baytar. 1874. *Kitab al-Jami li-Mufradat al-Adwiya wa-'l-Aghdhiya*. (in Arabic) (Compilation of simple drugs and foods). Bulaq, Cairo.

10. Lev, E. and Amar, Z. (in preparation). *Chomrei hamarpeh hamesoratim beEretz Israel beshilhei hamea ha-20*. (in Hebrew) (Traditional medical substances sold in markets of Jerusalem at the end of the 20th Century). Eretz, Tel-Aviv.

11. Levey, M. 1966. *The Medical Formulary of Aqrabadhin of al-Kindi*. The University of Wisconsin Press, Madison.

12. Manly, B.F.J. 1986. *Multivariate Statistical Analysis: A Primer*. Chapman and Hall, London.

13. Manniche, L. 1993. *An Ancient Egyptian Herbal*. British Museum Press, London.

14. Palevitch, D. and Yaniv, Z. 1991. *Tzimhei hamarpeh shel Eretz Israel* (in Hebrew) (Medicinal plants of the Holyland). Tamus Modan Press, Tel-Aviv.

15. Tzapar, S. 1991. *Refuot Teiman*. (in Hebrew) (Yemenite medicines). Tzapar Press, Jerusalem.

16. Yaniv, Z., Ranen, C., Levy, A. and Palevitch, D. 1989. Effect of temperature on the fatty acid composition and yield of evening primrose (*Oenothera lamarkiana*) seeds. *J. Exp. Bot*. 40:609-613.

17. Yaniv, Z., Schafferman, D., Elber, Y., Ben-Moshe, E. and Zur, M. 1994. Evaluation of *Sinapis alba*, native to Israel, as a rich source of erucic acid in seedoil. *J. of Ind. Crops*. 2:137-142.

18. Yu, K. and Pauls, K.P. 1992. Optimization of the PCR program for RAPD analysis. *Nucleic Acids Research*. 20:2606.

Variation of the Arbutin Content in Different Wild Populations of *Arctostaphylos uva-ursi* in Catalonia, Spain

Irene Parejo
Francesc Viladomat
Jaume Bastida
Carles Codina

SUMMARY. The crude drug of *Arctostaphylos uva-ursi* consists mainly of three groups of pharmaceutically relevant compounds, arbutin being the main phenolic constituent. In this study, four wild populations of *A. uva-ursi* were evaluated for their arbutin content in two different seasons, spring and autumn, coinciding with two periods of the vegetative cycle. The obtained results showed that autumn was a better period than spring to collect the plant material, and that leaves of *A. uva-ursi* from the site called Adraén Alt were found to contain the highest concentration of arbutin (9.1% of dry weight). No significant variations in

Irene Parejo, Francesc Viladomat, Jaume Bastida, and Carles Codina are affiliated with the Department of Natural Products, Plant Biology and Edaphology, Faculty of Pharmacy, University of Barcelona. Av. Diagonal, 643, 08028 Barcelona, Catalonia, Spain.

This work was financed by the LIFE programme of the European Commission (project ENV/E/000260).

[Haworth co-indexing entry note]: "Variation of the Arbutin Content in Different Wild Populations of *Arctostaphylos uva-ursi* in Catalonia, Spain." Parejo, Irene et al. Co-published simultaneously in *Journal of Herbs, Spices & Medicinal Plants* (The Haworth Herbal Press, an imprint of The Haworth Press, Inc.) Vol. 9, No. 4, 2002, pp. 329-333; and: *Breeding Research on Aromatic and Medicinal Plants* (ed: Christopher B. Johnson, and Chlodwig Franz) The Haworth Herbal Press, an imprint of The Haworth Press, Inc., 2002, pp. 329-333. Single or multiple copies of this article are available for a fee from The Haworth Document Delivery Service [1-800-HAWORTH 9:00 a.m. - 5:00 p.m. (EST). E-mail address: getinfo@haworthpressinc. com].

329

the arbutin content were observed within any of the studied natural sites.

KEYWORDS. Arbutin, *Arctostaphylos uva-ursi*, bearberry, Ericaceae, HPLC, geographical variations, seasonal variations

INTRODUCTION

Leaves of *Arctostaphylos uva-ursi* (L.) Sprengel (Ericaceae), also known as bearberry, are used as a urinary antiseptic and an astringent (1) because of their content of arbutin (hydroquinone-β-D-monogluco-pyranoside). This crude drug has been found to contain levels of arbutin of 5.95% (2) and 7.16% (4). The variations within and among natural populations of bearberry, however, have not been reported.

In this work several wild populations of bearberry have been evaluated for their arbutin content in two periods of the vegetative cycle, spring and autumn, to check possible variations depending on the collection time and the geographical location.

MATERIALS AND METHODS

Plant Material. Leaves of *Arctostaphylos uva-ursi* were collected in four natural populations growing at different altitudes in the Catalan Pyrenees of Spain. Table 1 shows geographical and some ecological data concerning the different characteristics of every natural site.

The ontogenic cycle of the bearberry is characterized by two growth periods interrupted by a summer dormancy. The plant begins to sprout from the end of April till the beginning of May, and the second growth period takes place from August to September (3). The plant material was collected in autumn 1997 and spring 1998. For each geographical site, samples from 5 different areas of the population were collected in order to check possible variations in the arbutin content.

Analytical Method. The levels of arbutine were determined by HPLC. Fifty milligram of dried powdered plant material was extracted with 25 ml of a mixture of water and methanol (95/5) at 25°C in an ultrasound bath for 30 minutes. After centrifugation at 10,000 rpm for 10 minutes, the supernatant was adjusted to 25 ml in a measuring flask. Every sample was extracted by triplicate,

TABLE 1. Geographical characteristics of the four studied wild populations of *Arctostaphylos uva-ursi.*

Population	Altitude (m)	Slope	Exposition	Subtrate	Tree occurrence	*A. uva-ursi* density
Adraén Alt	1740-1770	10-20%	NW	Calcareous	0-20% *Pinus sylvestris* *P. uncinata*	60-80%
Adraén Baix	1580-1640	10-40%	NW	Calcareous	20-30% *P. sylvestris*	80-100%
Cloterons	2000-2030	20-45%	S-SE	Calcareous	0-10% *P. uncinata*	80-100%
Guils de Cerdanya	1760-1770	0-15%	NE	Siliceous	50-60% *P. uncinata*	15-20%

and for each replicate two vials were filled, which were each injected twice. Therefore, the values obtained for each sample are the means of 12 analyses. Samples were quantified immediately after extraction to avoid possible chemical alterations. Blanks and standards of known arbutin concentrations were placed between the samples to monitor the quantification.

The quantitative analyses were carried out by RP-HPLC using a C8 column (20 $\times$ 3.9 mm) and a C8 pre-column (250 $\times$ 4 mm) packed with Nucleosil (5 µm; Teknokroma, Barcelona, Spain). Mobile phase was water/methanol (95/5), previously filtered through 0.45 µm filter (Millipore). Other parameters: flow rate: 1 ml/minute; UV detection: 280 nm; and injection volume: 10 µl. The retention time of arbutin was 3.7 minutes. The concentration of arbutin was determined using a calibration curve established with 6 dilutions of standard arbutin (cristalline arbutin $C_{12}H_{16}O_7$, Sigma-Aldrich), from 25 to 400 µg/ml. The correlation coefficient was r = 0.99997. Methanol and water used for both extraction and chromatographic separation were of HPLC grade (Scharlau). The method offers a good reproducibility ($\pm$2.8%, calculation based on the peak of arbutin during the period of analysis) and precision (RSD = 0.34%).

Statistical Analysis. The results were subjected to the statistical analysis of variance (Statistics for Windows 4.0, ANOVA two ways) to assess the observed differences in the arbutin levels between the collection period and the geographical sites. Differences were considered to be statistically significant when *p* value was less than 0.05. Then individual means were compared by using a Student-Newman-Keuls test. The coefficients of variations were also calculated to check the uniformity of the natural populations.

RESULTS AND DISCUSSION

Period of collection. The levels of arbutin corresponding to each wild population are shown in Table 2. In general, the arbutin content was higher in autumn than in spring, and the differences were found to be statistically significant. These differences were much higher in the sites called Adraén Alt and Adraén Baix, and lower in Cloterons. Comparing the levels of arbutin of each wild population between the two seasons, only the site called Guils de Cerdanya (the only site with a siliceous substart, and with the lowest bearberry density) showed no significant difference.

Differences among populations. The content of arbutin in the different sites was not the same in the two periods of collection (Table 2). Thus, while in spring the highest levels of arbutin were found in plants of the Cloterons site (8.05%), the lowest ones corresponded to the Adraén Baix site (6.30%). Nonetheless, in autumn the highest arbutin content was found in the site called Adraén Alt (9.10%) and the lowest in the Guils de Cerdanya one (6.97%).

After applying the Student-Newman-Keuls test, two groups of populations (Adraén Alt and Cloterons) can be statistically separated from the other two (Adraén Baix and Guils de Cerdanya) on the base of their arbutine content. According to these results, the Adraén Alt and Cloterons sites are the best to collect the bearberry leaves, independently of the period of collection.

Differences within populations. To evaluate the possible uniformity of every studied wild population in relation to the arbutin content, the coefficients of variation were calculated separately for the two periods of collection (Table 3).

TABLE 2. Levels of arbutin in leaves of bearberry collected in different time collections and wild populations. Values are expressed as percentage of dry weight, and represent the mean of 12 analyses ± SD.

Season	Sample	Adraén Alt	Adraén Baix	Cloterons	Guils de Cerdanya
Autumn					
	1	9.03 ± 0.16	7.49 ± 0.16	8.83 ± 0.12	6.58 ± 0.12
	2	8.84 ± 0.13	7.84 ± 0.06	8.62 ± 0.29	7.02 ± 0.13
	3	9.14 ± 0.06	8.25 ± 0.08	8.72 ± 0.22	6.91 ± 0.13
	4	9.11 ± 0.11	7.88 ± 0.12	8.69 ± 0.07	7.42 ± 0.09
	5	9.39 ± 0.12	8.60 ± 0.13	8.70 ± 0.11	6.93 ± 0.05
	$\bar{x}$	9.10 ± 0.20	7.99 ± 0.43	8.71 ± 0.07	6.97 ± 0.30
Spring					
	1	7.20 ± 0.20	6.19 ± 0.18	7.45 ± 0.20	6.57 ± 0.11
	2	6.00 ± 0.11	6.40 ± 0.12	7.87 ± 0.10	6.45 ± 0.07
	3	6.21 ± 0.04	6.25 ± 0.11	7.96 ± 0.07	7.24 ± 0.07
	4	7.48 ± 0.15	6.26 ± 0.08	8.41 ± 0.37	7.47 ± 0.08
	5	6.59 ± 0.11	6.39 ± 0.15	8.55 ± 0.21	6.61 ± 0.05
	$\bar{x}$	6.69 ± 0.63	6.30 ± 0.09	8.05 ± 0.44	6.87 ± 0.45

TABLE 3. Coefficients of variation (%) corresponding to the arbutin content within each natural site and collection time.

	Adraén Alt	Adraén Baix	Cloterons	Guils de Cerdanya
Autumn	2.40	4.96	1.73	3.86
Spring	9.93	1.43	5.47	6.55

These coefficients varied from 1.43% (Adraén Baix in Spring) to 9.42% (Adraén Alt in spring). In general, the different natural sites can be considered quite homogeneous, as the coefficients of variation were lower than 10%. The different populations seemed to be more uniform in autumn than in spring in what concerns the levels of arbutin (Table 3).

In summary, the analytical determination of arbutin in leaves of bearberry indicates that the arbutin content depends significantly on the season of the year. The levels of arbutin observed in the studied populations (from 6.3% to 9.1%) were found to be quite significant in relation to previous reports. On the basis of the obtained results, the collection of bearberry leaves should be carried out preferably during the autumn period because of their higher arbutin content. This concentration, however, can be different according to the characteristics of the sites where the plants grow.

REFERENCES

1. Karikas, G.A., M.R. Euerby, and R.D. Waigh. 1987. Isolation of piceoside from *Arctostaphylos uva-ursi*. *Planta Med.* 53: 307-308.

2. Matsuda, H., M. Higashino, Y. Nakai, M. Iinuma, M. Kubo, and F.A. Lang. 1996. Studies of cuticle drugs from natural sources. IV. Inhibitory effects of some *Arctostaphylos* plants on melanin biosynthesis. *Biol. Pharm. Bull.* 19: 153-156.

3. Jahodár, L., M. Sovová, and P. Klemera. 1986. The effect of ionizing irradiation on the growth of *Arctostaphylos uva-ursi* and the production of arbutin. *Folia Pharmaceutica* 10: 69-76.

4. Konndler, F., C. Schewer, D. Fritsche, and M. Pölin. 1990. Determination of arbutin in uvae-ursi folium (bearberry) by capillary zone electrophoresis. *J. Chromatogr.* 514: 383-388.

The Chemical Variability
of *Ocimum* Species

Hans Krüger
Sabine B. Wetzel
Bärbel Zeiger

SUMMARY. The composition of the essential oils of samples from the *Ocimum* gene bank collection (about 270 accessions) of the Institute of Plant Genetics and Crop Plant Research Gatersleben was investigated. The main components in the essential oils found were linalol (max. 71%), methyl chavicol (max. 92%), citral (max. 80%) and 1,8 cineole (max. 25%) as well as camphor (max. 63%), thymol (max. 35%), (E)-methyl cinnamate (max. 77%), eugenol (max. 80%), methyleugenol (max. 79%), methyl isoeugenol (max. 36%) and elemicin (max. 47%). *[Article copies available for a fee from The Haworth Document Delivery Service: 1-800-HAWORTH. E-mail address: <getinfo@haworthpressinc.com> Website: <http://www.HaworthPress.com> © 2002 by The Haworth Press, Inc. All rights reserved.]*

KEYWORDS. *Ocimum basilicum*, basil, chemotypes, essential oil, chemotaxonomy

Hans Krüger, and Bärbel Zeiger are affiliated with Bundesanstalt für Züchtungsforschung an Kulturpflanzen, Institut für Qualitätsanalytik, Neuer Weg 22-23, 06484 Quedlinburg, Germany.

Sabine B. Wetzel is affiliated with the Institut für Pflanzengenetik und Kulturpflanzenforschung, Corrensstr. 3, 06466 Gatersleben, Germany.

[Haworth co-indexing entry note]: "The Chemical Variability of *Ocimum* Species." Krüger, Hans, Sabine B. Wetzel, and Bärbel Zeiger. Co-published simultaneously in *Journal of Herbs, Spices & Medicinal Plants* (The Haworth Herbal Press, an imprint of The Haworth Press, Inc.) Vol. 9, No. 4, 2002, pp. 335-344; and: *Breeding Research on Aromatic and Medicinal Plants* (ed: Christopher B. Johnson, and Chlodwig Franz) The Haworth Herbal Press, an imprint of The Haworth Press, Inc., 2002, pp. 335-344. Single or multiple copies of this article are available for a fee from The Haworth Document Delivery Service [1-800-HAWORTH 9:00 a.m. - 5:00 p.m. (EST). E-mail address: getinfo@haworthpressinc.com].

It is known that there is a big morphological and biochemical variability in the genus *Ocimum* (1,2). Specially for *Ocimum basilicum* L. there have been many attempts to classify it into chemotypes. The following chemotypes have been described: linalol, methyl chavicol, citral and 1,8-cineole. But further work is required to describe the wide chemotaxonomical range. Material which is reproduced in gene banks under comparable conditions allows the separation of genetic factors from other influences such as production technologies, physiology of development, and environment. This method aims to better recognize genetic impacts, e.g., the composition of the essential oils. In addition, it excludes a possible hybridogenic nature of new types.

The *Ocimum* gene bank collection (about 270 accessions) of the Institute of Plant Genetics and Crop Plant Research Gatersleben was investigated. All samples originated from a cultivation in the experimental garden of the institute in 1999. Sample preparation and GC: 500 mg air-dried leaves of each were homogenized in 4 mL isooctane containing carvacrol as internal standard. After centrifugation the solutions were investigated by GC (Hewlett Packard Series II). GC conditions: column: 50 m × 0.32, HP 5; injector temperature: 250°C; detector temperature: 280°C; temperature program: 80° (1 min), 80°-220° (10°/min), 220° (9 min); carrier gas: nitrogen, 1 mL/min (constant flow); injection: 1 µL; split: 1:40.

The composition of the essential oils of all samples are shown in Appendix. The main components in the essential oils found were linalol (max. 71%), methyl chavicol (max. 92%), citral (max. 80%) and 1,8-cineole (max. 25%) as well as camphor (max. 63%), thymol (max. 35%), (E)-methyl cinnamate (max. 77%), eugenol (max. 80%), methyleugenol (max. 79%), methyl isoeugenol (max. 36%) and elemicin (max. 47%). The oil content in the dry leaves was found to vary from traces to 2.65%.

REFERENCES

1. Junghanns, W., Hammer, K. (1994). Vergleichende Untersuchungen bei Ocimum-Herkünften. *Herba Germanica*, 2, 92–94.

2. Svoboda, K.P., Kyle, S.K., Sinclair, W., Hampson, J.B., Asakawa, Y. (1999). Investigation of Essential Oil Composition and DNA Fingerprinting of *Ocimum Basilicum*. 30th International Symposium on Essential Oils, Leipzig 1999, Abstract 1-02.

APPENDIX. Composition of the essential oils of the Gatersleben *Ocimum* collection.

Gene bank number	[ml/100g air-dried leaves]	Ocimen	1,8-Cineol	γ-Terpinen	Linalol	Camphor	Methyl cavicol	Neral	Geraniol	Geranial	Thymol	(Z)-Methyl cinnamate	Eugenol	(E)-Methyl cinnamate	Methyl-eugenol	β-Elemen	α-Bergamoten	β-Caryophyllen	Methyl isoeugenol	β-Bisabolen	n.d. RT 17,484	Elemicin
												Oil composition (% in oil)										
Oa 1	0.17		17.03		11.67	1.47					00.00		36.43		12.33	1.06	3.92+					
Oa 8	0.64	0.20	23.40				7.47	·					40.65				1.09	1.59		7.35	7.15	
Oa 25	0.13		6.28		25.12	1.27	6.60						40.24		4.92	1.22	6.29					
Oa 29	0.37				25.48				0.93			7.77		49.42		1.99	2.21	6.57				
Oa 39	0.37		4.62		52.76	0.37	25.29								4.58	0.44	4.22	2.88			1.48	
Oa 58	0.41						87.89								0.90			2.33			2.01	
Oa 59	0.50						88.26								1.11			6.81				
Oa 65	0.56		0.38		46.01							5.04		34.75		1.65						
Oa 66	0.18		3.25		63.49	1.23							10.21		7.81		2.87+	7.29				
Oa 74	0.31		0.68		30.12		8.38					5.26	2.93	36.16		2.05	0.50	4.87				
Oa 75	0.34				29.36							7.90	1.78	46.81		1.65						
Oa 86	0.13		10.05		15.06		50.53								8.96		2.64+	2.31				
Oa 88	0.10						74.00								4.35	1.14	3.69+	1.51				
Oa 89	0.42						90.36								0.84						0.87	
Oa 91	0.35						91.76								1.18		1.27+	1.39			1.20	
Oa 92	0.28						89.96								1.15							
Oa 94	0.31		3.51		6.63		72.43					2.13				1.56	3.93+	3.31				
Oa 96	0.31		2.11		28.37		21.55					9.49		17.63		3.74	4.95	4.55				
Oa 97	0.26				10.62		0.80							66.96		1.31		5.43				
Oa 98	0.30		3.04		64.14		15.85						0.60			1.33		6.56				
Oa 99	0.28		4.93		39.38		31.77					0.50		8.00		1.50		7.49				
Oa 103	0.28		24.62				10.90						31.66		4.33		1.60	1.64		10.01	7.43	
Oa 104	0.28						88.45								4.13			1.64			1.42	
Oa 105	0.23						86.48								4.43			1.31			1.15	
Oa 108	0.36						89.60											1.28			1.22	
Oa 110	0.35		23.66				9.50						37.84				1.22	1.53		7.82	6.66	
Oa 117	0.16		14.99		26.08								11.38		29.65		6.03+					
Oa 123	0.17		5.42		2.88	1.93	72.10										6.27+	2.52				
Oa 125	0.31		3.93		9.90	1.49	3.59					4.37		59.90		1.20	3.56+	2.62				
Oa 126	0.13		17.86		3.52		10.39						14.40		31.86		7.16+					
Oa 127	0.22		17.83		1.24	0.52	33.44						5.73		16.35	0.80	2.58					
Oa 128	0.21		6.62		1.17	1.01	76.27								0.76		4.79+					
Oa 129	0.14		8.74		15.14	1.47	59.41						1.73		1.97		2.34					

APPENDIX (continued)

Oil composition (% in oil)

Gene bank number	[ml/100g air-dried leaves]	Ocimen	1,8-Cineol	γ-Terpinen	Linalol	Camphor	Methyl cavicol	Neral	Geraniol	Geranial	Thymol	(Z)-Methyl cinnamate	Eugenol	(E)-Methyl cinnamate	Methyl-eugenol	β-Elemen	α-Bergamoten	β-Caryophyllen	Methyl isoeugenol	β-Bisabolen	n.d. RT 17.484	Elemicin
Oci 131	0.68		0.56		20.07		71.55	0.44	0.47				0.23									
Oci 147	0.10		13.25		10.53	1.27							47.30		9.48		1.29+	1.13				2.83
Oci 155	0.45		24.65				9.15						36.64				1.36	1.68		8.23	7.46	
Oci 156	0.10		3.37		2.29	1.17	79.83										5.49					
Oci 157	0.12		5.67		16.87	3.27	62.90									0.99		4.09				
Oci 158	0.14		6.11		16.86	2.93	61.81											4.80				
Oci 162	0.18		5.84		45.34		2.99						23.19		4.50	1.49	7.45					
Oci 163	0.12		10.94		24.44		46.12						1.65		4.65	1.20	3.26+					
Oci 169	0.31		9.51		18.00	0.61						3.18	9.13	41.72	6.11	0.67	2.99+					
Oci 170	0.16		5.03		14.11	1.69	52.86					0.86		14.49			3.71+				1.77	
Oci 171	0.29		0.43		3.32			34.22		44.19							0.82	1.52				
Oci 174	0.19		17.21		26.16	0.91	1.73						33.58		2.29	0.96	6.65					
Oci 175	0.13		8.49		38.48								34.37		6.82		6.76					
Oci 177	0.35		10.19		28.76	0.41						2.78	17.88	26.31	0.69	1.00	1.63+					
Oci 178	0.31		24.81				9.40						36.91				1.25	1.53		7.99	6.59	
Oci 179	0.20				3.83			35.50		44.38							0.84	1.26			1.49	
Oci 184	0.20				3.93		2.14	34.73		44.02							0.77	1.21			1.52	
Oci 187	0.08		6.42		46.50		9.51						27.11			2.51	7.12					
Oci 191	0.31		23.57					34.01		43.15					15.15		1.35	1.61		8.54	7.15	
Oci 194	0.32		13.84		33.70	1.26	73.21						37.44			0.58	5.13					
Oci 195	0.12		6.36		5.39	3.09	38.15						23.23			1.12	4.5+					
Oci 196	0.25		3.14		35.69	0.69	37.71		0.46						0.79	1.77	0.92					
Oci 201	0.34		10.14		8.57		71.35						7.73	8.67	26.97	0.74	6.76					
Oci 202	0.08		2.62		6.43	1.35	13.89						0.81				3.93+					
Oci 203	0.36		10.52		21.91	0.65						2.96	2.20	34.24	0.97	0.91	1.22+					
Oci 212	0.60		4.09		0.37	1.64	84.35									0.49	5.37					
Oci 213	0.27		21.00				7.70						45.74		2.84		1.03	1.36		6.81	6.24	
Oci 214	0.13		2.84		14.31	1.38	66.38										4.03+					
Oci 215	0.24				3.25	0.50	1.99			43.15						0.78	0.83	1.66			1.69	
Oci 217	0.44		7.79		14.43	0.70	41.03	34.01				4.87	1.30	59.96	0.63		1.56+					
Oci 218	0.38		6.42		1.78		39.87								36.68	1.10	2.19	3.40			1.21	
Oci 219	0.13		6.88		21.53								3.26		16.27		7.07					
Oci 223	0.17		5.49		54.86	0.73	27.53		1.37						0.86	1.44						

APPENDIX. Composition of the essential oils of the Gatersleben *Ocimum* collection.

Gene bank number	ml/100g air-dried leaves	Ocimen	1.8-Cineol	γ-Terpinen	Linalol	Camphor	Methyl cavicol	Neral	Geraniol	Geranial	Thymol	(Z)-Methyl cinnamate	Eugenol	(E)-Methyl cinnamate	Methyl-eugenol	β-Elemen	α-Bergamoten	β-Caryophyllen	Methyl isoeugenol	β-Bisabolen	n.d. RT 17.484	Elemen
Oci 1	0.17		17.03		11.67	1.47					00.00		36.43		12.33	1.06	3.92+					
Oci 8	0.64	0.20	23.40										40.65		4.92		1.09	1.59		7.35	7.15	
Oci 25	0.13		6.28		25.12	1.27	6.60						40.24			1.22	6.29	6.57				
Oci 29	0.37				25.48				0.93			7.77		49.42		1.99	2.21					
Oci 39	0.37		4.62		52.76	0.37	25.29								4.58	0.44	4.22					
Oci 58	0.41						87.89								0.90			2.88			1.48	
Oci 59	0.50						88.26								1.11			2.33			2.01	
Oci 65	0.56				46.01							5.04		34.75		1.65		6.81				
Oci 66	0.18		3.25		63.49	1.23							10.21		7.81		2.87+					
Oci 74	0.31		0.68		30.12							5.26	2.93	36.16		2.05		7.29				
Oci 75	0.34				29.36		8.38					7.90		46.81		1.65	0.50	4.87				
Oci 86	0.13		10.05		15.05		50.53						1.78		8.96	1.14	2.64+					
Oci 88	0.10						74.00								4.35		3.69+					
Oci 89	0.42						90.36								0.84			2.31				
Oci 91	0.35						91.76								1.18		1.27+	1.51			0.87	
Oci 92	0.28						89.36								1.15			1.39			1.20	
Oci 94	0.31		3.51		6.63		72.43					2.13				1.56	3.93+	3.31				
Oci 96	0.31		2.11		28.37		21.55					9.49		17.63		3.74	4.95	4.55				
Oci 97	0.26				10.62		0.80							66.96		1.31		5.43				
Oci 98	0.30		3.04		64.14		15.85						0.60			1.33		6.56				
Oci 99	0.28		4.93		39.38		31.77					0.50		8.00		1.50		7.49				
Oci 103	0.28		24.62				10.90						31.66				1.60	1.64		10.01	7.43	
Oci 104	0.28						88.45								4.33			1.64			1.42	
Oci 105	0.23						86.48								4.13			1.31			1.15	
Oci 108	0.36						89.60								4.43			1.28			1.22	
Oci 110	0.35		23.66				9.50						37.84		29.65		1.22	1.53		7.82	6.66	
Oci 117	0.16		14.99		26.08								11.38				6.03+					
Oci 123	0.17		5.42		2.88	1.93	72.10					4.37					6.27+	2.52				
Oci 125	0.31		3.93		9.90	1.49	3.59							59.90		1.20	3.56+	2.62				
Oci 126	0.13		17.86		3.52		10.39						14.40		31.86		7.16+					
Oci 127	0.22		17.83		1.24	0.52	33.44						5.73		16.35		2.58					
Oci 128	0.21		6.62		1.17	1.01	76.27								0.76	0.80	4.79+					
Oci 129	0.14		8.74		15.14	1.47	59.41						1.73		1.97		2.34				

337

Oil composition (% in oil)

Gene bank number	[ml/100g air-dried leaves]	Ocimen	1,8-Cineol	γ-Terpinen	Linalol	Camphor	Methyl cavicol	Neral	Geraniol	Geraniol	Thymol	(Z)-Methyl cinnamate	Eugenol	(E)-Methyl cinnamate	Methyl-eugenol	β-Elemen	α-Bergamoten	β-Caryophyllen	Methyl isoeugenol	β-Bisabolen	n.d. RT 17.484	Elemicin
Oci 131	0.68		0.56		20.07		71.55	0.44		0.47			0.23				1.29+	1.13			2.83	
Oci 147	0.10		13.25		10.53	1.27							47.30		9.48		4.39+					
Oci 155	0.45		24.65				9.15						36.64				1.36	1.68		8.23	7.46	
Oci 156	0.10		3.37		2.29	1.17	79.83										5.49					
Oci 157	0.12		5.67		16.87	3.27	62.90									0.99		4.09				
Oci 158	0.14		6.11		16.86	2.93	61.81											4.80				
Oci 162	0.18		5.84		45.34		2.99						23.19		4.50	1.49	7.45					
Oci 163	0.12		10.94		24.44		46.12						1.65		4.65	1.20	3.26+					
Oci 169	0.31		9.51		18.00	0.61						3.18	9.13	41.72	6.11	0.67	2.99+					
Oci 170	0.16		5.03		14.11	1.69	52.86	34.22		44.19		0.86		14.49			3.71+	1.52			1.77	
Oci 171	0.29		0.43		3.32		1.73									0.96	0.82					
Oci 174	0.19		17.21		26.16	0.91							33.58		2.29		6.65					
Oci 175	0.13		8.49		38.48								34.37		6.82		6.76					
Oci 177	0.35		10.19		28.76	0.41						2.78	17.88	26.31	0.69	1.00	1.63+					
Oci 178	0.31		24.81				9.40	35.50		44.38			36.91				1.25	1.53		7.99	6.59	
Oci 179	0.20				3.83												0.84	1.26			1.49	
Oci 184	0.20				3.93		2.14	34.73		44.02							0.77	1.21			1.52	
Oci 187	0.08		6.42		46.50		9.51						27.11			2.51	7.12	1.61		8.54	7.15	
Oci 191	0.31		23.57										37.44		15.15	0.58	1.35					
Oci 194	0.32		13.84		33.70	1.26	73.21						23.23			1.12	5.13					
Oci 195	0.12		6.36		5.39	3.09	38.15									1.77	4.5+					
Oci 196	0.25		3.14		35.69	0.69	37.71		0.46				7.73		0.79	0.74	0.92					
Oci 201	0.34		10.14		8.57	1.35	71.35						0.81		26.97		6.76					
Oci 202	0.08		2.62		6.43	0.65	13.89							8.67			3.93+					
Oci 203	0.36		10.52		21.91	1.64	84.35					2.96	2.20	34.24	0.97	0.91	1.22+					
Oci 212	0.60		4.09		0.37		7.70						45.74			0.49	5.37	1.36		6.81	6.24	
Oci 213	0.27		21.00				66.38										1.03					
Oci 214	0.13		2.84		14.31	1.38	1.99								2.84		4.03+	1.66			1.69	
Oci 215	0.24				3.25	0.50		34.01		43.15		4.87		59.96	0.63	0.78	0.83					
Oci 217	0.44		7.79		14.43	0.70	41.03						1.30			1.10	1.56+					
Oci 218	0.38		6.42		1.78		39.87								36.68		2.19	3.40				
Oci 219	0.13		6.88		21.53		27.53						3.26		16.27		7.07				1.21	
Oci 223	0.17		5.49		54.86	0.73				1.37					0.86	1.44						

Sample	C1	C2	C2b	C3	C4	C5	C6	C7	C8	C9	C10	C11	C12	C13	C14	C15
Oc 231	0.15	5.11		47.24	1.30	34.58					1.70	1.02	2.94+			
Oc 233	0.12	8.96		29.52		28.90		6.80			12.63	0.98	7.21			
Oc 234	0.15	9.83		32.08	0.71	22.89		14.05			0.90	0.54	5.46			
Oc 237	0.35	7.14		56.27	1.17	19.46		0.52			3.41		4.74			
Oc 246	0.14	10.39		11.25	0.94	35.70		0.65			28.18		4.92			
Oc 247	0.13	2.36		51.70		15.35		5.06		5.90	7.06		3.44+			
Oc 243	0.13	3.43		23.78		60.63						0.80	5.48			
Oc 250	0.41	9.65		18.19	0.63			2.97	5.51	51.17	0.63	0.79	1.68+			
Oc 263	0.44	8.94		17.00	0.93	5.03		4.35	5.52	43.65	0.81		3.17+			
Oc 265	0.12	4.31		23.49	2.91	58.04							4.53			6.41
Oc 257	0.52	23.97				7.57		39.71					1.08	1.48	7.07	
D 7658	0.25	7.55	0.21	23.81	1.07	6.39		5.52	3.33	35.24	3.16	0.58	3.54+			
Feld 2	0.58	10.89		0.68		34.80					40.78		6.57	0.85		
Feld 3	0.15	11.82		33.38		9.19		32.11				2.54	0.98+			
Feld 4	0.12	5.93		38.34		3.62		17.29				5.67	1.76+			
Feld 5	0.14	15.24		52.52		12.99		14.46				1.16	3.22			
Feld 6	0.26	8.72		46.90	0.52			13.91				0.77	3.92			
Feld 7	0.28	11.16		50.60	0.51	31.41		17.57				1.01	4.97			
Feld 8	0.24	4.79		41.84		57.37		13.44				0.55	2.19			
Feld 9	0.46	4.47		28.92		3.48		0.62				0.87	1.37			
Feld 10	0.30	6.89		49.19		49.28		20.25				1.40	6.91			
Feld 11	0.51	9.77		0.84		30.86		0.33			28.06	0.21	5.90			
Feld 12	0.34	4.92		40.25	0.94	55.00		5.91				1.23	4.93			
Feld 13	0.32	5.46		26.02		1.28		1.89				0.71	1.80			
Feld 14	0.15	5.90		53.15		57.43		29.46				0.94	2.27			
Feld 15	0.25	10.28		15.60	1.00	0.85						1.15	0.73+			
Feld 16	0.22	4.67		51.89			1.39	28.53				1.31	2.68			
Feld 17	0.24	7.85		51.82	0.98	11.87		14.91				2.24	3.41			
Feld 18	0.33	4.56		50.22	0.45			30.16				0.80	5.98			
Feld 19	0.33	6.29		46.70	0.36	38.81		20.82				1.06	2.15			
Feld 20	1.37	3.23	0.24	39.83		56.42	0.52					1.16	2.68			
Feld 21	0.38	11.75		18.51		20.79						1.36				
Feld 22	0.12	4.88		44.02	0.46	48.72		14.95			26.50	0.98	4.10			
Feld 23	0.64	10.62		1.08		55.00						0.50	6.05			
Feld 24	0.46	6.26		40.87		43.29	1.59					1.10	2.14			
Feld 25?	0.17	3.79		40.68		27.83						1.02	1.34			
Feld 25?	0.32	8.60		37.40		38.48		11.97				0.58	3.75			
Feld 26	0.20	6.33		36.79		42.12		7.11					3.35			
Feld 27	0.28	5.43		34.24				5.00				0.99	2.55			
Feld 28	0.28	13.81		33.00		61.85		27.58				1.20	6.24	0.73		
Feld 29	0.33	3.55		24.75								0.85	2.31			

APPENDIX (continued)

Oil composition (% in oil)

Gene bank number	[ml/100g air-dried leaves]	Ocimen	1,8-Cineol	γ-Terpinen	Linalol	Camphor	Methyl cavicol	Neral	Geraniol	Geranial	Thymol	(Z)-Methyl cinnamate	Eugenol	(E)-Methyl cinnamate	Methyl-eugenol	β-Elemen	α-Bergamoten	β-Caryophyllen	Methyl isoeugenol	β-Bisabolen	n.d. RT 17.484	Elemicin
Feld 30	0.38		8.76		43.58	0.30	12.11						21.46			0.80	2.71					
Feld 31	0.51		9.97		0.61		47.38						0.35		29.86		5.93					
Feld 32	0.09		6.02		67.15		1.99						6.89				4.52					
Feld 33	0.46		3.35		25.64	0.46	60.40						0.33			0.51	4.12					
Feld 34	0.17		4.95		40.68	0.66	1.36					4.29	6.98	27.92		1.51	2.57					
Feld 36	0.33		9.60		26.69	1.00	30.80						9.08			0.80	7.78					
Feld 37	0.27		4.37		34.80		49.03									0.66	3.09					
Feld 38	0.32		7.78		53.18								23.21			0.71	6.11					
Feld 39	0.26		3.79		24.19		62.46									0.60	2.56					
Feld 41	0.27		5.50		53.09		2.83						20.97			1.22	3.42					
Feld 42	0.67		8.49		0.58		65.44								14.78	0.23	5.67					
Feld 43	0.16		4.71		29.77		57.75						0.73			0.83	2.81					
Feld 44	0.28		5.25		35.53		44.02						2.66			0.53	3.46					
Feld 45	0.43		5.73		29.49		54.68						0.30			1.40	4.61					
Feld 46	0.25		9.79		35.34	0.87	20.44						13.70			0.93	3.47					
Feld 47	0.51		8.27		18.67	0.72	57.36									1.77	11.26					
Feld 48	0.23		4.80		27.54	1.57	22.33						11.13			1.52	11.27					
Feld 49	0.27		6.81		30.58	1.44	14.41						12.49			0.68	2.80					
Feld 50	0.39		3.86		28.44		54.48						3.03				2.40					
Feld 52	0.14		5.17		32.52	0.56	57.21						0.80									
Feld 53	0.25		3.01		42.78	1.79							38.64			2.23		1.69				
Feld 54	0.07		15.26		36.12	0.85	62.43						22.57			1.42	6.03					
Feld 55	0.41		7.63		21.33	0.53										0.92						
Feld 56	0.20		12.06		40.64			0.45	15.48	0.34			26.47			1.32	4.80	0.77				
Feld 57	0.25		9.94		56.00				8.46	0.75						1.78	1.15	0.69				
Feld 59	0.34		15.78		58.14	0.34										1.04	1.08					
Feld 60	0.17		10.46		47.64		47.81						24.62		28.34	0.69	6.37					
Feld 61	0.51		9.86		0.84		5.17						0.33			0.28	6.61					
Feld 62	0.31		7.70		45.97	0.78							25.62			1.01	2.87					
Feld 63	0.30		11.89		40.55	0.41							26.41			0.78	7.30					
Feld 64	0.22		8.04		50.70								19.74			1.21	6.32					
Feld 65	0.22		4.48		52.69								28.37			1.27	2.66					
Feld 66	0.22		14.37		35.51	1.57							25.93			0.75	8.11					

Feld	(1)	(2)	(3)	(4)	(5)	(6)	(7)	(8)	(9)	(10)	(11)	(12)	(13)
Feld 67	0.17		1.03	68.96		8.57		7.29				1.44		4.13
Feld 68	0.19		12.61	35.62				32.14						5.15
Feld 69	0.36		9.60	39.53	0.81			32.28				0.75		4.89
Feld 70	0.44		11.19	18.87	0.68	51.93						0.96		4.80
Feld 73	0.17		15.91	37.04	0.60			25.87			21.27	1.15		6.15
Feld 75	0.54		9.41	0.64		56.57		0.31				0.22		4.42
Feld 79	0.18		1.91	42.42		47.53						0.72		6.37
Feld 80	0.12		9.29	38.09				34.06				1.11		5.19
Feld 81	0.08		16.84	41.09				24.72						
Feld 82	0.15		10.77	51.11				18.82				1.97		3.04
Feld 83	0.37		12.27	39.82				28.11				0.61		5.48
Feld 85	0.21		1.70	56.45		13.48	1.14					0.89		8.60
Feld 86	0.15		13.27	41.31				22.81				0.76		9.17
Feld 87	0.20		7.82	33.93	0.68	46.32						0.62		3.80
Feld 88	0.16		11.29	43.35				27.27				0.70		6.60
Feld 89	0.38		1.11	60.67		0.48	13.16					5.65		
Feld 90	0.14		3.74	28.87	1.82			37.11				4.03		6.78
Feld 91	0.38		2.09	61.72	1.64		8.86	1.02				1.43		1.11
Feld 92	0.24		4.97	65.74				15.07				0.80		4.09
Feld 93	0.44		10.48	1.12		40.86					35.56	0.36		5.34
Feld 94	0.22		9.47	35.22	0.72	43.19								3.75
Feld 95	0.50		7.16	38.10	0.48	43.19		1.57				0.61		2.29
Feld 96	0.18		7.42	53.47				22.30				0.66		4.86
Feld 97	0.11		9.81	41.17	0.29			27.75						8.69
Feld 98	0.13		11.02	39.02	2.18			29.27				0.88		9.56
Feld 99	0.21		11.36	43.08	0.58		1.15	22.58				1.75		4.48+
Feld 100	0.19		2.52	52.02				29.47				0.92		6.50
Feld 105	0.48		10.01	0.71		46.12		0.38	5.00		30.39	0.36		5.98
Feld 108	0.41		9.82	14.97	1.96			15.28		34.76	0.51		1.56	5.61
Feld 109	0.35		4.94	23.16	1.10	32.67		17.07			4.79			5.63
Feld 110	0.64		4.35	0.35	1.47	83.33					0.84	0.59		5.34
Feld 112	0.36	0.21	12.67	1.00	0.31	9.17		1.44	10.00		61.50	0.37		7.88
Feld 113	0.57		10.67	3.47		0.50			9.02	62.55				1.32
Feld 115	0.53		11.76	5.28	0.26	0.83			10.57	58.65				0.92+
Feld 116	0.40		8.83	1.29	0.29	0.88			9.79	65.74				0.79+
Feld 117	0.62		12.23	5.58		0.23			10.90	58.29				1.64
Feld 118	0.37		8.89	1.08		0.82			10.20	65.33				0.93+
Feld 119	0.71		12.57	9.55					9.88	55.49				2.01
Feld 120	0.70		9.37	3.12	0.32	0.55				64.11				1.24
Feld 121	0.47		12.54	6.70	0.24	0.68		0.97	9.89	54.23		0.22		0.66+
Feld 122	0.44		6.70	0.60		2.39			11.36	66.72				2.98

APPENDIX (continued)

Gene bank number	[ml/100g air-dried leaves]	Ocimen	γ-Terpinen	1,8-Cineol	Linalol	Camphor	Methyl cavicol	Neral	Geraniol	Geranial	Thymol	(Z)-Methyl cinnamate	Eugenol	(E)-Methyl cinnamate	Methyl-eugenol	β-Elemen	α-Bergamoten	β-Caryophyllen	Methyl isoeugenol	β-Bisabolen	n.d. RT 17.484	Elemicin
											Oil composition (% in oil)											
Feld 124	0.66			13.32	10.33	0.23						9.62		53.67	0.59		2.00					
Feld 125	0.52			9.65	0.75		57.38							0.54	19.37		6.02					
Feld 126	0.66			9.76	7.04	0.33	1.13					9.86	0.51	57.52		0.26	2.73					
Feld 127	0.56			10.43	2.16	0.29	0.76					10.01		61.92			0.72+					
Feld 130	0.13			2.20	17.47	2.04	21.59					6.18		38.92			1.33+					
Feld 131	0.34			1.13	35.52	0.44	43.52		1.21				2.33				0.92	1.06				
Feld 132	0.15			1.94	19.55	1.85	32.60					4.74		29.62		2.45	1.75					
Feld 135	0.40			7.80	1.37		1.74					12.17		64.92		0.28	2.72					
Feld 136	0.21			18.04	15.78	1.54			1.14				41.30		1.66		7.02					
Feld 139	0.46			10.55	0.85		41.74										6.76					
Feld 141	0.18			7.31	23.85		51.14								33.21	0.98						
Feld 147	0.36			0.84	3.94			32.33		46.12							1.25	2.98				
Feld 148	0.18			2.74	71.02	0.77	12.66						1.12				3.98					
Feld 149	0.44			0.87	3.85	1.94		32.77		46.46							1.21	2.88				
Feld 150	0.16			2.71	29.57		60.05					8.23	3.63	39.34			3.31+					
Feld 152	0.52			9.66	0.47										18.42		5.90					
Feld 155	0.64			1.15	34.97			17.66		25.90							4.66	2.99				
Feld 157	0.43			10.71	0.77		36.07								39.36		7.17					
Feld 158	0.13			10.54	40.10								24.20		2.03	1.79	4.37+					
Feld 159	0.09			4.31	16.04	2.69	63.23									1.18	2.91+					
Feld 160	0.17			5.06	11.39	1.97	67.23									1.30	3.65					
Feld 161	0.12			3.82	50.42		22.91									1.20	2.08					
Feld 162	0.07			6.09	34.35		33.72						3.53		4.05							
Feld 163	0.11			6.01	42.95	1.29	8.60						5.26			1.49	5.35					
Feld 164	0.07			7.64	4.37		66.58						25.73				6.07	3.92				
Feld 165	0.12			6.02	10.21	2.49	64.81									1.32	5.25					
Feld 167	0.10			6.82	8.81	1.32	68.11									2.09	0.81					
Feld 169	0.18			2.38	2.69		78.29								1.07	1.19	4.84+					
Feld 176	0.58			9.11	0.98		59.52								19.66		5.21					
Feld 180	0.09			4.24	34.24	1.41	46.55										4.40					
Feld 187	0.26			7.05	22.05		55.88									0.92	6.84					
Feld 188	0.17			9.04	42.88	0.92			0.81				23.75		2.03	1.19	3.54+					
Feld 189	0.13			9.80	37.69		19.59						12.61			1.54	7.34					

Feld 190	0.44		9.65		0.78		44.81						0.32		32.99		5.69					
Feld 191	0.24		7.15		39.68	0.89			1.09				31.36		1.05	1.55	3.53+					
Feld 192	0.26		9.46		38.44				0.87				31.04		0.90	1.52	3.26+					
Feld 193	0.17		8.59		43.48	1.16			0.79				24.04		0.96	1.65	4.08+					
Feld 194	0.16		10.79		42.18	0.90			1.11				20.56		2.09	1.47	3.94+					
Feld 196	0.51		10.16		0.72		42.16						0.29		33.65	0.25	6.49					
Feld 197	0.30		15.75				8.77						46.80				1.41	1.59		8.14	7.54	
Feld 205	0.28		14.98				8.95						48.30				1.35	1.61		7.70	7.51	
Feld 206	0.55				0.54										55.19	4.34		29.58				
Feld 209	0.56		8.81		0.82		61.93								18.72		4.43+					
Feld 213	0.39				47.87							5.53	0.49	31.75		1.09	0.56	6.46				
Feld 216	0.45				48.48							6.34	0.45	33.46		1.34	0.49	2.98				
Feld 218	1.23		0.86		14.78		73.49						0.23				1.96+	1.50			4.95	
Feld 219	0.74				0.88							8.66		77.02			5.69	5.72				
Feld 222	0.31		2.71		44.98		2.48					4.48	0.50	28.84		1.55	5.26	3.39				
Feld 229	0.28						91.81								0.96			1.94				
Feld 230	0.49		10.33		0.97		37.82						0.45		38.36		5.81+					
Feld 234	0.37				3.79			32.79		46.81							1.20	2.93				
Feld 128	0.15		10.20		39.89				1.90				26.95				4.64					
Feld 236	0.38		4.05	0.89	19.46	2.71							4.60			7.65	8.87					
Feld 237	0.46		4.19	0.81	20.02	2.56							4.69			6.80	10.05	0.51				
Feld 246	1.61		2.12		67.21		6.06						3.20			3.94	3.09					
Feld 251	0.32				0.51										59.48	4.04		25.05				
Feld 252	0.29	4.81	18.21		5.43								13.65		4.05	8.99		17.18				
Feld 254	0.27										1.50		54.55		1.54	12.97		27.50				
Feld 257	0.82			28.52	0.85						33.44		0.27			0.35	0.58	3.20				
Feld 258	0.70				2.04							7.38	0.22	75.66			5.86	5.86				
Feld 260	0.26										1.02		54.54	0.45		12.70		26.76				
Feld 262	0.96	6.24	2.34										46.76		12.58	0.89		12.55		13.59		
Feld 270	0.52	0.22	9.57		0.86		51.57						0.71		25.18		6.61+					
Feld 273	0.52				0.58		0.28								56.10	5.58		25.14				
Feld 274	0.19	4.60	19.86		3.96								10.45		2.85	12.90		16.19				
Feld 275	2.65		11.03	0.39		62.86												0.67				
Feld 277	0.41		7.65		0.56		74.96								7.81		3.76+					
Oci 38	0.50	17.06			0.69								0.35		11.90	1.43	0.44	4.94				47.51
Oci 145	0.07				1.63						7.29		62.55			9.09		17.73				
Oci 42	1.43			34.34	0.61						31.42					0.28	0.38	3.44				
Oci 53	0.72		4.20	0.31	44.31	0.56		0.24	0.33	0.43	1.04		19.33			3.10	2.01					
Oci 60	0.44		4.31		28.38		55.37				2.74		0.43			0.92	1.37					
Oci 73	0.47		3.73	0.26	41.89	0.53		0.22	0.73	0.37			31.52			1.32	4.89					
Oci 100	1.11		2.30	0.46	35.69	0.80							4.22			6.00	16.12	0.44				

Oil composition (% in oil)

Gene bank number	[ml/100g air-dried leaves]	Ocimen	1.8-Cineol	γ-Terpinen	Linalol	Camphor	Methyl cavicol	Neral	Geraniol	Geranial	Thymol	(Z)-Methyl cinnamate	Eugenol	(E)-Methyl cinnamate	Methyl-eugenol	β-Elemen	α-Bergamoten	β-Caryophyllen	Methyl isoeugenol	β-Bisabolen	n.d. RT 17.484	Elemicin
Ods 115	1.71	9.80											78.57									
Ods 151	0.33	26.30	1.07		0.67								0.86					1.95				
Ods 154	0.58	26.60	1.09		0.61												10.07	3.30	34.54			
Ods 166	0.28		6.62		33.59	0.96				0.46			31.60				10.84	2.67	33.23			
Ods 167	0.74		2.16	0.25	56.00			0.24	0.80		5.51		23.11			2.42	8.57					
Ods 176	0.36		9.65		40.48	1.04			0.87				23.91		0.55	1.69	2.26+	0.90				
Ods 182	1.36			28.21	1.20						35.43		0.25			0.31	1.81+	3.36				
Ods 183	0.38		4.23		53.33	0.39					1.65		17.87			2.48	0.21					
Ods 222	1.30			29.66	1.34						34.08		0.24			0.28	2.46+	1.85				
Ods 256	0.31		6.40		46.63	0.58				1.08	1.15		14.22			3.92	4.25+					
Ods 258	0.78		2.75	0.45	46.47								13.06			4.99	2.7+					
Ods 101	1.27		2.34	0.46	41.26	0.90							3.68			7.64	14.04					
Ods 111	0.43												51.54			15.14		28.63				
Ods 122	1.86	11.48											75.92					2.27				
Ods 137	0.17												2.00		79.29	3.39	2.55	11.18				
Ods 139	0.24												0.76		77.12	6.66		13.59				
Ods 152	0.06										2.53		61.15			12.14		22.30				
Ods 153	0.40	8.96			0.43								0.32			0.73	11.59	3.46	36.06			
Ods 189	1.95	25.83											80.38					2.27				
Ods 211	0.53	25.34	1.12		0.41								0.52			1.29	10.89	3.15				
Ods 257	0.39		19.31		6.57	1.47			0.71				18.92		25.04		5.04		34.05			
Ods 269	0.90												62.10			11.82		22.90				
D 7654	1.63	9.93											78.45					2.29				

Phenotypic Variation
in *Hypericum perforatum* L.
and *H. maculatum* Crantz
Wild Populations in Lithuania

Jolita Radusiene
Edita Bagdonaite

SUMMARY. The investigations described the variation in morphological and chemical characters of *Hypericum* spp. wild populations. The data was subjected to multivariate statistical analysis to determine inter-population variability. The results of quantitative analysis (HPLC) of phenolic compounds in leaves and flowers are presented. The results indicated that wild populations of both *Hypericum* species are potentially important sources for breeding and improvement of the cultivated varieties. *[Article copies available for a fee from The Haworth Document Delivery Service: 1-800-HAWORTH. E-mail address: <getinfo@haworthpressinc.com> Website: <http://www.HaworthPress.com> © 2002 by The Haworth Press, Inc. All rights reserved.]*

Jolita Radusiene and Edita Bagdonaite are affiliated with the Laboratory of Economic Botany, Institute of Botany, Zaliuju ezeru 49, LT-2021 Vilnius, Lithuania.

The authors are very grateful to Professor Tur-Henning Iversen from the Biocentre of Trondheim University in Norway for granting permission for the performance of HPLC analysis.

[Haworth co-indexing entry note]: "Phenotypic Variation in *Hypericum perforatum* L. and *H. maculatum* Crantz Wild Populations in Lithuania." Radusiene, Jolita, and Edita Bagdonaite. Co-published simultaneously in *Journal of Herbs, Spices & Medicinal Plants* (The Haworth Herbal Press, an imprint of The Haworth Press, Inc.) Vol. 9, No. 4, 2002, pp. 345-351; and: *Breeding Research on Aromatic and Medicinal Plants* (ed: Christopher B. Johnson, and Chlodwig Franz) The Haworth Herbal Press, an imprint of The Haworth Press, Inc., 2002, pp. 345-351. Single or multiple copies of this article are available for a fee from The Haworth Document Delivery Service [1-800-HAWORTH 9:00 a.m. - 5:00 p.m. (EST). E-mail address: getinfo@haworthpressinc.com].

345

KEYWORDS. *Hypericum perforatum, Hypericum maculatum,* wild populations, phenotypic variation, phenolic compounds

INTRODUCTION

Hypericum perforatum L. and *H. maculatum* Crantz are common species in Lithuania. In Lithuanian folk medicine, *H. perforatum* is known as a remedy against 99 diseases. Worldwide, *H. perforatum* is the best known herb being touted as an antidepressant. Meanwhile, *H. maculatum* has been known and used less as raw material; there have been fewer research studies of it. Studies of both species of *Hypericum* could help reshape the antidepressant market by providing herbal non-prescription remedies. According to the World Health Organization, Lithuania tops Europe's suicide rate (7). Depression could be among the main reasons for this phenomenon.

Only one Russian variety, 'Zolotodolinskaja', has been known so far in Lithuania. In order to make successful use of wild material in breeding, it is important to study the interpopulation variation. The object of the present work is to investigate the phenotypic variation of *Hypericum* based on morphological and reproductive traits as well as on quantity composition of phenolic compounds. The study is conducted under the framework of the National Plant Genetic Resources Conservation Programme (1998-2002).

MATERIALS AND METHODS

In 1998-1999, 15 populations of *H. perforatum* and 10 of *H. maculatum* were investigated in the phase of full flowering. A total of 20 morphological characters were measured in 30 randomly selected individuals from the wild. The spectrophotometric method was used for the determination of the total amount (percentage) of flavonoids. Quantitative analysis of phenolic compounds in the metanolic extracts of flowers and leaves was carried out by modified HPLC gradient elution method (2).

The data was analyzed using the SPSS/PC+ computer package (2). Differences among populations were tested by one-way analysis of variance (ANOVA). The Scheffe's test was employed to identify significantly homogeneous groups among populations. The dendrograms of hierarchical cluster analysis of the populations were constructed on the grounds of agglomerative grouping by the method of the average linkage between groups using squared Euclidean distances. The relationship between morphological characters and quantity of chemical compounds was assessed using regression analysis.

RESULTS AND DISCUSSION

Morphological variation. The one-way analysis of variance (ANOVA) revealed highly significant differences ($p < 0.001$) among *Hypericum* populations within all measured traits. Peak values of the F statistic in *H. perforatum* were observed for the height of plants, length of sepals, weight of raw material, width of raceme, parameters of leaves, and density of dark glands; in *H. maculatum*, for the height of plants and the parameters of petals.

A dendrogram of hierarchical cluster analysis provides evidence for the existence of two groups among the studied *H. perforatum* populations on the basis of morphological characters (Figure 1). The dendrogram manifested the distinct character of certain populations, too. The special case of the 99EB07 population could be explained by the largest weight of plants, their parts and the largest production of raw material. The examined populations of *H. maculatum* can also be divided into two groups (Figure 2). Highest plants and biggest production of biomass mark the individual character of population 99EB10.

The most differentiating character in the populations of both species was productivity of biomass. The germplasm of these populations will be under research and conservation *in situ* and *ex situ*.

According to the previous data (6), *H. perforatum* exhibits a wide range on its ecological adaptation scale. All populations of *H. perforatum* were grouped according to their growing habitats. On the basis of cluster analysis, the similarity of populations in different habitats according to their morphological and productivity characters was defined. The dendrogram shows great pheno-

FIGURE 1. Average linkage (between groups) cluster analysis dendrogram of *H. perforatum* populations based on the mean phenotypic characters.

```
                        Rescaled Distance Cluster Combine
                0         5        10        15        20        25
                +---------+---------+---------+---------+---------+
Population Case
98JR10       2   -+-------------+
98JR13       3   -+             +---------------+
99EB13      13   ---------------+               I
99EB01       6   ---+-------+                   I
99EB03       8   ---+       +-------+           +-----+
99EB09      11   -----------+       +---+       I     I
99EB05       9   -----------+-------+   I       I     I
99EB14      14   -----------+           +-------+     I
99JR22       4   ---------+---+         I             +---+
99EB15      15   ---------+   +-------+ I             I   I
98JR24       5   -------------+       +-+             I   +-------+
99EB11      12   ---------------------+               I   I       I
99EB02       7   -------------------------------------+   I       I
99EB04       1   -----------------------------------------+       I
99EB07      10   -------------------------------------------------+
```

typical similarity of populations in different communities of *Agropyretalia repentis* and *Trifolio-Geranietea sanguinei* (Figure 3). The vitality of plants and the production of raw material is highest in these communities.

May the phenotypic plasticity of *Hypericum* be understood as adaptation to the different conditions? We treat the modifying effect of the environment as an important issue. Therefore, it can be expected that the differences observed among populations have a genetic character, as the populations from different habitats have indicated great similarity. The genetic character of phenotypic variation was noted in previous works (4). The obtained results pointed to the importance of investigations and selection of populations under uniform conditions in the field.

Chemical variation. Analysis of variance was performed on phenolic compounds in four groups: leaves and flowers of *H. perforatum* as well as leaves and flowers of *H. maculatum*. ANOVA revealed that all compounds differed significantly among groups, except for total amount of flavonoids (Table 1).

FIGURE 2. Average linkage (between groups) cluster analysis dendrogram of *H. maculatum* populations based on the mean phenotypic characters.

```
                        Rescaled Distance Cluster Combine
                 0         5        10        15        20        25
                 +---------+---------+---------+---------+---------+
Population  Case
99JR15       2   -+---------+
99EB12      10   -+         I
99EB04       6   -+-+       +-----------------+
99EB08       8   -+ +---+   I                 I
99EB06       7   ---+   +---+                 +-------------------+
98EB03       1   -------+                     I                   I
98JR17       3   -------+---------+           I                   I
98JR30       5   -------+         +-----------+                   I
98JR22       4   -----------------+                               I
99EB10       9   -------------------------------------------------+
```

FIGURE 3. Average linkage (between groups) cluster analysis dendrogram of *H. perforatum* habitat types based on the mean phenotypic characters.

```
                                  Rescaled Distance Cluster Combine
                               0         5        10        15        20        25
                               +---------+---------+---------+---------+---------+
Habitat type             Case

Trifolio-Geranietea        3   -+---+

Agropyretalia repentis     5   -+    +---+

Festuco-Brometea           4   -----+   +---------------------------------------+

Epilobium angustifolii     1   ---------+                                       I

Festuco-Sedetalia          2   -------------------------------------------------+
```

TABLE 1. One-way ANOVA for the chemical compounds in leaves and flowers in *H. perforatum* and *H. maculatum*.

Character	Mean square		F	Significant level
	between groups	within groups		
Caffeic acid	0.725	0.053	13.676	0.000
Chlorogenic acid	8.143	2.066	3.942	0.019
Hyperoside	113.519	24.843	4.569	0.010
Quercetin	5.000	0.376	13.282	0.000
Quercetrin	1.680	0.262	6.421	0.002
Rutin	1155.878	121.904	9.480	0.000
Total flavonoids	4.435	3.673	1.207	0.325

Degrees of freedom (df) between groups equals 3; within groups equals 27.

A post hoc test allowed us to specify differences of compounds among certain groups. Scheffé's range tests allowed the selection of homogeneous groups within the data set analyzed. Mean quantities of caffeic acid and quercetin in the flowers of *H. maculatum* stood apart from other groups ($p < 0.001$), forming a separate group. Amount of chlorogenic acid and total concentration of flavonoids did not differ significantly between species or between plant parts, and created a single homogeneous group. The amount of hyperoside and quercetrin in *H. maculatum* leaves and *H. perforatum* flowers differed significantly at $p < 0.05$. The amount of rutin depended on the species and the plant part ($p < 0.05$). Rutin concentration of both species in either leaves or flowers formed separate homogeneous groups (Figure 4).

Wild populations of *H. perforatum* are more efficient in accumulating secondary metabolites compared with cultivated 'Zolotodolinskaja'. A broad range of constituents is totally consistent with the previous reports on *H. perforatum* (2). The data presented by other authors (1,3) is quite discrepant. The disparity of the metabolite concentrations in published data to a great extent could be caused by the mismatch of the blossoming phase, extraction method, modification of HPLC method, and equipment.

Two stepwise regression analysis procedures showed that the length of leaves was the main factor influencing the variation of flavonoid concentration in both species (Figure 5). Relationship between flavonoid concentration and other morphological characters was quite weak. Qualitative bivariate regression model showed that the flavonoid concentration was discriminated mostly by the population affiliation of the particular *Hypericum* species. On average,

FIGURE 4. Mean values of compound concentration in leaves and flowers of *H. perforatum and H. maculatum* wild populations.

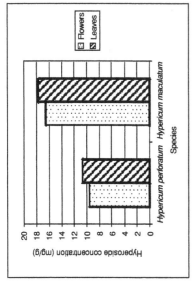

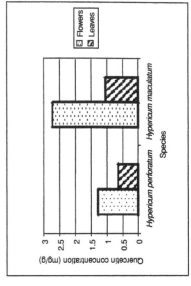

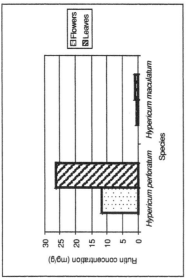

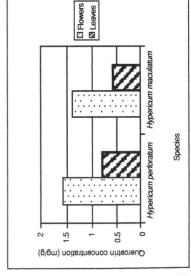

FIGURE 5. Regression of flavonoid concentration (percentage) against leaf length of *Hypericum*.

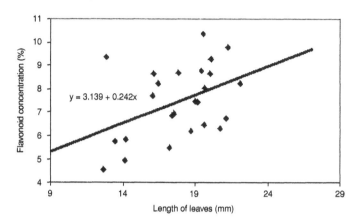

the flavonoid concentration was 2.76 percentage points higher in *H. maculatum* populations than in *H. perforatum* ones.

One can expect that wild populations of both species of *Hypericum* are a potentially important source of genetic variation for the improvement of the cultivated material.

REFERENCES

1. Bombardelli, E. and P. Morazzoni. 1994. *Hypericum peforatum. Fitoterapia* 66:43-68.

2. Hölzl, J. and E. Ostrowski. 1987. Johanniskraut *(Hypericum perforatum* L.). HPLC–Analyse der wichtgen Inhaltsstoffe und deren Variabilität in einer Population. *Deutsche Apotheker Zeitung* 127:1227-1130.

3. Martonfi, P. and M. Repcak. 1994. Secondary metabolites during flower ontogenesis of *Hypericum peforatum* L. *Zahradnictvi* 21:7-44.

4. Martonfi, P., R. Brutovska, E. Cellarova and M. Repcak. 1996. Apomixis and hybridity in *Hypericum perforatum. Folia Geobotanica Phytotaxonomia* 31:389-396.

5. Norusis, M. 1989. SPSS/PC+ V2.0. Base manual. Chicago.

6. Radusiene, J. and E. Bagdonaite. 2000. *Hypericum perforatum* L. augimviecių ekologine bei fitocenologine charakteristika (Ecological and phytocenological characterization of *Hypericum perforatum* L. habitats). *Botanica Lithuanica* (in print).

7. *The Economist.* 1998. The pain of being set free, August 29, p. 30.

Variation for Agronomic and Essential Oil Traits Among Wild Populations of *Chamomilla recutita* (L.) Rauschert from Central Italy

Paola Taviani
Daniele Rosellini
Fabio Veronesi

SUMMARY. In Italy, Chamomile is widely used for flower-head infusion but is almost totally imported. Wild populations are easily found, especially in disturbed soils and as weeds of cereal crops. Eleven chamomile populations were collected in Central Italy with the purpose of estimating the diversity and potential economic value of wild germplasm from this area. The eleven populations, together with two previously col-

Paola Taviani, Daniele Rosellini, and Fabio Veronesi are affiliated with the Department of Plant Biology and Agro-Environmental Biotechnology–Section Genetics and Breeding, Borgo XX giugno, 74, 06121 Perugia, Italy.

The authors are grateful to Dr. F. Scartezzini (Istituto Sperimentale per l'Assestamento Forestale e l'Alpicoltura, Trento, Italy) for his advice on essential oil distillation, and Prof. C. Bicchi (Dipartimento Scienza e Tecnologia del Farmaco, University of Torino, Italy) for gas chromatography analysis.

Funding was provided by the Italian Ministry of Agricultural and Forestry Policies, project "Incremento della Produzione di Piante Officinali," 1999.

[Haworth co-indexing entry note]: "Variation for Agronomic and Essential Oil Traits Among Wild Populations of *Chamomilla recutita* (L.) Rauschert from Central Italy." Taviani, Paola, Daniele Rosellini, and Fabio Veronesi. Co-published simultaneously in *Journal of Herbs, Spices & Medicinal Plants* (The Haworth Herbal Press, an imprint of The Haworth Press, Inc.) Vol. 9, No. 4, 2002, pp. 353-358; and: *Breeding Research on Aromatic and Medicinal Plants* (ed: Christopher B. Johnson, and Chlodwig Franz) The Haworth Herbal Press, an imprint of The Haworth Press, Inc., 2002, pp. 353-358. Single or multiple copies of this article are available for a fee from The Haworth Document Delivery Service [1-800-HAWORTH 9:00 a.m. - 5:00 p.m. (EST). E-mail address: getinfo@haworthpressinc.com].

353

lected wild populations from North Italy, the Slovak variety Bona and an Italian selection (Syn1) were compared in a spaced-plant trial in 1999. High diversity was found for both agronomic and quality traits, and some wild populations appeared to be better than the best check (Bona) for flower head and essential oil yield. As for the essential oil composition, the populations were tentatively grouped into four previously defined chemotypes and a new one. Four wild populations appeared equal or better than Bona for α-bisabolol (43-55%) or chamazulene (16.3-22.5%) content, and can be useful for developing high oil quality varieties. A significant negative correlation was found between flower head weight and α-bisabolol content. *[Article copies available for a fee from The Haworth Document Delivery Service: 1-800-HAWORTH. E-mail address: <getinfo@ haworthpressinc.com> Website: <http://www.HaworthPress.com> © 2002 by The Haworth Press, Inc. All rights reserved.]*

KEYWORDS. Chamomile, biodiversity, medicinal plants

INTRODUCTION

Chamomilla recutita (L.) Rauschert (Compositae) ($2n = 2x = 18$) is an important species for medicinal and cosmetic use. In Italy it is widely used for flower-head infusion but is almost totally imported from East Europe, Egypt and Argentina. Wild populations are easily found, especially in disturbed soils and as weeds of cereal crops. Previous research conducted in North Italy has indicated that valuable materials can be found in the wild (1). As part of a national research project on aromatic and medicinal plants supported by the Italian Ministry of Agricultural and Forestry Policies, wild chamomile populations were collected in Central Italy with the purpose of estimating the diversity and potential economic value of wild germplasm from this area.

MATERIALS AND METHODS

Eleven chamomile populations were collected in 1997-98 (Figure 1). Four other populations were compared, two previously collected wild populations from North Italy (1), the Slovak variety Bona and an Italian selection (Syn1) (2). Bona was used as a check because it is considered one of the best cultivars for the Italian conditions.

In January 1999, seeds of all populations were sown in petri dishes and transplanted to small pots in a greenhouse. In April, eighty plants per *population* were transplanted in the field at S. Martino in Campo, Perugia, 220 m.a.s.l.

FIGURE 1. Collection sites of wild chamomile populations and their altitudes.

Collection place	Altitude m.a.s.l.
Casalina	168
S. Romualdo	380
Bosco	207
Senigallia	147
Abbadia di Fiastra	186
Pollenza	341
Sanguineto	300
Ciampino	125
Arpino	450
Castelvecchio	490
Furci	550
Moretta	262
Fossalon	1

Each plot was made up of 20 plants 10 cm apart in 2 rows 40 cm apart; plots were arranged in a randomized block design with four replications. Yield and quality traits were evaluated (Table 1). In June, the flower heads were harvested by hand, *dried* at 45°C for 72 hours, and distilled (2 hours hydrodistillation with pentan). For each population, samples of dry flower heads from the first and second replication, and from the third and fourth replication, were bulked before essential oil distillation. Essential oil composition was determined by *gas chromatography* (two analyses per sample, total of four analyses per population). Field data were subjected to ANOVA using a fixed model (3).

RESULTS

Large diversity for both yield and quality traits was observed (Table 1). The commercial variety Bona was out-performed by Bosco, Casalina and S. Romualdo for dry flower head weight, by Abbadia di Fiastra for fresh 100 flower head weight and by Pollenza, Bosco, S. Romualdo, Sanguineto, Abbadia di Fiastra, Senigallia, Casalina and Ciampino for essential oil yield.

One ecotype, Arpino, had essential oil content (0.29% w/w, that is 0.32% v/w) lower than the minimum value established by the Italian pharmacopoeia (0.4% v/w) (4). A population from North Italy showed higher exterior flower head diameter than Bona (17.35 vs. 16.9 mm). For essential oil composition,

TABLE 1. Means of flower head traits, plant height, essential oil yield and its components in chamomile populations.

Population	Dry flower head weight (kg/ha)	Fresh 100 flower head weight (g)	Exterior diameter (mm)	Interior diameter (mm)	Height of plant (m)	Essential oil yield (% w/w)	α-bisabolol (%)	Chamazulene (%)
Casalina	440.00 ab	8.95 abc	15.99 abcd	6.01	32.29 abc	0.47	4.37	12.98
S. Romualdo	437.81 ab	10.40 a	15.85 abcd	6.35	32.09 abc	0.66	18.40	14.54
Bosco	485.00 a	8.95 abc	16.54 abc	6.25	30.33 abc	0.83	5.23	12.76
Senigallia	333.70 bcd	9.00 abc	15.51 abcd	5.51	27.44 cde	0.51	44.63	6.70
Abbadia di Fiastra	367.80 abc	10.85 a	15.85 abcd	6.26	27.36 cde	0.58	17.03	5.06
Pollenza	338.10 abcd	9.05 abc	16.89 abc	6.14	25.33 de	0.89	34.42	16.68
Sanguineto	385.00 ab	10.45 a	16.86 abc	6.38	33.96 ab	0.60	17.79	14.08
Ciampino	233.70 cde	7.20 c	15.31 bcd	5.42	24.96 de	0.47	50.58	16.60
Arpino	218.70 de	8.35 bc	14.24 d	6.18	28.50 bcd	0.29	15.08	3.52
Castelvecchio	324.10 bcd	8.10 c	14.94 cd	5.91	22.50 e	0.42	51.82	20.57
Furci	161.60 e	7.80 c	14.35 d	5.77	21.83 e	0.39	43.12	11.92
Moretta	374.40 abc	10.65 a	17.35 a	6.18	31.67 abc	0.61	11.15	9.75
Fossalon	484.70 a	10.05 abc	15.68 abcd	5.95	34.73 a	0.54	4.61	10.72
Bona	400.90 ab	10.50 a	16.91 ab	6.90	31.38 abc	0.45	35.91	13.41
Syn1	403.40 ab	10.75 a	16.70 abc	6.47	29.43 abcd	0.58	10.26	5.06

Means followed by the same letter are not different at $P < 0.05$.

the populations were tentatively grouped into four previously defined chemotypes (5) and a new one (Figure 2). Chemotype C, characterized by high α-bisabolol and chamazulene content, is represented by five wild populations. In particular, Ciampino (50.58% α-bisabolol and 16.6% chamazulene) and Castelvecchio (51.82% and 20.57%) appeared better than Bona (35.91% and 13.41%). It should be noted that Bona had an unusually low α-bisabolol content in this trial, but even considering the normal average value of about 40-45% for Bona, the value of the two best wild populations remains high.

A significant, positive correlation (0.76, $P \leq 0.01$, 13 d.f.) between dry flower head weight and plant height, and a significant negative correlation between α-bisabolol content and dry flower head weight ($-0.67, P < 0.01$, 13 d.f.) and plant height ($-0.82, P < 0.01$, 13 d.f.) were evidenced. Chamazulene concentration showed no correlation with yield traits.

DISCUSSION

High diversity was found among the populations studied for all the traits examined and some of them performed well compared to the variety Bona. The

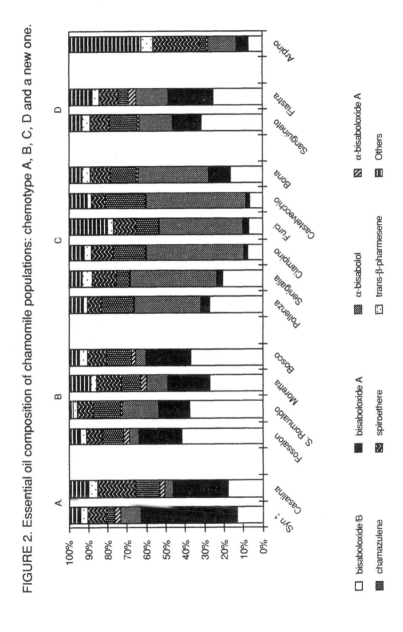

FIGURE 2. Essential oil composition of chamomile populations: chemotype A, B, C, D and a new one.

357

same materials are under evaluation in the current year in an Autumn-transplanted field test to confirm the results here presented and to assess other important traits such as plant growth habit and the thickness of the flower-head layer.

α-Bisabolol and chamazulene content are important components of chamomile quality, and their evaluation is time consuming and costly. Therefore, selection of new, locally adapted, cultivated varieties should start from materials with preliminarily ascertained good oil quality. Fortunately, such populations can be found in the wild: Pollenza, Senigallia, Ciampino, Furci and Castelvecchio appear promising. Possible negative correlations between yield and oil quality need to be taken into account, but the high within-population variability should permit the reconciliation of productivity and oil quality in varieties adapted to the Italian conditions.

REFERENCES

1. Aiello, N., L. D'Andrea, F. Scartezzini and C. Vender. Bio-agronomic evaluation of wild chamomile populations (*Chamomilla recutita* (L.) Rauschert). Proceedings of 5° Convegno Nazionale Biodiversità "Biodiversità e sistemi ecocompatibili," September 9-10, 1999, Caserta, Italy, in press.

2. Tavoletti, S., F. Veronesi, N. Aiello, A. Bezzi, F. Scartezzini and C. Vender. 1996. Variability for agronomic and quality traits in *Chamomilla recutita* (L.) Rausch. Proceedings of International Symposium "Breeding Research on Medicinal and Aromatic Plants," June 30-July 4, 1996, Quedlinburg, Germany. pp. 244-246.

3. SAS Institute Inc. 1985. *SAS User's Guide: Statistics, Version 5 Edition*. SAS Institute Inc., Cary, NC, USA. 437 pp.

4. Farmacopea Ufficiale Italiana. 1991. *Droghe vegetali e preparazioni (camomilla comune)*. Istituto Poligrafico Zecca dello Strato, Rome, Italy.

5. Schillcher, H. 1987. Die Kamille. *Wiss. Verlagseges.*, Stuttgart: 97-98.

Variation of Morphology, Yield and Essential Oil Components in Common Chamomile (*Chamomilla recutita* (L.) Rauschert) Cultivars Grown in Southern Italy

Laura D'Andrea

SUMMARY. The common chamomile (*Chamomilla recutita* (L.) Rauschert) is one of the oldest and most popular medicinal plants, used for its numerous properties. Two diploid and two tetraploid cultivars, sown and transplanted in Southern Italy, were compared to determine the variation in morphology, yield and essential oil components at two harvest times. An analysis of variance was performed, and means were compared using Duncan's test. The greatest height, maximum number of flowers per flower head, and number of main stems of the plants were obtained with transplanted plants. The *larger* diameter and highest weight of flowers were obtained from tetraploid cultivars. The greater fresh and dry flower *head* yields were obtained from diploid cultivars. There were no statistical differences on oil percentage among the four

Laura D'Andrea is affiliated with the Dipartimento di Scienze delle Produzioni Vegetali-Facoltà di Agraria-University of Bari, via Amendola 165/A, 70126 Bari, Italy (E-mail: laura.dandrea@libero.it).

This research was supported by "Ministero delle Politiche Agricole e forestali" (MiPA) in Italy, within the Project "Incromento della Produzione di Piante Officinali" (I.P.P.O.), Publication n. 22.

[Haworth co-indexing entry note]: "Variation of Morphology, Yield and Essential Oil Components in Common Chamomile (*Chamomilla recutita* (L.) Rauschert) Cultivars Grown in Southern Italy." D'Andrea, Laura. Co-published simultaneously in *Journal of Herbs, Spices & Medicinal Plants* (The Haworth Herbal Press, an imprint of The Haworth Press, Inc.) Vol. 9, No. 4, 2002, pp. 359-365; and: *Breeding Research on Aromatic and Medicinal Plants* (ed: Christopher B. Johnson, and Chlodwig Franz) The Haworth Herbal Press, an imprint of The Haworth Press, Inc., 2002, pp. 359-365. Single or multiple copies of this article are available for a fee from The Haworth Document Delivery Service [1-800-HAWORTH 9:00 a.m. - 5:00 p.m. (EST). E-mail address: getinfo@haworthpressinc.com].

cultivars and between the harvest times. The chemical type classification was determined, as suggested by Schilcher. *[Article copies available for a fee from The Haworth Document Delivery Service: 1-800-HAWORTH. E-mail address: <getinfo@haworthpressinc.com> Website: <http://www.HaworthPress. com> © 2002 by The Haworth Press, Inc. All rights reserved.]*

KEYWORDS. Chamomile, diploid cultivars, tetraploid cultivars, morphology, yield, essential oil

INTRODUCTION

The common chamomile (*Chamomilla recutita* (L.) Rauschert) is one of the oldest and most popular medicinal plants, used for its numerous properties (1). Its inflorescences and essential oil are used as anti-inflammatory, antispasmodic, antiseptic, antibacterial and antifungal treatments. "Camomile tea," made from the inflorescence, is used very much in Italy, where few hectares are cultivated in chamomile and the demand is *very high*.

The aim of this study was to determine the variation of morphology, yield and essential oil components of two diploid and two tetraploid cultivars of chamomile between two cultivation techniques and two harvest times.

MATERIALS AND METHODS

Two diploid (Bona and Dotto) and two tetraploid (Kor and Bk2) cultivars (cv.) (2), were examined. The field trials were carried out at the experimental station "E. Pantanelli" of University of Bari, in the area of Policoro (MT) in Southern Italy. There were two experiments, referred to as *experiment 1* and *experiment 2*. The seeds were sown at the end of winter with an inter-row spacing of 40 cm. The seedlings, propagated by seeds in December in the greenhouse, were transplanted into the field in February, with a distance of planting 40 cm × 10 cm. A two factor randomized block design with four replications was used for each experiment. The flowers were harvested by hand and were dried at 40°C in an oven. The essential oils were obtained by steam distillation in a Clevenger-type apparatus and the main essential oil components were determined by gas chromatography.

Experiment 1. A comparison between two cultivation techniques, sowing and transplanting, was carried out. The following morphological and productive characteristics were determined at full flowering: height and number of main stems of the plants; height and number of flowers in the flower head; diameter and weight of 100 fresh flowers; fresh and dry flower head yield and essential oil yield.

Experiment 2. A comparison between two harvest times was carried out: 27 May (first cutting, beginning of flowering, mostly flowers at II development stage (3)) and 10 June (second cutting, full flowering, mostly flowers at III development stage (3)). The following parameters were determined: fresh and dry flower head yield, main essential oil components: trans-β farnesene, bisabololoxide B, bisabololoxide A, α-bisabolonoxide A, α-bisabolol, chamazulene, and spiro-ether. The chemical types were determined as suggested by Schilcher (5).

In both experimentals an analysis of variance was performed, and means were compared using Duncan's test.

RESULTS

Experiment 1. The data concerning the morphological and yield characteristics are shown in Table 1. Differences are observed among cultivars, between genotypes and between cultivation techniques. The height of the plants is high in transplants (75.5 cm) and in diploid cultivars (78.7 cm) whereas the tetraploid cultivars (Kor and Bk2) have the wider flower diameter (17.0 cm) and the highest weight of 100 fresh flowers (20.6 g).

Experiment 2. The fresh flower heads yield is shown in Figure 1: better yield was obtained from Dotto, among the cultivars, and from the diploid, between genotypes. The harvest time (Table 2) influences the drug yield of chamomile with an increase from 16.7 q/ha (in first cutting) to 19.3 q/ha (in second cutting). The results of the chemical analyses of essential oils (Table 3) revealed that the major constituents were found to be α-bisabolol and α-bisaboloxide A. The chemical types classification is shown in Figure 2, where chemotype A (high content of bisaboloxide A) and chemotype C (high content of bisabolol) were determined, as suggested by Schilcher (5). Dotto (diploid cv.) and Bk2 (tetraploid cv.) were classified as Chemotype A; Bona (diploid cv.) and Kor (tetraploid cv.) were classified as Chemotype C.

There were no statistically significant differences in oil percentage among the four cultivars and between the harvest times. These results agreed with the experiments carried out in Germany by Franz (3,4).

DISCUSSION

The results show that both chemotype and yield differences can be observed among cultivars of the same genotype. Tetraploid cultivars have wider diame-

TABLE 1. Main characteristics of chamomile in two cultivation techniques.

technique of cultivation	genetic	cultivars	plant		full layer of flowers		100 fresh flower		flower heads yield	
			height (cm)	stems (n)	height (cm)	flowers (n)	diameter (mm)	weight (g)	fresh (q/ha)	dry (q/ha)
sowing	diploid	Bona	67.4	1.8	14.3 C	28.7 D	14.4	13.8 b	80.2 b	14.1
		Dotto	75.0	1.3	16.2 C	34.6 CD	13.7	12.5 b	77.5 b	16.6
	tetraploid	Kor	46.2	4.8	15.4 C	52.0 CD	16.2	21.3 a	44.2 e	7.8
		Bk2	65.0	2.1	15.7 C	31.7 CD	14.7	21.3 a	90.4 ab	16.4
transplant	diploid	Bona	75.7	13.1	28.2 B	279.8 A	16.5	11.3 b	79.0 b	16.8
	tetraploid	Kor	59.2	14.8	23.7 B	148.0 BC	17.3	25.0a	66.4 be	11.5
		Bk2	71.2	11.9	25.4 B	227.8 AB	19.9	15.0 b	74.6 b	16.3
		Bona	71.6 B	7.4 B	21.2 AB	154.3 ab	15.5 ab	12.5 C	79.6 A	15.5 A
		Dotto	85.9 A	6.4B	24.8 AB	183.1 ab	14.0 b	12.5 C	91.9 A	19.5 A
		Kor	52.7 C	9.8 A	19.5 B	100.0 b	16.7 a	23.1 A	55.3 B	9.6 B
		Bk2	68.1 B	7.0 B	20.5 B	129.7 ab	17.3 a	18.1 B	82.5 A	16.3 A
	diploid		78.7 A	6.9 B	23.0A	168.7 a	14.7 B	12.5 B	85.8 a	17.5 A
	tetraploid		60.4 B	8.4 A	20.0 B	114.9 b	17.0 A	20.6 A	68.9 b	13.0 B
sowing			63.4 B	2.5 B	15.4 B	36.8 B	14.7 B	17.2	73.1	13.7 B
transplant			75.7 A	12.8 A	27.7 A	246.8 A	17.0 A	15.9	81.6	16.7 A
sowing	diploid		71.2	1.5	15.2 C	31.7 C	14.0	13.1	78.9	15.4
	tetraploid		55.6	3.4	15.5 C	41.8 C	15.4	21.3	67.3	12.1
transplant	diploid		86.2	12.3	30.8 A	305.7 A	15.4	11.9	92.7	19.6
	tetraploid		65.2	13.3	24.6 B	187.9 B	18.6	20.0	70.5	13.9
means			69.5	7.6	21.5	141.8	15.9	16.6	77.3	15.2

Means with the same capital letter are not significantly different at P = 0.01.
Means with the same small letter are not significantly different at P = 0.05.

FIGURE 1. Fresh flower yield (q/ha) of chamomile in two harvest times.

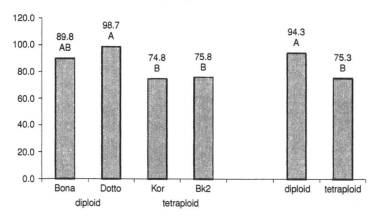

TABLE 2. Drug yield (q/ha) of chamomile in two harvest times.

genetic	cultivars	harvest time		harvest
		27 May	10 June	mean
diploid	Bona	16.8 b	23.9 a	20.3 AB
	Dotto	22.3 a	20.7 ab	21.5 A
tetraploid	Kor	11.5 c	15.9 bc	13.7 C
	Bk2	16.3 b	16.6 b	16.4 BC
diploid		19.6	22.3	20.9 A
tetraploid		13.9	16.2	15.0 B
means		16.7 b	19.3 a	18.0

Means with the same capital letter are not significantly different at P = 0.01.
Means with the same small letter are not significantly different at P = 0.05.

TABLE 3. Main essential oil components (%) of four chamomile cultivars in two harvest times.

components	diploid				tetraploid			
	Dotto		Bona		Bk2		Kor	
	27 May	10 June	27 May	10 June	27 May	10 June	27 May	10 June
trans-β farnesene	6.2	11.1	6.6	7.3	5.4	10.7	3.0	4.4
bisaloloxide B	6.4	7.9	2.4	4.5	15.1	10.9	1.0	1.4
α-bisabololoxide A	4.4	13.3	1.6	0.3	9.9	6.5	0.0	0.0
α-bisabolol	28.8	1.7	**63.9**	**67.6**	6.0	9.8	**71.3**	**71.9**
chamazulene	3.3	0.8	9.3	10.4	11.5	12.6	16.3	14.1
bisaboloxide A	**42.5**	**57.1**	6.9	1.0	**41.0**	**35.3**	0.3	0.4
spiroethers	1.3	1.9	1.4	1.2	1.7	3.0	0.9	1.2
others	7.3	6.4	8.0	7.7	9.4	11.3	7.2	6.7

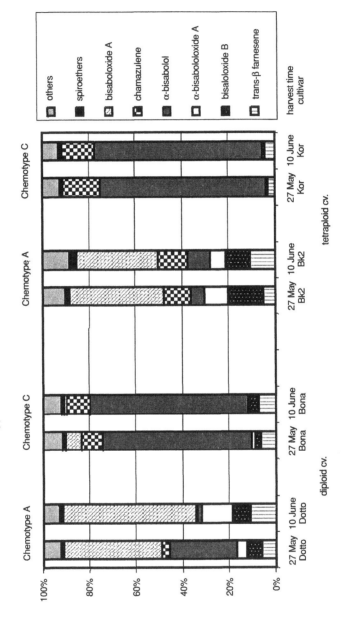

FIGURE 2. Chemical type classification of four chamomile cultivars in two harvest times

ter flowers but a significantly lower number of flowers in the flower head as well as higher content of chamazulene (important for the oil market price) than diploid cultivars. All the characteristics are influenced by the development stage of the flowers. The variability of chamomile observed in this study indicates that the best cultivation choice depends on the expected use.

REFERENCES

1. Hornok, L. (1992). Cultivation and Processing of Medicinal Plants. John Wiley & Sons.

2. Aiello, N. (1997). Varietà diploidi e tetraploidi di camomilla comune. *Erboristeria Domani*, 9, 71-72.

3. Franz, C. (1981). Actual viewpoints on the cultivation of medicinal and aromatic plants. *Atti del II seminario internazionale sulle piante medicinali ed aromatiche. Città di Castello*, 11-13 settembre 1981, 299-317.

4. Franz, C. (1992). Genetica biochimica e coltivazione della camomilla (*Chamomilla recutita* (L.) Rauschert). *Agricoltura Ricerca*, 131, 87-96.

5. Schilcher, H. (1987). Die Kamille. Wiss. Verlagsges., Stuttgart.

Evaluation and Preservation of Genetic Resources of Carob (*Ceratonia siliqua* L.) in Southern Italy for Pharmaceutical Use

Girolamo Russo
Laura D'Andrea

SUMMARY. The carob tree (*Ceratonia siliqua* L.) is widely used in the Mediterranean countries for ornamental and industrial purposes. The carob is used in the pharmaceutical industries for therapeutical treatment and in the liquor industry. The area devoted to carob cultivation in Italy is now restricted mostly to marginal areas. The aim of this study was to contribute to the knowledge of the carob genetic resources at a time of increasing risk of its genetic erosion with an investigation into the morphological variability of *in situ* collected plants in different regions of Southern Italy. The morphology (length width and thickness, apical, basal and middle) of seeds and fruits was determined, using a descriptor list. The results showed considerable variation for most of the examined characters. *[Article copies available for a fee from The Haworth Document Delivery Service: 1-800-HAWORTH. E-mail address: <getinfo@haworthpressinc. com> Website: <http://www.HaworthPress.com> © 2002 by The Haworth Press, Inc. All rights reserved.]*

Girolamo Russo and Laura D'Andrea are affiliated with the Dipartimento di Scienze delle Produzioni Vegetali-Facoltà di Agraria-Università di Bari, Via Amendola, 165/A-70126 Bari, Italy (E-mail: girolamo.russo@agr.uniba.it or laura.dandrea@libero.it).
The authors have contributed equally to this study.
This research was supported by "C.N.R." within the project "Biodiversità."

[Haworth co-indexing entry note]: "Evaluation and Preservation of Genetic Resources of Carob (*Ceratonia siliqua* L.) in Southern Italy for Pharmaceutical Use." Russo, Girolamo, and Laura D'Andrea. Co-published simultaneously in *Journal of Herbs, Spices & Medicinal Plants* (The Haworth Herbal Press, an imprint of The Haworth Press, Inc.) Vol. 9, No. 4, 2002, pp. 367-372; and: *Breeding Research on Aromatic and Medicinal Plants* (ed: Christopher B. Johnson, and Chlodwig Franz) The Haworth Herbal Press, an imprint of The Haworth Press, Inc., 2002, pp. 367-372. Single or multiple copies of this article are available for a fee from The Haworth Document Delivery Service [1-800-HAWORTH 9:00 a.m. - 5:00 p.m. (EST). E-mail address: getinfo@haworthpressinc.com].

KEYWORDS. *Ceratonia siliqua* L., carob tree, genetic erosion, morphology of seeds and fruits

INTRODUCTION

The carob tree (*Ceratonia siliqua* L.), belonging to the family Leguminosae, is widespread in the Mediterranean Countries (Spain, Italy, Portugal, Cyprus, Greece, Israel, etc.) (1). It is cultivated for ornamental and industrial (food, pharmaceutical, chemical) purposes (2-4). The edible part is usually the processed seed and legume (5). The extracts and meal of carob are used in the pharmaceutical industries for the therapeutical treatment of diarrhoea enteritis in infancy (6). The ground legumes, without seeds, are used in the liquor industry (7). The area devoted to carob cultivation in Italy declined steadily after World War II, and is now restricted mostly to the marginal areas in a limited number of farms (8). It was necessary to describe the actual characters of carob populations because of a wide dispersion throughout the country of the carob genetic resources and the increasing risk of its genetic erosion. The aim of this study was to contribute to the knowledge of morphological variability of plants collected *in situ* in different regions of Southern Italy.

MATERIALS AND METHODS

The research was carried out on seeds and fruits of plants collected *in situ* in different regions of Southern Italy: 40 from Puglia, 8 from Calabria, 3 from Basilicata and 3 from Campania.

The morphology of seeds and fruits was determined using a descriptor list (9). Length, width and thickness of ventral rib; length, width and thickness of dorsal rib; thickness and width of inter-rib zone; length, apical width, basal width and middle width of fruit; length and width of false peduncle of fruit; length and width of true peduncle of fruit; height, width, depth and spacing of mesocap loculus of seeds; length, width and thickness of seeds; length/width ratio of seeds; width/thickness ratio of seeds and length/thickness ratio of seeds were observed. An analysis of variance was performed, and means were compared using Duncan's test.

RESULTS

The characters observed are reported in Tables 1, 2 and 3. The results showed considerable variation for most of the examined characters. Cultivar

TABLE 1. Rib characters of cultivars.

Cultivars	ventral rib			dorsal rib			inter-rib zone	
	length (mm)	width (mm)	thickness (mm)	length (mm)	width (mm)	thickness (mm)	length (mm)	width (mm)
1	16.5 GO	6.2 JO	10.2 IM	17.4 JQ	6.2 NR	9.8 IP	6.1 JQ	8.8 FI
2	17.0 DM	5.8 MQ	10.2 IN	17.4 JQ	6.5 LR	9.2 LR	6.2 HQ	9.1 DH
3	18.0 BG	5.6 NR	8.9 OU	19.6 CG	7.1 GQ	10.6 FK	4.9 ST	8.7 FJ
4	17.3 DK	6.6 FN	9.8 JQ	18.1 EN	6.4 MR	10.5 GK	7.4 BD	11.6 A
5	14.8 P	5.0 PS	7.6 VW	17.1 KQ	4.7 UV	7.1 TU	5.9 LR	10.1 BE
6	16.9 EN	7.6 AF	12.4 BE	17.9 FN	8.8 AC	11.3 BG	5.8 NR	7.6 IO
7	15.9 JP	8.1 AD	13.0 BC	18.2 EM	8.2 BI	10.7 EK	6.0 KQ	8.0 HN
8	16.6 GN	8.4 AB	13.3 AB	18.2 EM	9.1 AB	11.8 BE	6.2 IQ	8.7 FJ
9	17.6 CJ	4.8 QS	9.0 NU	17.9 FN	5.4 RU	8.7 PS	6.5 FM	10.4 AC
10	17.0 EM	8.5 A	12.1 CG	18.3 EM	8.3 BF	11.8 BD	6.8 DJ	6.9 NP
11	16.2 IP	6.1 KO	10.0 IP	17.0 LQ	6.1 PT	9.6 KQ	6.2 IQ	8.9 EH
12	17.2 DK	8.5 A	8.3 SW	18.0 EN	7.9 BK	8.2 RS	7.0 CH	6.9 NP
13	16.1 JP	3.6 T	5.3 Y	17.3 JQ	4.1 V	5.0 W	4.2 U	11.5 A
14	15.5 LP	8.0 AD	12.3 BE	16.3 NQ	8.7 AC	10.5 FK	6.5 FM	6.8 NP
15	16.4 GP	7.5 AG	10.9 GJ	17.7 HP	9.1 AB	11.6 BF	7.0 CF	5.9 P
16	16.2 IP	7.1 DL	10.3 IL	17.5 JQ	8.2 BH	10.9 DJ	6.9 CI	8.4 GL
17	15.2 NP	8.2 AC	9.5 KS	15.7 Q	8.6 AD	10.4 GK	6.5 FM	7.3 KO
18	17.9 BI	6.7 FN	11.3 EI	18.2 EM	8.1 BI	10.7 DK	7.3 CE	6.7 OP
19	18.7 AD	7.3 BI	10.9 HJ	19.8 CE	6.3 MR	10.1 HN	6.8 DJ	7.0 MP
20	16.4 GO	6.1 LP	10.4 IL	18.3 EM	6.4 MR	9.9 IP	5.9 LR	8.2 GM
21	19.4 AB	7.6 AG	12.1 BF	21.4 AB	9.7 A	11.0 DI	6.6 EK	7.1 LP
22	17.4 DK	7.2 DK	11.1 FI	17.8 GO	8.1 BI	11.2 CH	6.7 EK	6.8 NP
23	11.3 Q	6.5 GO	11.7 DH	14.1 R	8.5 BE	10.3 GM	7.3 CE	6.8 NP
24	17.9 BH	7.9 AE	12.4 BE	19.1 CJ	7.9 BJ	12.2 BC	6.6 EL	7.5 IO
25	16.3 HP	4.5 ST	6.4 XY	17.1 KQ	5.0 SV	6.7 UV	4.6 TU	9.0 DH
26	17.1 DL	7.2 CJ	11.8 CH	17.9 FN	7.3 FP	10.4 GK	6.0 KQ	8.6 GK
27	16.8 FN	6.9 EM	10.3 IL	17.3 JQ	6.5 LR	9.9 IP	6.3 GQ	8.4 GL
28	17.3 DK	6.2 JO	9.0 MT	22.6 A	7.5 DM	10.5 FK	5.6 PR	11.0 AB
29	15.4 LP	3.6 T	6.1 Y	17.8 GN	5.0 TV	6.4 UV	5.8 MR	10.9 AB
30	14.9 OP	4.6 RS	7.9 TW	17.0 LQ	9.0 AB	9.9 IP	5.8 MR	9.1 DH
31	14.9 OP	6.1 KO	10.2 IN	16.0 OQ	7.6 CL	10.4 GL	7.0 CG	8.0 HN
32	18.4 AF	6.3 IO	10.0 IP	20.8 BC	6.7 LQ	9.8 IP	6.8 DI	8.5 GK
33	17.9 BI	8.1 AD	14.4 A	19.3 CI	8.4 BF	13.3 A	6.3 GQ	6.9 NP
34	17.1 DL	6.4 HO	9.5 KR	17.9 FN	6.0 QT	10.3 GM	7.4 BD	8.9 EH
35	17.3 DK	6.5 FO	10.1 IN	18.2 EM	6.8 KQ	9.9 IO	7.6 BC	10.7 AB
36	18.4 AF	7.4 BH	12.6 BD	19.1 CJ	8.6 AD	12.4 AB	6.4 FO	9.3 CH
37	15.9 KP	7.1 DL	12.7 BD	16.5 MQ	8.3 BG	12.2 BC	7.0 CF	8.5 GK
38	19.2 AB	6.2 JO	10.7 HK	19.7 CF	7.0 HQ	9.1 MR	5.8 MR	8.0 HN
39	19.8 A	6.0 LP	9.7 JQ	20.5 BD	6.1 OS	7.8 ST	5.2 RS	8.9 EH
40	19.1 AC	6.8 FM	11.8 CH	19.6 CG	8.0 BJ	10.3 GM	6.3 GQ	8.4 GL
41	18.3 AF	5.5 OS	8.6 QV	19.5 CI	7.0 IQ	9.7 JQ	6.4 FP	8.7 FJ
42	19.2 AB	6.8 FM	10.1 IO	19.6 CH	8.4 BF	10.4 GL	6.5 FN	9.0 DH
43	16.7 GN	6.6 FN	10.6 HK	17.4 JQ	6.9 JQ	10.1 HN	8.0 AB	8.6 GK
44	18.5 AE	6.1 KO	10.7 HK	19.6 CG	6.5 LR	9.1 MR	7.2 CE	10.2 BD
45	17.4 DK	5.9 MP	9.3 LS	19.0 CJ	6.5 LR	8.9 NS	7.0 CF	10.7 AB
46	19.7 A	4.8 RS	7.3 WX	22.9 A	6.7 LQ	6.6 UV	4.6 TU	8.3 GL
47	16.2 IP	4.8 RS	8.5 RW	17.0 LQ	5.5 HU	8.1 HI	5.7 OH	9.4 CG
48	16.3 HP	4.6 ST	5.7 Y	18.0 EN	4.7 UV	5.1 W	5.6 QR	9.2 CH
49	18.5 AF	4.9 QS	5.8 Y	18.8 DL	4.8 UV	5.7 VW	4.9 ST	8.7 FJ
50	16.8 FN	5.7 NR	8.9 PU	17.7 IP	5.5 RU	8.9 NS	6.6 EK	10.1 BE
51	15.4 MP	6.8 FM	8.8 QV	15.9 PQ	7.3 EO	9.0 NS	6.4 FO	9.9 BF
52	16.6 GN	6.6 FM	7.7 UW	17.4 JQ	7.0 HQ	8.8 OS	5.9 LR	9.1 DH
53	16.3 HP	6.9 EM	10.6 HK	19.4 CI	7.4 EN	9.5 KQ	8.3 A	11.1 AB
54	16.8 EN	5.5 OS	8.7 QV	18.9 DK	5.4 RU	8.6 QS	6.2 IQ	7.5 JO
maximum	19.8	8.5	14.4	22.9	9.7	13.3	8.3	11.6
minimum	11.3	3.6	5.3	14.1	4.1	5.0	4.2	5.9
means	17.0	6.4	10.0	18.3	7.1	9.6	6.3	8.7
CV %	7.0	12.2	9.2	7.2	12.2	9.2	8.1	10.5

Means with the same letter are not significantly different at P = 0.01.

TABLE 2. Fruit characters of cultivars.

Cultivars	fruit				peduncle			
	length	width			false		true	
		apical (mm)	basal (mm)	middle (mm)	length (mm)	width (mm)	length (mm)	width (mm)
1	16.4 KS	15.9 SW	16.8 PR	21.2 OS	6.3 OS	4.4 BF	2.7 HR	4.6 CG
2	16.6 IS	17.6 NS	17.2 OR	22.4 IP	6.4 OS	4.5 BE	2.8 FR	4.6 CG
3	17.5 EP	14.9 VX	19.5 FM	23.0 EO	7.4 LS	4.2 BI	3.0 DO	3.9 GM
4	16.9 GR	20.6 BH	20.2 CK	23.9 DJ	6.7 NS	3.8 GN	2.2 RV	4.0 GM
5	15.9 OS	14.1 X	17.3 NQ	19.5 SU	8.9 FL	2.8 ST	2.3 OU	2.5 R
6	16.5 JS	16.5 RV	18.5 JP	23.9 DJ	7.5 KR	4.5 BE	4.6 A	5.4 AB
7	15.6 PS	18.2 KR	20.1 CL	24.9 BD	9.3 DL	4.3 BH	2.9 EP	5.7 A
8	16.5 KS	16.7 QU	18.3 LP	22.4 IP	8.9 FL	5.1 A	4.2 AB	5.1 BC
9	17.5 EP	19.8 EL	19.5 EM	21.8 LR	7.8 JQ	3.7 GP	2.4 NU	3.8 IN
10	17.4 EP	19.8 DL	20.6 CI	24.8 CE	8.7 GM	4.2 BJ	3.7 BD	3.8 IN
11	16.5 JS	15.8 TW	16.7 PR	21.2 OS	6.2 PS	4.4 BF	2.6 JS	4.5 CH
12	17.5 EP	19.9 DK	21.0 CH	24.6 CF	8.4 HN	4.6 AC	4.1 AC	4.5 CH
13	16.3 LS	16.4 RW	16.1 QR	20.1 RT	6.4 OS	3.3 NS	4.6 A	2.7 QR
14	15.3 QS	20.4 CI	20.0 CL	23.2 DN	9.0 FL	4.0 CL	3.1 DN	4.5 CH
15	17.0 GQ	20.0 DJ	20.9 CI	24.0 CJ	7.8 JQ	3.6 KQ	3.2 DM	3.7 JO
16	16.3 MS	20.7 BG	21.6 BD	24.2 CI	6.8 MS	3.4 MR	3.2 DK	4.0 FL
17	14.7 ST	20.5 BH	20.1 CL	24.0 CJ	7.6 KR	4.5 BE	3.3 DJ	4.6 CG
18	17.5 DO	18.6 IP	20.7 CI	23.7 DK	8.2 HO	3.8 FN	3.1 DN	3.8 HN
19	19.2 BE	16.9 PU	17.9 MQ	21.7 MR	7.9 IQ	4.3 BG	3.2 DL	3.7 JO
20	16.3 MS	19.2 FN	19.7 DL	21.9 KQ	9.9 DH	3.9 FN	2.5 MT	3.7 KO
21	19.3 BD	20.0 DJ	21.0 CH	24.7 CF	9.0 FL	4.5 BE	3.2 DK	4.9 BE
22	17.2 FP	20.6 BH	20.4 CJ	23.6 DL	7.8 JQ	3.8 GN	2.9 FQ	4.0 FL
23	13.3 T	20.8 BG	19.1 HN	22.7 GO	7.9 IQ	3.8 GO	2.8 HR	4.0 FL
24	17.8 CN	21.4 AE	21.1 CG	24.8 CE	9.8 DI	4.2 BJ	2.7 HR	4.3 DK
25	16.1 NS	18.1 LR	17.6 MQ	22.0 KQ	11.0 BD	2.7 TU	3.5 CG	3.5 LP
26	17.0 GQ	21.4 AE	21.2 CG	24.4 CH	9.5 DK	4.0 EM	2.5 LS	4.4 DI
27	16.4 KS	19.7 EL	19.3 GM	22.4 IP	8.1 HP	3.8 GO	2.2 PU	4.1 FL
28	20.2 AB	22.8 A	24.3 A	26.7 A	10.4 CG	4.2 BJ	3.6 BE	4.3 DK
29	16.6 HR	20.6 BG	19.2 HN	22.8 FO	6.3 OS	3.7 IP	2.8 GR	3.2 NQ
30	15.8 OS	17.8 MR	18.4 KP	20.4 QT	9.1 EL	3.7 HP	3.3 DJ	3.9 HM
31	15.1 RS	20.2 CJ	20.4 CJ	23.4 DM	6.7 NS	3.4 MR	2.5 KS	4.0 FM
32	18.9 BF	19.9 DL	20.0 CL	23.4 DM	7.7 JQ	3.4 LR	3.5 CG	4.1 FL
33	18.2 CK	22.2 AB	21.3 CF	24.7 CE	9.5 DJ	4.6 AD	3.5 CF	4.6 CF
34	16.6 HR	21.5 AE	20.1 CL	24.5 CG	6.0 QS	3.8 FN	2.6 IS	3.8 HN
35	17.0 GQ	21.9 AC	21.3 CF	24.6 CF	6.6 NS	3.7 HP	2.2 RV	4.5 CH
36	18.1 CM	16.4 RW	17.4 NQ	22.3 JP	6.8 MS	4.6 AB	2.9 EP	4.9 BD
37	15.0 RS	22.7 A	21.8 BC	25.0 BD	6.7 NS	3.6 KQ	2.2 RV	4.5 CH
38	18.7 BG	19.4 FM	20.8 CI	22.6 GO	11.0 BD	4.0 DM	3.4 DH	4.1 FL
39	19.5 BC	18.7 HO	19.9 CL	21.5 NR	12.7 AB	3.7 IQ	3.0 DO	3.8 IN
40	18.9 BF	20.3 CI	21.4 CE	24.3 CI	11.8 BC	3.7 IP	2.2 QV	4.2 EK
41	18.4 CJ	18.6 IP	20.8 CI	23.1 DN	10.9 BE	2.8 ST	2.6 JS	2.9 PR
42	18.5 CH	20.8 BG	21.1 CG	25.8 AC	9.0 FL	3.1 QT	2.4 NU	3.3 MQ
43	15.9 OS	21.0 BF	21.3 CF	23.9 CJ	5.7 RS	3.5 KQ	2.7 HS	3.9 HM
44	18.2 CL	21.7 AD	23.3 AB	24.9 BE	6.0 QS	3.8 FN	2.1 RV	4.4 DJ
45	17.0 GQ	19.8 EL	21.5 CD	24.2 CI	6.2 PS	3.7 HP	2.0 SV	4.4 CI
46	21.1 A	16.5 RV	17.6 MQ	20.7 PT	8.5 HN	3.2 OT	3.3 DI	3.3 MQ
47	16.1 NS	18.9 GO	19.3 GM	22.6 HO	6.3 OS	3.2 PT	2.8 HR	3.1 OR
48	17.6 DO	15.2 UX	16.4 QR	19.4 TU	14.3 AB	2.9 RT	1.8 TV	3.5 LP
49	18.4 CI	14.7 WX	15.5 R	18.2 U	12.1 BC	2.3 U	2.1 RV	2.6 R
50	17.0 GQ	19.6 EL	20.2 CL	23.5 DM	5.6 S	3.5 KQ	1.7 UV	4.0 GM
51	15.3 QS	18.4 JQ	19.1 HN	22.5 IP	10.7 CF	3.4 MR	2.3 PU	4.2 EK
52	16.5 IS	17.4 OT	18.9 IO	22.5 IP	11.1 BD	3.5 KR	1.5 V	4.5 CH
53	17.9 CN	20.2 CI	20.9 CH	26.5 AB	8.2 HO	3.6 JQ	2.4 NU	3.7 JO
54	15.7 OS	13.7 X	13.1 S	15.0 V	10.6 CG	4.1 BK	3.3 DJ	3.3 MQ
maximum	21.1	22.8	24.3	26.7	14.3	5.1	4.6	5.7
minimum	13.3	13.7	13.1	15.0	5.6	2.3	1.5	2.5
means	17.1	19.0	19.6	23.0	8.4	3.8	2.9	4.0
CV %	7.7	7.1	7.0	5.8	16.5	11.1	17.6	11.9

Means with the same letter are not significanlty different at P = 0.01.

TABLE 3. Seed characters of cultivars.

Cultivars	mesocarp loculus				seed			seed ratio		
	height (mm)	width (mm)	depth (mm)	spacing (mm)	length (mm)	width (mm)	thickness (mm)	length/ width (mm)	width/ thickness (mm)	length/ thickness (mm)
1	15.1 EN	8.6 LN	4.0 DM	1.8 NQ	8.1 S	6.7 IR	4.1 CJ	1.2 NP	1.6 OR	2.0 YZ
2	15.2 EN	8.8 HN	4.1 CK	1.9 IP	9.6 CK	7.3 BC	4.6 AB	1.3 GN	1.6 PR	2.1 VY
3	14.9 JO	9.5 CJ	3.5 LT	2.2 FK	11.2 A	7.8 A	3.8 HR	1.4 AE	2.1 AB	2.9 A
4	14.6 KR	10.0 BD	4.6 AC	2.8 CD	9.2 IO	7.0 CK	4.4 AE	1.3 GN	1.6 PR	2.1 UY
5	17.0 AB	11.0 A	4.5 AD	2.1 GN	9.0 KP	6.8 EO	4.1 CK	1.3 FM	1.7 IR	2.2 PY
6	14.3 MS	8.9 GM	3.7 GR	2.5 DF	9.5 EK	6.9 DM	3.9 FP	1.4 AJ	1.8 EQ	2.4 GR
7	15.2 EN	8.6 KN	3.3 QT	2.1 GN	9.9 BF	7.1 BG	3.5 QT	1.4 AJ	2.1 A	2.9 AB
8	17.2 A	9.2 DM	3.6 IR	2.4 EG	9.7 BH	6.7 GQ	3.4 RT	1.4 AC	2.0 AF	2.8 AD
9	14.8 KP	9.6 CJ	4.0 DL	2.0 HO	9.0 KP	7.0 CK	4.2 CI	1.3 IN	1.7 IR	2.2 QY
10	15.2 EN	9.3 CM	4.1 CI	2.1 GM	10.1 BC	7.2 BE	4.7 A	1.4 AH	1.5 R	2.2 QY
11	15.2 EN	8.5 MN	4.1 CJ	1.7 OR	8.5 PS	6.7 GQ	4.0 CM	1.3 KO	1.7 IR	2.1 TY
12	15.2 EN	9.2 DM	4.1 CJ	1.8 MQ	9.8 BG	7.2 BD	4.7 A	1.4 BL	1.5 R	2.1 VY
13	13.6 QS	8.1 N	3.0 ST	1.7 PS	9.2 HO	6.3 RS	3.5 QT	1.5 AB	1.8 CK	2.7 AI
14	14.4 LS	8.6 LN	4.0 EN	1.9 JP	9.2 HO	6.4 PS	4.0 DN	1.4 AD	1.6 PR	2.3 MW
15	14.7 KQ	8.9 GM	4.1 CI	2.1 GN	9.5 DK	6.8 EO	4.2 CG	1.4 AI	1.6 NR	2.3 NX
16	15.1 EN	9.2 DM	4.0 DM	2.3 EH	9.4 FM	7.1 BI	4.3 BF	1.3 EL	1.7 JR	2.2 PY
17	14.4 LS	9.1 DM	3.9 EP	1.9 LQ	8.3 RS	6.2 S	3.9 GQ	1.3 CL	1.6 OR	2.2 QY
18	14.3 MS	9.1 EM	4.5 AD	2.0 IO	9.7 BH	6.8 EO	4.2 BG	1.4 AE	1.6 OR	2.3 LW
19	13.5 RS	9.0 FM	3.6 HR	1.4 RS	9.2 HO	6.7 GR	4.1 CJ	1.4 BK	1.6 KR	2.2 OY
20	14.2 MS	8.8 JN	3.7 HR	1.9 KQ	8.9 LQ	6.4 OS	3.2 T	1.4 AJ	2.0 AD	2.8 AE
21	16.1 AH	9.4 CL	3.9 EP	2.1 GN	9.5 FL	6.7 HR	3.8 JR	1.4 AH	1.8 EQ	2.5 EO
22	15.6 DL	9.3 CM	4.7 AB	2.5 DF	9.8 BG	6.9 DN	4.1 CK	1.4 AF	1.7 IR	2.4 IU
23	13.6 QS	9.4 CL	3.2 RT	1.9 IP	9.1 IO	6.6 JS	3.9 FP	1.4 AJ	1.7 HR	2.4 KV
24	15.3 DN	9.1 FM	3.5 LT	2.2 FJ	9.2 HO	6.7 FQ	3.7 KR	1.4 BL	1.8 DN	2.5 FP
25	14.9 IO	9.1 DM	3.5 NT	2.0 HO	9.8 BG	7.3 BD	3.9 FO	1.4 BL	1.9 BJ	2.5 FP
26	16.0 BJ	8.8 IN	4.2 BG	2.2 FL	9.2 HO	6.5 LS	3.4 RT	1.4 AH	1.9 AH	2.7 AJ
27	13.8 OS	8.6 LN	3.4 OT	2.0 HN	9.6 CK	6.8 EP	3.9 FP	1.4 AH	1.7 GR	2.5 GQ
28	17.1 AB	9.9 BE	3.4 OT	1.6 QS	10.2 B	6.8 EO	3.6 NS	1.5 A	1.9 AI	2.8 AE
29	16.8 AC	10.1 BC	4.4 AE	1.9 IP	10.1 BD	7.1 BG	3.9 FO	1.4 AG	1.8 DO	2.6 CM
30	14.7 KQ	9.4 CL	3.5 KS	2.3 EH	9.0 KP	6.5 MS	3.5 PT	1.4 AJ	1.8 CL	2.6 DN
31	14.8 JO	9.1 DM	4.3 AF	2.1 GN	9.6 CJ	6.8 EO	4.1 CJ	1.4 AH	1.7 JR	2.3 LW
32	16.3 AE	9.6 CH	4.2 CH	2.0 HN	10.1 BE	7.0 CJ	3.6 NS	1.4 AD	1.9 AG	2.8 AF
33	15.0 GO	9.5 CK	3.3 PT	3.4 A	9.5 FL	7.0 CJ	3.7 LS	1.4 BL	1.9 AH	2.6 CM
34	14.2 MS	9.1 FM	4.4 AE	1.8 MQ	9.8 BG	7.2 BE	4.4 AD	1.4 BK	1.6 KR	2.2 OY
35	16.2 AF	9.6 CJ	4.1 CK	2.1 GN	9.1 JP	6.9 CM	4.4 AC	1.3 HN	1.6 QR	2.0 WY
36	15.6 DL	8.9 GM	4.5 AD	1.9 JP	9.5 EK	6.7 GR	3.7 MS	1.4 AH	1.8 CM	2.6 BL
37	15.4 DM	8.7 KN	3.9 EO	2.3 EH	9.2 IO	6.5 LS	3.8 GQ	1.4 AH	1.7 HR	2.4 JU
38	16.4 AD	9.7 CG	3.8 FQ	3.2 AB	9.1 IO	6.3 QS	3.3 ST	1.4 AE	1.9 AG	2.8 AF
39	15.3 DN	9.6 CJ	3.3 QT	3.3 AB	9.3 GN	6.4 PS	3.2 T	1.5 AC	2.0 AE	2.9 AC
40	15.7 CK	9.3 CM	4.3 AF	3.1 AB	9.3 GN	6.4 PS	3.5 PT	1.5 AB	1.8 DP	2.6 BK
41	16.1 AI	10.6 AB	3.8 FQ	2.7 D	9.0 KP	7.4 B	4.0 DN	1.2 NP	1.9 BJ	2.2 OY
42	16.4 AD	9.7 CG	3.9 FP	2.7 D	8.8 MR	6.6 JR	4.1 CJ	1.3 DL	1.6 OR	2.1 RY
43	15.1 FN	10.1 BC	4.6 AC	2.7 D	8.8 NR	7.2 BF	4.2 BG	1.2 MP	1.7 HR	2.1 VY
44	15.2 EN	9.7 CG	4.1 CH	2.4 EG	9.0 JP	6.9 DM	4.2 CG	1.3 GN	1.8 CL	2.1 SY
45	15.6 DL	10.1 BC	4.8 A	3.0 BC	8.9 LQ	6.9 CL	4.3 BF	1.3 JN	1.6 MR	2.1 VY
46	15.0 HO	9.9 BF	3.8 FQ	3.0 BC	9.7 BH	7.3 BD	3.6 NS	1.3 CL	2.0 AD	2.7 AH
47	13.7 PS	9.3 CM	3.5 MT	2.6 DE	8.4 QS	7.1 BH	4.1 CL	1.2 OP	1.8 FR	2.1 VY
48	13.3 S	9.6 CI	3.9 EO	2.0 HO	8.9 MQ	7.0 CJ	3.9 FP	1.3 KO	1.8 EP	2.3 NX
49	14.2 NS	8.7 KN	3.0 T	2.3 EH	8.8 MR	6.5 KS	3.6 OT	1.4 BL	1.8 DN	2.5 FP
50	15.4 DN	9.7 CH	4.1 CK	1.6 PS	8.7 OR	6.9 CL	4.4 AE	1.3 LO	1.6 PR	2.0 XZ
51	14.8 KP	9.5 CJ	4.2 BG	1.5 RS	9.6 BI	7.0 BI	4.0 EN	1.4 BK	1.8 EQ	2.4 GS
52	13.9 OS	9.3 CM	3.6 JR	1.9 IP	9.1 IO	6.8 EO	3.8 IR	1.3 DL	1.8 DP	2.4 HT
53	16.2 AG	9.6 CH	3.5 NT	2.3 FI	9.3 GN	7.0 CJ	3.4 RT	1.3 EL	2.0 AC	2.7 AG
54	9.2 T	8.7 KN	3.6 JR	1.4 S	7.3 T	6.4 NS	4.2 CH	1.1 P	1.5 R	1.7 Z
maximum	17.2	11.0	4.8	3.4	11.2	7.8	4.7	1.5	2.1	2.9
minimum	9.2	8.1	3.0	1.4	7.3	6.2	3.2	1.1	1.5	1.7
means	15.0	9.3	3.9	2.2	9.3	6.8	3.9	1.4	1.8	2.4
CV %	5.6	6.4	9.9	10.5	4.4	4.4	7.2	5.7	8.7	8.9

Means with the same letter are significantly different at P = 0.01.

'46' has the biggest length ventral and dorsal rib. Width mean of ventral rib was 6.4 mm, with values ranging between 8.5 and 3.6 mm. Length fruit ranged from 13.3 mm to 21.1 mm. Cultivar '28' has the wider apical, basal and middle width of fruit. Length means of false and true peduncles were 8.4 mm and 2.9 mm; width means of false and true peduncles were 3.8 mm and 4.0 mm. Seed length ranged from 11.2 to 7.3 mm with a cultivar average of 9.3 mm. The mean width of seed was 6.8 mm and the mean thickness of seed was 3.9 mm.

DISCUSSION

The data obtained in this study have shown the great variability of the carob cultivars collected in several areas of Southern Italy. These results suggest the importance of preserving the genetic resources of carob and could be a starting point for further studies, with the aim of the selection of cultivars with the highest yields of seeds and fruits for intensive cultivation or with particular qualitative characteristics for improved industrial uses.

REFERENCES

1. Spina, P. (1986). Il carrubo. Edagricole, Bologna.

2. Battle, I. and Tous, J. (1985). Algarrobo: situaction, avances en su conocimiento y posibilidades de futuro. *I Congreso Espanol dee Frutos secos*. Reus: 357-380.

3. Crescimanno, F. G. et al. (1987). Aspetti morfologici e carpologici di cultivar di carrubo (Ceratonia siliqua L.). In: Book of abstracts. *II International carob symposium. Valencia, Spain*, Sept 29-Oct 1, A-9: 13.

4. Rotundo, A. et al. (1992). La coltura del carrubo in Basilicata. Indagini su cultivar e/o cloni di particolare interesse. In: *Atti del Convegno su Germoplasma Frutticolo. Salvaguardia e Valorizzazione delle risorse genetiche*. Alghero, Italy, 21-25 Settembre 1992, Ed. Delfino: 611-614.

5. Tutin, T. G. et al. (1978). *Flora Europea*, Vol 2: 83-84.

6. Benigni R., Capra C., Cattorini P. E. (1965). *Piante Medicinali-Chimica Farmacologia e Terapia*. Inverni & Della Beffa-Milano.

7. Fenaroli, G. (1963). *Le sostanze aromatiche naturali*. Rd. Hoepli Milano

8. Pesce, S. (1985). Aspetti economici della produzione del carrubo in Italia. *Tecnica Agricola* 37 (1): 33-61.

9. Donno, G., Panaro, A. D. (1965). Su alcune cultivar di carrubo della provincia di Bari e proposta di una scheda descrittiva (Primo contributo). *Scienza e Tecnica Agraria* Vol V, N. 4-5; 3-39.

ECONOMIC, ETHICAL AND LEGAL ASPECTS

Stakes in the Evolutionary Race: The Economic Value of Plants for Medicinal Applications

Timo Goeschl

SUMMARY. According to estimates, biologically active compounds derived from plants are–directly or indirectly–at the basis of between a quarter and half of all prescription drugs sold in the world. This implies that medicinal plants have generated considerable economic value for

Timo Goeschl is affiliated with the Department of Land Economy, University of Cambridge, Cambridge, UK (E-mail: tg203@cam.ac.uk).

This paper arose thanks to the invitation of Melpo Skoula and Christopher Johnson. It owes a lot in spirit and content to the joint work that the author has undertaken with Tim Swanson of University College, London, on the valuation and management of biodiversity. Mr. Swanson's contribution is herewith acknowledged, without implicating him in any errors contained in this paper. The author welcomes comments and suggestions.

[Haworth co-indexing entry note]: "Stakes in the Evolutionary Race: The Economic Value of Plants for Medicinal Applications." Goeschl, Timo. Co-published simultaneously in *Journal of Herbs, Spices & Medicinal Plants* (The Haworth Herbal Press, an imprint of The Haworth Press, Inc.) Vol. 9, No. 4, 2002, pp. 373-388; and: *Breeding Research on Aromatic and Medicinal Plants* (ed: Christopher B. Johnson, and Chlodwig Franz) The Haworth Herbal Press, an imprint of The Haworth Press, Inc., 2002, pp. 373-388. Single or multiple copies of this article are available for a fee from The Haworth Document Delivery Service [1-800-HAWORTH 9:00 a.m. - 5:00 p.m. (EST). E-mail address: getinfo@haworthpressinc.com].

society in the past. The theme of this paper is to explore the likely economic values of plants for medicinal applications in the future. These values arise predominantly from the potential of plants to provide templates for solutions to medical problems not yet or imperfectly solved at present. This situates the question of economic values of plants for medicinal purposes in the context of pharmaceutical R&D. Studies that have attempted to ascertain these values in R&D are then surveyed and the significant diversity in their results is discussed. What gives the paper its title is that these studies generally assume that solutions once found will work forever. For one of the most important areas of medicinal applications, infectious diseases, this is not true. There, humanity is engaged in an evolutionary race with pathogens that render solutions ineffective over time. Plant-derived biologically active compounds are stakes in this race, promising to provide solutions for problems whose nature is unknown at present. The evolutionary character of this race has significant implications for the valuation of plants for medicinal applications by both the pharmaceutical industry and society at large. *[Article copies available for a fee from The Haworth Document Delivery Service: 1-800-HAWORTH. E-mail address: <getinfo@haworthpressinc.com> Website: <http://www.HaworthPress.com> © 2002 by The Haworth Press, Inc. All rights reserved.]*

KEYWORDS. Economic value of plants, evolution, bioprospecting

INTRODUCTION

One of the undisputed achievements of the Jesuit order has been that they made available to the wider world what constituted for a long time the only effective treatment of malaria. It was one of their cardinals, Johannes de Lugo, who popularized the use of a powder, *Jesuit's powder,* that had been derived from the bark of a tree growing in present-day Ecuador. Jesuit missionaries had noticed the fever-suppressing properties[1] of the *quina* tree and subsequently developed the logistics for delivering the product from its South American origins to the European markets, where the drug was widely marketed from the middle of the seventeenth century onwards, and further east. The active compound within the powder was not identified until 1820 when two French chemists, Pelletier and Caventou, isolated what they called "*quinine*" from the bark of the *quina* tree. Quinine and the cultivation of its source, the *cinchona* tree (as it was called by then) became a business operation on a global scale, pitching the British and the Dutch against each other in fierce commercial competition over the control of quinine production. Its economic effects, particularly through its inexpensive availability to colonial powers expanding into malaria-infested corners of the globe in the 19th and early 20th century, can only guessed.

Quinine is of course only one example of the ubiquitous use of plants for medicinal purposes. Before the rise of the synthetic drug, the plant kingdom was the predominant supplier of solutions to physical ailments in human communities.[2] Ancient sources, such as the Egyptian Ebers papyrus written around 1500 BC, provide ample proof of the systematic study and use of plants for curing diseases and improving the quality of life. The link between drugs and plants was a very material one in that drugs consisted of one or several processed forms of the plant itself. In this, plants served as a direct input into the production of pharmaceuticals and their availability with respect to quantity and quality at any given point in time often determined the supply of pharmaceuticals at the end of the production chain.

The possibility of producing drugs synthetically severed the input dependence of drug production on the availability of plants as a raw material. *Aspirin*, a drug introduced by the German company *Bayer* in 1899, was probably the first synthetic drug. Although based on properties of the willow bark known for centuries, the drug itself contains a human-made compound, *acetylsalicylic acid*, as the active compound. This compound could be inexpensively produced without relying on a supply from a biological source. Its success relied on *Bayer*'s discovery that the combination of *salicylic acid*, the fundamental compound in willow bark, with *acetyl*, another chemical, resulted in a drug with greatly reduced side effects. The combination of organic chemistry with pharmacological knowledge thus did two things: it freed the supply of drugs from the often highly volatile and expensive supply of natural compounds and allowed the creation of products of superior quality (2).

The rise of synthetic drugs may be assumed to have diminished the role of plants in medicinal applications. This is true with respect to the role of plants as raw material in the drug production process. With respect to the role of plants as the source of information about possible solutions to medicinal problems, however, it is certainly wrong. *Aspirin* may be synthetic, but it could not have been created without previous knowledge about the properties of *salicin*, the major active compound in the willow bark. Likewise, the synthetic substitutes that were created for substituting quinine[3] were essentially modelled on the chemical structure of quinine itself and are therefore aptly called "analogues." This implies that the value of plants for medicinal purposes shifted increasingly from their use as an input into drug production to their use as an input into drug discovery.[4] In terms of magnitude, this means that plants play an important, and perhaps increasing, role in the age of synthetic drugs. Today, an estimated 25 percent of prescription drugs in the US are based on plants (6), possibly 50 percent world-wide (3), and for cancer drugs, the share is estimated to be in the region of 40 percent (4).

The purpose of this paper is to review the economic value of plants such as the cinchona tree or the willow for human health. For this purpose, we first need to understand the methodology underlying economic studies of the value of medicinal plants before presenting the results of selected studies. Although interesting, these studies are predominantly based on the present. The far more policy-relevant question is what the value of medicinal plants is likely to be in the future.[5] This is because the decision about the socially desirable amount of conservation efforts with regard to plants depends on assessing their future contributions to human well-being. We will therefore review some of the theoretical models used to forecast these contributions. In the last part, we will consider the implications of evolution in these estimates. If plants are stakes in the evolutionary race between people and their pathogens, their economic value to society is likely to exceed previous valuations by magnitudes.

THE ASSESSMENT OF ECONOMIC VALUES OF PLANTS

Methodology

There have been several efforts to ascertain the economic value of plants in the context of their pharmaceutical use. Since these efforts were often motivated by different problems, it is not surprising that they differ quite considerably with respect to the type of value that they attempt to measure and with respect to the assumptions that inform the numerical estimation.

With regard to the type of value, the two most common models fall in the categories of utility models and rent models. Utility models attempt to ascertain the total economic value of a resource to society. Rent models attempt to estimate the value that one of the recipients can appropriate from the resource. Utility models are therefore well suited to express society's interest in a resource while rent models will deliver good estimates of the individual economic interest of private agents (firms, consumers) in a resource. It is easy to illustrate the difference with an example. When the British company *Glaxo* undertook the R&D that would eventually result in its anti-ulcer drug *Zantac,* its valuation of the product (and–by implication–of the plant-based input, a distant relative of *liquorice*) consisted of the expected value of twelve years of revenue stream from being able to sell the drug under a patent-protected monopoly. This is the rent that *Glaxo* took into consideration for making its investment decision because it could assume to be able to appropriate these rents from the market. Even at the monopolistic market price that the company charged under its patent, however, there were presumably many customers that would have paid more for the drug than what they had to pay for it in actual fact.[6] In sum, therefore, society received more utility out of the introduction of the drug than the *Glaxo* could appropriate. This also means that *Glaxo*'s and society's valuation of the plant-based source of the drug must have differed as their valuation of the drug differed. This is because *Glaxo* looked at the patent rent created by the plant-based source while society looked at the social utility created by the plant-based source.[7] When estimating the value of plants, employing one of these two concepts will have significant effects on the result. Also, each concept faces its particular empirical difficulties with respect to data requirements, accuracy and reliability. This is also true for the distinction of gross and net values so derived. The gross value approach does not take into account the search costs, R&D costs and conservation costs that society at large or individual parties may have incurred as a consequence of drug discovery and production.[8] The results contained in studies are often complicated further by reporting the figures thus estimated in the form of a royalty to a region/community on whose territory the plants exist.

Results on the Current Economic Value of Medicinal Plants

Table 1 reports the results of two seminal studies on the current value of plants to the pharmaceutical industry. As an example of the rent based approach, Farnsworth and Soejarto (6) estimate the value of plants used in the pharmaceutical industry based on US drug sales figures. They conclude that the average value of each of the approximately 40 plant species used in the pharmaceutical industry is around $203 million US dollars per year, i.e., the equivalent of what they generate in terms of turn-over in the marketplace.

TABLE 1. Current economic value of medicinal plants.

Authors	Objective of Study	Valuation Result
Farnsworth and Soejarto (6)	Value of Plants used in Pharmaceutical Industry	$203 million per successful species per year. Based on 1980 US gross drug sales, survey data showing that 25% of all prescriptions contain one or more active plant-based agents and 40 plants account for those active agents.
Principe 1989b (3)	Value of Plants used in Pharmaceutical Industry	$1.5 trillion/year total value of plant-based drugs (US and OECD). Based on value of statistical life ($8 million in 1983 USD), percentage of lives saved by anti-cancer drugs (15%), and percentage of plant-based anti-cancer drugs (40%).

Source: Adapted from Cartier and Ruitenbeek (28).

From this we conclude that the total value of plant species within the US pharmaceutical industry is $8.12 billion US dollars per year.

Principe (3) starts from a very different angle by looking at the utility generated by plants used in the pharmaceutical industry. He examines the impact of pharmaceuticals in terms of lives saved in OECD countries by recourse to the statistical value of a human life. Using this methodology, Principe arrives at a total value of $1,500 billion US dollars, almost two hundred times the value established by Farnsworth and Soejarto (6). This starkly illustrates the impact of different value concepts and the resultant choice of methodologies on the estimation result.

Despite the methodological differences, both studies highlight the volume of benefits that plants are currently generating in the pharmaceutical industry. As the health sectors in the developed economies often account for a significant share of economic activity, this should not be too surprising. In fact, these studies are likely to underestimate their real value (cf. 8). Since they are limited both geographically and institutionally,[9] the global values of current medicinal plants may exceed their estimates by magnitudes. On the other hand, both studies only look at the gross figures rather than at the underlying costs incurred during R&D and production of the drug. This means that they are a poor measure of the efficiency of plant use in the pharmaceutical industry.

Results on the Future Values of Plants for Medical Applications

The estimates on the current value of medicinal plants are informative, but are of little policy relevance in the age of synthetic drug production. To the extent that the output of pharmaceuticals relies on the physical input of plant material, we can expect the market mechanism to ensure an efficient supply of these inputs into drug production. To the extent that the values reflect the informational input of plants into R&D, it merely informs us that society has profited from the diversity of plants in the past. The crucial question, however, is to what extent society is likely to benefit in the future. These benefits must guide the allocation of resources to the protection of plants now on the basis of their value potential for future generations. This leads us into the economic theory of bio-prospecting, i.e., the assignment of values to the diversity of plants on the basis of discovery of future drugs in a process of R&D.

Table 2 reports the findings of three studies on the values generated by bio-prospecting of plants. Here the emphasis is on how the value of future plant-based drugs can be translated into the value of the underlying resource, the diversity of plants that will be screened in the hope of finding useful substances.

Mendelsohn and Balick (9) and Aylward (10) examine the average value of a plant species resulting from a new discovery while Simpson et al. (11) look at

TABLE 2. Economic value of medicinal plants in R&D.

Authors	Objective of Study	Valuation Result
Mendelsohn and Balick (9)	Value of Plants and Land to Pharmaceutical Industry	If ALL potential drugs are discovered: $449 million NPV per successful species for a total $147 billion NPV or $48 per ha tropical rain forest. Based on NPV of new drug source and 328 plant-based drugs yet to be discovered and developed.
Aylward (10)	Net Private and Social Returns to Biotic Samples in Costa Rica	Net return per biotic sample: $21.23 (priv.) and $33.91 (social). Based on NPV of a new drug source. Social cost inclusive of costs for biodiversity protection and publicly provided taxonomic information. Success rate 1:10,000.
Simpson et al. (11)	Value of the Marginal Species to the Pharmaceutical Industry	Value of marginal species: $9,000 based on 250,000 species to test at success rate of 1:83,333 and revenue to cost ratio of 1.5. Sensitivity analysis: at success rate of 1:12,500, value is zero (redundancy); revenue cost ratio at 1.1 reduces value to $2.20.

Source: Adapted from Cartier and Ruitenbeek (28).

the marginal change in the value of plant diversity when a species that is a potential source of a new drug is added to the diverse pool. All these studies benefit from taking into account the cost of R&D, i.e., they examine "net" values. Efficiency is therefore included in the value analysis. Not surprisingly, the models are very sensitive to their parameter specification. Particularly the probability of discovery, also referred to as the "hit rate," plays a central role.

The first study rests on two important estimations, one regarding the net present value of drug discovery, estimated at around $450 million, and the other regarding the pool of undiscovered drugs left in the plant kingdom, estimated at 328 (see Table 2). For the tropical rainforest, Mendelsohn and Balick thus arrive at an estimate for the net present value of bio-prospecting on the order of $147 billion US dollars.

The second study by Aylward (10) relies on a slightly different approach to estimate the social and private rents from bio-prospecting. It explicitly takes into account the social input costs into R&D through the provision of conserved areas that provide the samples to be screened and the provision of taxonomic information (out of public research) required to identify samples. The "hit rate" is in the order of 1:10,000 for each sample and thus results in estimates of between $21 for private industries and about $34 for society at large for each biotic sample.

In the economic literature, the currently most influential article on the valuation of genetic resources was published in 1996 by Simpson et al. (11). The authors developed a search-theoretic perspective on the problem that was inspired by Brown and Swierzbinski (12). The aim of their work is to quantify the willingness of private firms to invest in the conservation of biodiversity when the value of each sample is the outcome of a Bernoulli trial (the screen). In other words, they evaluate genetic resources from the vantage point of expected private profits from research.

The typical model features a fixed probability p of identifying a valuable trait in a sample where valuable traits give rise to a product with fixed revenue R through a process of further R&D. The cost of screening a sample is fixed at level c. The expected value of a search over n samples can then be expressed as $V(n)$ which is

$$V(n) = pR - c + (1 - p)(pR - c) + (1 - p)^2(pR - c) + \ldots \tag{1}$$

The marginal value of the nth sample is then

$$v(n) = (pR - c)(1 - p)^n \tag{2}$$

The empirical problem with the formulation in equation (2) is that the probability of a "hit," p, is the most important parameter for estimating $v(n)$, but that data on p is notoriously difficult to obtain. Simpson et al. (11) solve this dilemma by evaluating the expected value of the marginal species under the most optimistic conditions. One interesting finding is that the function mapping the probability of success in any single trial to the value of the marginal species is single-peaked and strongly skewed to the right. This means that once the probability of a successful trial is such that the expected marginal value of a trial exceeds the cost of the trial, the value will rise very rapidly to its maximum value and then decrease again rapidly. This observation is crucial as it shows several points: sampling costs are an essential determinant of the marginal value, and studies that do not take these costs into account (such as 6,13) are bound to overestimate the marginal value significantly. Secondly, the fact that the marginal value of the species is not a monotonously increasing function of the probability of success in the Bernoulli trial brings an issue to the fore that had previously been overlooked by many researchers, namely the presence of substitutability between species.

The degree of relative scarcity of "successful" traits is one of the key elements in the search-theoretic perspective: more than one sample can be a "success" in the Bernoulli trial, such that once a trial has been successful, there is no further need for sampling.[10] If substitutability is very scarce, i.e., the probability of success is very low, then the marginal value is depressed since the expected revenue from the marginal trial is too low to warrant a high volume of trials. If substitutability is not scarce, then the expected revenue from the marginal trial is too low to warrant a high volume since it is very likely that a success has occurred already. In other words, if there is much redundancy within the stock of samples, a significant proportion of the samples can be discarded prior to screening with little loss of expected revenue since it is very likely that a success will be found within the remaining portion.

Based on a number of reasonable assumptions regarding the market value of a product and other parameters, Simpson et al. (11) derive an upper boundary for the willingness to pay for the marginal sample at around $9,000.[11] Their cogent conclusion is that there is little reason to expect that the industrial use of genetic resources will result in their preservation by private investors.[12] This conclusion has been confirmed in its essence by further work on optimal bio-prospecting (14).[13] To challenge these results, a very different perspective on the role of plants for human well-being is required.

THE EVOLUTIONARY RACE BETWEEN HUMANS AND PATHOGENS

Plasmodium, the pathogen that induces malaria, is one of many pathogens that pose ecological challenges to the human population. The most notorious among these pathogens are those that cause infectious diseases. On a global basis, infectious diseases remain the leading cause of death and disability (15).[14] Without doubt, their occurrence is strongly correlated with per-capita income which leads to insufficient levels of resources to be spent on medical care. But even where public health standards are in the top echelon of the world, infectious diseases remain among the leading causes of death (30). As put so aptly by MacNeill in his book "Plagues and Peoples," ". . . ingenuity, knowledge, and organization alter but cannot cancel humanity's vulnerability to invasion by parasitic forms of life. Infectious disease which antedated the emergence of humankind will last as long as humanity itself, and will surely remain, as it has been hitherto, one of the fundamental parameters and determinants of human history" (16). The message is that medicine should be conceived of as living defense system rather than a static technology. That is, this field of human activity consists of continuing efforts to combat the erosion of human-erected defenses against a hostile biological world. The defenses erected are neither absolute nor perpetual; they are constantly eroding under the pressure of the forces of nature (17). The welfare losses that society incurs due to the presence of sizeable pathogen populations are substantial. More disturbingly, these losses appear not to be decreasing over time despite considerable economic resources spent on R&D in the pharmaceutical sector.

These trends over time deserve particular attention. As explained in the introduction, *quinine* offered the first effective non-genetic response of humans to *Plasmodium*'s ecological challenge.[15] However, *quinine* lost its curative properties over time as cumulative use increased. Most strains of *Plasmodium* have become resistant against quinine and many of its synthetic successor drugs (18). Very similarly, the effectiveness of many antibiotics is decreasing over time. Figure 1 shows a particularly telling example of the occurrence of a class of *vancomycin*-resistant microbes. Use of this specific antibiotic is particular high in hospital intensive care units because of the general immune weakness of the patients concerned. It is much less frequently used in non-intensive care settings.[16] Figure 1 presents the visible correlation between the intensity of use and the speed at which resistance develops in the pathogen population. It is just one particularly striking example of the mounting evidence for the immediate relevance of the evolutionary perspective on drug development.[17]

As evolutionary theorists do not tire of telling us, ecological relationships like the one between humans and their pathogens are dynamic over time (19). Very often, they involve a rapid succession of different "strategies" of attack (pathogen) or defense (host). Ecological relationships of this kind have been termed "Red Queen" races when a constant change of strategy is necessary in order to prevent an increase of losses to the opponent in this ecological race (20).[18] The interaction between humans and their pathogens carries all the characteristics of such a Red Queen race.

FIGURE 1. Percentage of nosocomial enterococci reported as resistant to vancomycin isolated from infections in patients in intensive-care units (ICUs) and non-ICUs, by year–National Nosocomial Infections Surveillance System, 1989-March 31, 1993.*

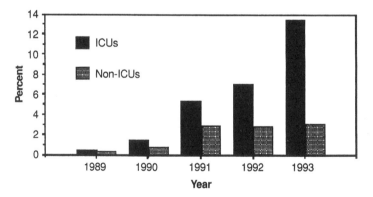

* Treatment options for patients with nosocomial infections associated with vancomycin-resistant enterococci are limited, often to unproven combinations of antimicrobials or experimental compounds.
Source: Centers for Disease Control (27).

What does the presence of a Red Queen race imply for the valuation of plants for medicinal purposes? Solutions against pathogens, once found, do not remain effective forever and new strategies to fend off pathogens have to be devised. This means that our search for active compounds is likely to be a continuous process without a known time horizon. This changes our modelling of the search process significantly because it implies the screening for active compounds on a continuous basis in order to find solutions to problems that do not exist at present and are not foreseeable in their specific form prior to their emergence. In order to carry out this continuous screening process, we need to be able to have permanent access to a library of solution concepts. The reason why plants are likely to continue to offer such a library of templates for solutions to pathogen problems is apparent: plants have co-evolved with pathogens over millions of years.[19] For evolutionary biologists, it is therefore no surprise that the genetic make-up of plant populations contains the solutions to many problems posed by pathogens to higher organisms (21-23).[20]

In a recent paper, Goeschl and Swanson offer a first approach to modelling a situation where biological innovations, such as drugs or new crop varieties, induce a response from pathogens in relation to the scale of their application (24). The aim of the paper is to explain the investment in biodiversity preservation of industry in such a situation and to contrast it with society's optimal choice in the same setting. Their approach, based on a modified version of the Aghion and Howitt (25) model of endogenous growth through creative destruction, highlights the fundamental divergence of the public and the private

FIGURE 2. Investment in conservation at different rates of biological innovation.

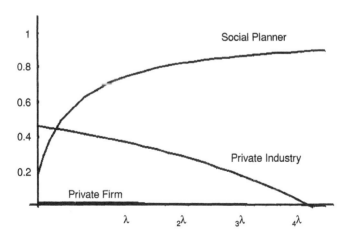

valuations of plant diversity as an input into R&D for medicinal purposes. The results differ radically from the previous search-theoretic approaches.

Figure 2 shows one of the results of this paper, namely the optimal investment size of society and private industry as a function of the speed at which pathogens react to each unit of drug being applied. It reveals two fundamental insights. The first is that, for most parameter specifications, social valuation of plant diversity for R&D is significantly above the valuation of the industry and far above the valuation of the individual firm. This means that in settings of this type, estimates about private valuations may be very bad proxies for society's valuation of this resource. The second insight is that society and industry react exactly opposite to an increase in the response rate of pathogens. Society increases its valuation in situations where the response rate of pathogens increases, i.e., where resistance develops more rapidly. Industry, however, decreases its valuation of the resource when evolution speeds up.

This means that the economic value of plants with potential medicinal properties against pathogens will increase for society if the evolutionary race "heats up," but the we will observe industry investing less and less in their conservation.[21]

CONCLUSION

This paper has offered an overview of some of the economic literature that tries to assess the economic value of medicinal plants. These values have certainly been considerable in the past, in terms of plants as both a physical and as an informational input into drug production, although the comparison of estimates is complicated through methodological complexities. Since the start of synthetic drug production, the value of plants for medicinal purposes has shifted increasingly from their use as an input into drug production to their use as an input into drug discovery. For these values to result in a conservation of plant diversity, however, we need to demonstrate its value as a source of new drugs in the future. If the assumption is that solutions once found remain effective forever, this value is very much in question as the recent bio-prospecting literature aims to demonstrate. In the opposite case, theoretical models clearly indicate that the continuing need of humanity to discover new solutions to problems whose nature is unknown at present results in a drastic increase of society's valuation of plant diversity. This contrasts with the reaction of industry whose valuation of plants as an input into R&D is likely to decrease. In all areas where humanity is engaged with pathogens in some form of an evolutionary race, we can therefore expect that private industry valuation of plant diversity is a poor guide to the social values contained in them. The challenge for society in managing its natural environment is to carefully manage its plant-based stakes in its evolutionary race with pathogens that will–in the words of William McNeill–continue to shape "human history."

NOTES

1. For lack of space, the question of precedence of discovery by the Incas is not discussed here although it would conveniently lead to the interesting question of the role of ethnobotany in drug discovery.

2. In fact, the plant kingdom supplied solutions to quite different problems as well, but the prolific use of, e.g., belladonna to dispose of undesirable members of the imperial family in Ancient Rome is not the topic of this paper. An excellent coverage of these aspects is contained in (1).

3. The impetus for finding a substitute was triggered by the abuse of monopoly power over quinine production by the Dutch at first and by the drying up of supplies of quinine after the destruction of the Dutch plantations in Java by the Japanese later.

4. On the implications of this shift for the relationship between the North and the South, see (5).

5. The curious implication of taking this perspective is that many of the plants to be used in the future are presently not subsumed under the heading "medicinal plants" since their medicinal properties have not yet been identified.

6. Let us assume for the sake of the argument that patients would have had to purchase the drug at their expense, rather through some other institution such as a national health service or insurance.

7. This is not to say that the utility-based model is more accurate. Depending on the question, one or the other of these concepts will be more applicable.

8. The assignment of the correct property rights structure is crucial for the efficiency of the drug discovery process. This is discussed in the context of genetic resources in (7).

9. These studies do not take into account the use of plants that is not accounted for in market transactions (6) or lies outside a disease-specific interest (3).

10. The biological equivalent is that there may be abundance of species with very similar genetic make-up and that the same bio-active compound (that results in a "success" in the screen) can be produced by species of completely different genetic structure.

11. Simpson et al. (11) translate this estimate into an per-area willingness-to-pay for conservation using the common MacArthur-Wilson approach of relating habitat size to the extant stock of biodiversity. Based on this computation, the maximal willingness to pay for a hectare of biodiverse lands in Western Ecuador, one of the "biodiversity hot spots," is US$20.63. The rainforests of the Amazon elicit only US$2.59 per hectare. This implies that most areas with even extraordinary biodiversity do not justify significant payments from the pharmaceutical industry for their preservation.

12. This search-theoretic approach has been considerably refined in order to include differences in the value of individual hits or situations in which the assumption of independence between the probability distribution of individual traits is violated.

13. Rausser and Small (14) successfully criticize the brute-force approach to search modeled in the paper by Simpson et al. (11) and propose an alternative search specification. However, this essentially leads to little more than a redistribution of the total value of bio-prospecting among different habitats, favoring those with higher endemic diversity.

14. Infectious diseases were responsible for 56.3% of premature deaths world-wide in 1990, measured in disability-adjusted life years (non-communicable diseases: 26.5%) and for 45.8% of the total loss of disability-adjusted life years in 1990 (non-communicable diseases: 42.2%).

15. A previous genetic response was the development of sickle-cells in human bodies. This genetic mechanism prevents the spread of malaria cells in the body through a self-destruction mechanism in red blood cell that gets activated in response to infection.

16. Vancomycin is a very powerful antibiotic that has customarily been regarded as the "last line of defense" in the present array of antibiotics.

17. This development of resistance has immediate economic effects. In a study detailing per patient costs, the hospital stay of a patient with methicillin-susceptible *S. aureus* (MSSA) was $24,280, compared with $64,370 for a patient with methicillin-resistant *S. aureus* (MRSA) (27).

18. The term originates from Lewis Carroll's "Alice in Wonderland" where a similar setting is positively and unambiguously identified, probably for the first time.

19. Plants are not mobile and can therefore not evade their predators. This has led to the speculation that plants are more chemically creative than animals.

20. Goeschl (29) develops the implication of the Red Queen races for genetic resource management in agriculture. Many of the results derived there also apply to pharmaceuticals, particularly the need for diversified strategies for pathogen management [see also the optimal rule of application developed in (26)].

21. The reasons for the differences are manifold and are discussed in detail in (24).

REFERENCES

1. Mann, J. (1994): *Magic, Murder and Medicine*. Oxford: Oxford University Press.

2. Schoenberg-Albers, A. (1995): The pharmaceutical discovery process. In: Swanson, T. (ed.): *Intellectual property rights and biodiversity conservation: an interdisciplinary analysis of the values of medicinal plants*. Cambridge: Cambridge University Press.

3. Principe, P. (1989b): *The economic value of biological diversity among medicinal plants*. Paris: OECD.

4. Principe, P. (1989a): The economic significance of plants and their constituents as drugs. In: Wagner, H., H. Hikino and N. Farnsworth (eds.): *Economics and medicinal plant research 3*. London: Academic Press, 655-6.

5. Swanson, T. (1996): The Reliance of Northern Economics on Southern Biodiversity: Biodiversity as Information. *Ecological Economics*, 1-6.

6. Farnsworth, N. and D. Soejarto (1985): Potential consequences of plant extinction in the United States on the current and future availability of prescription drugs. *Economic Botany* 39(2), 231-240.

7. Swanson, T. and T. Goeschl (2000): Property rights issues involving plant genetic resources: implications of ownership for economic efficiency. *Ecological Economics* 32(1), 75-92.

8. Lambert, J., J. Srivastava, and N. Vietmeyer (1997): *Medicinal plants: Rescuing a global heritage*. Technical Paper, no. 355. Washington, DC: World Bank.

9. Mendelsohn, R. and M. Balick (1985): The value of undiscovered pharmaceuticals in tropical forests. *Economic Botany* 51(3), 328.

10. Aylward, B. (1993): *The economic value of pharmaceutical prospecting and its role in biodiversity conservation*. Discussion Paper 93-05. International Institute for Environment and Development. London: IIED.

11. Simpson, R.D., R.A. Sedjo, J.W. Reid (1996): Valuing biodiversity for use in pharmaceutical research. *Journal of Political Economy* 104(1), 163-185.

12. Brown, G. and J. Swierzbinski (1988): Optimal genetic resources in the context of asymmetric public goods. In: Smith,V.K. (ed.): *Environmental Resources and Applied Welfare Economics*. Washington: RFF, 293-312.

13. Pearce, D.W. and S. Puroshothaman (1992): The economic value of plant-based pharmaceuticals. In: T. Swanson (ed.): *Intellectual Property Rights and Biodiversity Conservation*. Cambridge: Cambridge University Press.

14. Rausser, G. and A. Small (2000): Valuing research leads: Bioprospecting and the conservation of genetic resources. *Journal of Political Economy* 108(1), 173-206.

15. World Bank (1993): World development report 1993: Investing in health. New York: Published for the World Bank by Oxford University Press.

16. McNeill, W.H. (1994): *Plagues and peoples*. Harmondsworth: Penguin Books.

17. Swanson, T. and T. Goeschl (1999): Ecology, information, externalities and policies: The optimal management of biodiversity for agriculture. In: Peters, G.H. and J. von Braun (eds.): *Food Security, Diversification and Resource Management: Refocusing the Role of Agriculture*. Brookfield: Ashgate.

18. Cornes, R., N.V. Long, S. Shigemura (1995): Drugs, Pest and Resistance. *Mimeo*. Department of Economics. McGill University.

19. Anderson, R.M. and R.M. May (1991): *Infectious Diseases of Humans. Dynamics and Control*. Oxford Science Publications. Oxford: Oxford University Press.

20. Maynard Smith, J. (1976): A comment on the Red Queen. *The American Naturalist* 110, 325-330.

21. Myers, N. (1997): Biodiversity's genetic library. In: Daily, G.C. (ed.): *Nature's Services: Societal Dependence on Natural Ecosystems*. Washington: Island Press.

22. Frank, S. (1994): Recognition and polymorphism in host-parasite genetics. *Philosophical Transactions of the Royal Society, London, Series B* 346, 191-197.

23. May, R.M. and R. Anderson (1983): Epidemiology and genetics in the coevolution of parasites and hosts. *Proceedings of the Royal Society London B* 219, 281-313.

24. Goeschl, T. and T. Swanson (1999): *Endogenous Growth and Biodiversity: The Social Value of Biodiversity for R&D*. GEC 99-12. Centre for Social and Economic Research on the Global Environment. Norwich and London: CSERGE.

25. Aghion, P. and P. Howitt (1992): A model of growth through creative destruction. *Econometrica* 60, 323-51.

26. Goeschl, T. and T. Swanson (2000): *Lost Horizons: The Interaction of IPR Regimes and Resistance Management.* Invited Paper to the Harvard University Centre for International Development's Symposium on Antibiotic Resistance, February 28, 2000.

27. Centers for Disease Control (1994): *Addressing Emerging Infectious Disease Threats: A Prevention Strategy for the United States.* MMWR 43(RR-5). Atlanta: CDC.

28. Cartier, C.M. and H. Ruitenbeek (1998): *Preliminary Review of Studies and Models for Marine Biodiversity Valuation.* Study prepared for the World Bank. Washington, DC: World Bank.

29. Goeschl, T. (1998): *The Economics of the Red Queen: Biodiversity in Agriculture and Long-Term Food Supply.* Paper presented at the 2nd International Conference of the European Society for Ecological Economics, Geneva.

30. McGinnis, M. and W.H. Foege (1993): Actual causes of death in the United States. *Journal of the American Medical Association* 270, 2207-12.

European Plant Intellectual Property

Margaret Llewelyn

SUMMARY. This paper outlines the current level of intellectual property protection for plant material within Europe. In particular, it focuses on specific areas of concern for the plant breeder such as access to plant material for research purposes and the extension of protection to derived material. *[Article copies available for a fee from The Haworth Document Delivery Service: 1-800-HAWORTH. E-mail address: <getinfo@haworthpressinc. com> Website: <http://www.HaworthPress.com> © 2002 by The Haworth Press, Inc. All rights reserved.]*

KEYWORDS. Intellectual property rights, patents, plant variety rights

INTRODUCTION

In recognition of the economic importance of biotechnological innovation, the World Trade Organization requires member states to provide patent protection for all inventions irrespective of the field of technology (Article 27(1)). Members are given the option of excluding plants, but they must provide a form of intellectual property protection. The option to exclude plants from patent protection has not been exercised in Europe with the result that plant varieties are protectable by plant variety rights and all other plant material is patentable. The concern raised by the availability of private rights over plant

---

Margaret Llewelyn is affiliated with Sheffield Institute for Biotechnology Law and Ethics (SIBLE), University of Sheffield, UK.

[Haworth co-indexing entry note]: "European Plant Intellectual Property." Llewelyn, Margaret. Co-published simultaneously in *Journal of Herbs, Spices & Medicinal Plants* (The Haworth Herbal Press, an imprint of The Haworth Press, Inc.) Vol. 9, No. 4, 2002, pp. 389-397; and: *Breeding Research on Aromatic and Medicinal Plants* (ed: Christopher B. Johnson, and Chlodwig Franz) The Haworth Herbal Press, an imprint of The Haworth Press, Inc., 2002, pp. 389-397. Single or multiple copies of this article are available for a fee from The Haworth Document Delivery Service [1-800-HAWORTH 9:00 a.m. - 5:00 p.m. (EST). E-mail address: getinfo@haworthpressinc.com].

material is that they will inhibit research and deny those engaging in plant research access to essential genetic information.

Whilst this paper concentrates on European plant intellectual property protection, this protection must be looked at in the context of global trends in intellectual property provision. Of particular relevance are:

1. The International Convention on the Protection of New Plant Varieties (UPOV) which sets down the minimum standard required for the grant of a plant variety right. The grant, and enforcement of the right, is left to member states. The UPOV Convention was last revised in 1991. The European Union (EU) is a member of the 1991 Convention through the adoption of the Community Regulation on Plant Variety Rights,[1] as are all EU member states, although many are still in the process of amending their national plant variety laws.[2]

2. The Agreement on Trade Related Aspects of Intellectual Property Rights (TRIPS) which is overseen by the World Trade Organization (WTO). As already stated, members are required to provide patent protection for *all* types of inventions irrespective of the field of technology, with certain optional exceptions. These relate to inventions (a) the exploitation of which would be contrary to morality, which includes harm to human, animal, or plant life or health and to the environment (Article 27(2)); (b) diagnostic, therapeutic and surgical methods for the treatment of humans or animals (Article 27(3)(a)); and (c) plants and animals, provided that either patent protection and/or an effective *sui generis* system of protection is available for plant varieties. European patent law currently contains all three exclusions.

If a member state fails to comply with the provisions of the TRIPS Agreements then action will be taken by the WTO. Decisions over granting patents are taken at the local level as is any enforcement action. This gives rise to some disparity in the interpretation and application of the rights.

Article 30 of the TRIPS Agreement permits limited exceptions to the right conferred by a patent; however, it is not clear whether derogations, such as the free use of patented material for research purposes and the practice of farm saved seed, would come within the scope of this provision. Article 31 of the Agreement also allows member states to grant compulsory licenses over patented material. It is not clear, however, if this is a general provision which can be exercised whenever the member state thinks necessary, or if it only applies in exceptional circumstances. These will be matters to be decided by the WTO.

Both the TRIPS Agreement and UPOV Conventions are ostensibly intended to define when private rights should be granted. However, there is a clear presumption in favour of granting rights and any limitations on such rights are invariably restrictively applied. Together they form the legislative framework within which recent European plant intellectual property has developed.

EUROPEAN PLANT INTELLECTUAL PROPERTY

The two main types of protection are the plant variety right and the patent.

Plant Variety Protection. Plant variety rights can obtained either at the national or Community level but not both. The main differences between the two are: (a) national rights are purely local in effect whilst a Community right is enforceable, via the national courts, in all EU member states; (b) the Community right is based on the 1991 UPOV Convention whereas some EU member states have yet to bring their national legislation into line with the revised Convention; and (c) certain new concepts have been introduced in the Community right, such as a system of additional payments or remuneration for farm saved seed, which have not yet been introduced fully at the national level.

For the purposes of this paper I shall concentrate on Community plant variety rights.

Plant variety rights are available over plant groupings which collectively are distinct from other varieties which are common knowledge and are uniform and stable following repeated reproduction.[3] To qualify for protection the variety must also be new in that it must not have been sold in the Community more than one year before the date of filing the application or more than four years outside the Community.

The right is granted following two years of trials during which the variety is assessed against other control varieties and lasts for 30 years for trees and shrubs and 25 years for all other varieties. No species of plants are excluded nor is there an option to exclude varieties on the grounds of morality.

If granted, the right extends over the variety and, in particular, over the reproductive material of the variety. In certain circumstances it extends to material directly obtained from the variety, provided this material is (i) capable of producing whole plants of the variety and (ii) it has been obtained through an authorized use. The right does not extend to component parts of the variety which do not produce whole plants nor to processes for producing plant material.

The holder is entitled to prevent all others from producing or reproducing the protected variety, conditioning the variety for the purposes of propagation, offering to sell or selling, exporting or importing the variety.

The rationale behind the grant of a plant variety right is the protection of the time invested in producing new varieties. It is not necessary to show that the results were particularly novel, or non-obvious. This can be contrasted with the patent system which protects non-obvious novel results.

Patent Protection. European patent law is, at present, a mix of EU and non-EU legislation. The governing Convention is the European Patent Convention (EPC) which is a non-EU instrument and not subject to EU dictate or legislation. All existing EU member states are parties to the European Patent Convention. The EPC has had two key effects on European patent law. Firstly, a single application results in a patent which is enforceable in as many member states of the EPC as the applicant designates. The effect of such a grant is to give the European patent the same national status as a patent granted by a national office. Secondly, therefore, to ensure that members provide the same type of right, national laws must conform to the provisions of the EPC.

Patents are available over inventions (products or processes) which are novel (not previously known in that form anywhere in the world), involve an inventive step (not obvious to a person skilled in the art), are capable of industrial application and which are sufficiently disclosed in the patent specification. The term "invention" is not defined; however, the Convention does state what will not be regarded as an invention. This includes discoveries, and inventions which are excluded from patent protection for reasons of public interest. This last includes inventions which would be contrary to morality, plant and animal varieties and essentially biological processes for the production of plants and animals.

As the granting criteria are not defined in any detail there has been considerable national variation in the interpretation and application of patent protection, particularly in respect of inventions involving genetic material. In an attempt to clarify the situation, in 1998 the European Union introduced a Directive on the Legal Protection of Biotechnological Inventions.[4] The Directive is currently being implemented into the national laws of the EU.[5]

The Directive has not met with universal approval. The Commission was constrained in what it could do, as EU legislation has no direct effect on the operation or application of the EPC. If the Directive had introduced new principles of patent protection, then member states could have been in violation of their obligations under the EPC, not to mention the creation of possible conflict and contradiction between EU patent law and the EPC.[6] The Directive remains the subject of intense controversy and it is the subject of a challenge, brought by the Netherlands and Italy, before the European Court of Justice.[7] At best the Directive merely provides guidance on how existing patent law should be applied and this function has been recognized by the European Patent Office (EPO) which in June 1999 adopted the Directive into its Implementing Rules for the purposes of supplementary interpretation.

Both the Community Regulation and the Directive are very new pieces of legislation and their exact scope is unclear. This paper will, therefore, simply set out those aspects of the rights which are likely to have the most impact on plant research. It will be for the courts, however, to determine the full extent of the protection which is available.

KEY CONCERNS

Protectable Subject Matter

Plant variety rights protect plant varieties and, in particular, those parts of the plant which constitute the reproductive material of the variety. Protection does not extend to processes for producing new plant varieties. As the right is intended to protect the time invested in breeding new varieties rather than rewarding innovation, no distinction is drawn between a variety which results from genetic engineering and one which results from the cultivation of a discovery. All that the breeder has to show is that the variety is distinct, uniform and stable in its essential characteristics following repeated reproduction.

Patent protection, on the other hand, can extend to *any* material, product or process, which is novel, inventive, capable of industrial application (these are broadly defined) provided it is not specifically excluded.

The categories of excluded material are of particular significance, as it is clear both from the case law of the European Patent Office[8] and from the wording of the EU Directive[9] that these are to be restrictively applied. The four which are relevant here are discoveries, plant varieties, essentially biological processes and inventions which would be contrary to morality.

Discoveries. In patent law a discovery is information for which a use has yet to be found. Once a use is found, provided that use is both novel and inventive, then it is patentable. For example, naturally occurring material, which has been isolated from its natural environment and for which a novel use has been found will, provided it can be shown that the isolation and novel use are inventive, be patentable. Equally where commonly known genetic information is used in a novel and inventive way, then it is possible to obtain patent protection over the material in that novel and inventive form. The EU Directive is quite specific that it is not possible to obtain patent protection for genetic material for which a function or use has yet to be found.[10] This is intended to prevent patents being filed over basic genetic information.

Plant Varieties. Both the EPC and the EU Directive specifically exclude plant varieties from patent protection[11] because it was thought that plant variety rights were more appropriate. For many years the exclusion appeared to apply to *all* plant material; however, as a result of the increased development of GM, crops there has been pressure for the exclusion to be more restrictively applied. Recent case law of the EPO and Article 4(2) of the EU Directive state quite clearly that the exclusion only applies to plant varieties as such.[12] This means that patent claims can be made over genes, individual plants, and plant species even where this encompasses a number of plant varieties.

It should also be noted that Article 64(2) of the EPC and Article 8(2) of the Directive state that where a patent has been granted over a process, the protection conferred by that patent also extends to any product made as a direct result of using the protected process. This means that where a patent covers a process of producing plant varieties and a plant variety is made using that process, the patent will extend to the variety as well as to the process.

Patent protection is, therefore, available over *any* plant material provided it is novel, inventive, has an industrial application and does not take the specific form of a plant variety.

Essentially Biological Processes (EBP). EPBs were traditionally excluded from patent protection as they were not regarded as technical. The degree of technical intervention is critical. Both the Directive and EPC state that a biological process is only excluded where it consists entirely of natural phenomena such as crossing.[13] It would seem that where there is any technical intervention by man, no matter how small this might be, then, provided the resulting process is novel, inventive and capable of industrial application, it will be patentable.

Morality. Article 53(a) of the EPC excludes inventions, the publication or exploitation of which would be contrary to morality. As with the other categories of excluded material this is restrictively applied. The EPO case law shows clearly that a patent will only be rejected on the grounds of morality if the sole purpose of the invention is to cause harm which would be regarded as abhorrent.[14] Where an invention is capable of providing a benefit then, even if it is also capable of causing harm, the patent will be granted, providing the benefit outweighs the harm. Article 6 of the Directive slightly refines Article 53(a). Article 53(a) states that where "publication or exploitation" of the patented invention would be contrary to morality then a patent will be refused. Article 6 states that it is the "commercial exploitation" which must be contrary to morality. This means that a product resulting from unethical research work would not be unpatentable under the Directive whereas it might be under the EPC.

The second key issue is the extent of protection granted by a patent.

Scope of the Right

Research and Extension of the Right. Protection granted under plant variety rights extends only to that material of the variety which is capable of producing whole plants. The right does not prevent others from using the protected material for research or breeding purposes. However, where the result of a breeding programme is a variety which can be shown to be essentially derived from the original protected variety, then the consent of the plant variety rights holder will be necessary before the essentially derived material can itself be exploited, or rights granted over it. What precisely constitutes an "essentially derived" variety is at present unclear.

A patent allows the patent holder to prevent an unauthorized use of the protected material with only very limited exceptions. Whilst it is ostensibly possible to use patented material for research purposes, this use is restricted in that it must be for non-commercial purposes (normally taken to mean it must be for experimental purposes) only. Protection extends to any material within which the protected invention can be found or which results from the direct use of a patented process. Article 30 of the TRIPS Agreement does permit certain limitations on the right provided by a patent but it is not clear whether the WTO will accept the research exemption as one of them.

Even where protected material is freely available for research purposes this does not mean that the results of that research will be free from the patent once the research has been completed. The Directive clearly states in Articles 8 and 9 that any material within which the protected material is found will be covered by the patent even where the material has been made available for use in research. Put more simply, patent protection extends to any material within which the protected material can be found, no matter how that material has been derived. In determining the scope of protection the sole deciding factor is what the patent holder has claimed within the patent specification. Clever patent drafting will enable the holder to claim rights over most uses, and forms, of the protected material.

Farmers' Privilege. The right of farmers to retain seed from one year to the next without having to pay an additional remuneration to the plant breeder has been regarded as fundamental to the relationship between the breeder and the farmer. However, the right has been criticized. As a result, the Community Plant Variety Right now permits only a limited number of crop types to be subject to the farm saved seed exemption.[15] The farmer is still required to pay an equitable remuneration, sensibly lower than the price originally paid (this is commonly taken to means between 50-90% of the original price). Only "small farmers"[16] are exempt. Any other retention of harvested material will be an infringing act.

Until the adoption of the EU Directive there was no equivalent provision in patent law. Article 11(1), however, introduces the concept and this states that a farmer may retain seed from protected plant material on the same basis as under the Community plant variety right.

It is not certain if the WTO will accept the farm saved seed exemption as an acceptable limitation under Article 30 of TRIPS.

Compulsory Licensing

Under Community plant variety rights the Community Office can grant a compulsory exploitation right to a person other than the plant variety rights holder if the Office considers it to be in the public interest. It is not clear when, and in what circumstances, this right will be granted.

Ostensibly the same provision applies within European patent law. However, as patent offices see the protection of the interests of the patent holder as their primary obligation, compulsory licenses are very rarely granted as they are regarded as diluting the right. Also, a compulsory license will only be granted if the patentee has already rejected an application for a license. Patent offices are very unwilling to force a patent holder to do that which he has already refused to do. The Directive, like the EPC, is silent on the matter of compulsory licensing but it does introduce the concept of compulsory cross license where a patent holder cannot exploit his patent invention without infringing a plant variety right and vice versa. However, the license is only available if the person seeking it can show that their invention or plant variety constitutes significant technical progress of considerable economic interest.

Article 31 of the TRIPS Agreement permits the granting of compulsory licenses in limited circumstances based on proving public interest. It is unclear as to the extent to which member states of TRIPS have the freedom to decide when they will grant such licenses. This will be a matter for the WTO.

CONCLUSION

It can be seen that, as a result of recent legislative activity in Europe, protection is available for all types of plant material. However, increased protection for rights holders is likely to come at a price-reduced access for third parties. It is questionable whether this is a price which those engaged in plant research will wish to pay.

NOTES

1. *Council Regulation No 2100/94 on Community Plant Variety Rights.*

2. For example, Austria, Finland, France, Ireland, Italy and Portugal are members of the 1978 Act whereas Denmark, Germany, the Netherlands, Sweden and the UK have ratified the 1991 Act. Greece is currently not a member of UPOV.

3. Article 6.

4. Directive 98/44/EC.

5. For example, Denmark and Finland have already implemented the Directive, the UK hopes to implement by the end of July 2000, the German implementation is in doubt due to the Green Party and France is refusing to implement the Directive as it contradicts the French bioethics laws which forbid the patenting of any part of the human body.

6. For a discussion of this, see Llewelyn, M. *The Legal Protection of Biotechnological Inventions: An Alternative Approach* (1997) 3 European Intellectual Property

Review 115-127; Llewelyn, M. *The Legal Protection of Plant Varieties in the European Union: A Policy of Consensus, or Confusion?* (1997) 2 Bioscience Law Review 50-61.

7. *Case C-377/98.*

8. See in particular *Novartis* (2000) European Patent Office Reports 303.

9. See Articles 2,3,4, and 5.

10. Article 5(3).

11. Articles 53(b) and 4(1)(a), respectively.

12. In the *Novartis* decision, note 8, it was held that the exclusion only applied to plant material which could be protected by a plant variety right and that it did not apply to any other plant material. This is reiterated in Article 4(2) of the Directive and in Recital 29.

13. Article 2(2) and *Novartis* note 8.

14. See, for example, *Plant Genetic Systems/Glutamine synthetase inhibitors* (1995) European Patent Office Report 357.

15. Article 14. The crop types are fodder plants, cereals, potatoes and oil and fibre plants.

16. This is defined according to Council Regulation (EEC) No 1765/92 of 30 June 1992.

The Effects of Different Nitrogen Doses on *Artemisia annua* L.

Filiz Ayanoğlu
Ahmet Mert
Saliha Kırıcı

SUMMARY. The purpose of this study was to determine the effects of different nitrogen doses (0, 6, 12 and 18 kg/da) on the plant parameters, yields and quality of *Artemisia annua* L. ecotype from Adana, Turkey. A two-year experiment was conducted between 1996-1998 under Adana ecological conditions. During this study, plant height (cm), number of branches per plant, fresh herb yield (kg/da), dry herb yield (kg/da), dry leaf yield (kg/da), essential oil content (%) essential oil yield (l/da), artemisinin content (%), and artemisinin yield (kg/da) were investigated. The results indicated that application of different N doses statistically affected the essential oil content and oil yield only the first year of the study. But artemisinin content and yield were not significantly affected. *[Article copies available for a fee from The Haworth Document Delivery Service: 1-800-HAWORTH. E-mail address: <getinfo@haworthpressinc.com> Website: <http://www.HaworthPress.com> © 2002 by The Haworth Press, Inc. All rights reserved.]*

KEYWORDS. *Artemisia annua* L., annual wormwood, nitrogen, artemisinin, essential oil

Filiz Ayanoğlu and Ahmet Mert are affiliated with the Department of Field Crops, Faculty of Agriculture, Mustafa Kemal University, 31034 Antakya, Hatay, Turkey.

Saliha Kırıcı is affiliated with the Department of Field Crops, Faculty of Agriculture, University of Çukurova, 01330 Adana, Turkey.

[Haworth co-indexing entry note]: "The Effects of Different Nitrogen Doses on *Artemisia annua* L." Ayanoğlu, Filiz, Ahmet Mert, and Saliha Kırıcı. Co-published simultaneously in *Journal of Herbs, Spices & Medicinal Plants* (The Haworth Herbal Press, an imprint of The Haworth Press, Inc.) Vol. 9, No. 4, 2002, pp. 399-404; and: *Breeding Research on Aromatic and Medicinal Plants* (ed: Christopher B. Johnson, and Chlodwig Franz) The Haworth Herbal Press, an imprint of The Haworth Press, Inc., 2002, pp. 399-404. Single or multiple copies of this article are available for a fee from The Haworth Document Delivery Service [1-800-HAWORTH 9:00 a.m. - 5:00 p.m. (EST). E-mail address: getinfo@haworthpressinc.com].

INTRODUCTION

There are 22 *Artemisia* species native to Turkey which are annual, biennial and perennial (2). *Artemisia annua* L. (Compositae) (annual wormwood) is a herbaceous annual, approximately 0.5 to 2 m high (1,7). Artemisinin which is gathered from *A. annua* L. has special value because of its antimalarial effect (10). As the compounding of artemisinin is complex and has a very low yield, only extraction from cultivated plant material is economically viable (3). In addition to artemisinin, *A. annua* L. also produces an essential oil in its leaves and flowers which can be used in perfumery and as an antimicrobial agent (6,11).

MATERIALS AND METHODS

The purpose of this study was to determine the effects of different nitrogen doses (0, 6, 12 and 18 kg/da) on the plant parameters, yields and quality of *A. annua* L. ecotype from Adana, Turkey. A two-year experiment was conducted between 1996-1998 under Adana ecological conditions. *Artemisia* seeds were sown in January and transplanted to the field in March. In each treatment, half of the N dose was applied at planting, the other half was applied one month after planting. Also, 6 kg/da P_2O_5 was applied in each plots at planting. The experimental design was a completely randomized block with three replications. Plots were four rows of 3 m length with inter and intra row spacings of 30 and 15 cm, respectively. The plots were harvested at full flowering period (in September). The essential oil content in plant material was obtained by a Neo-Clavenger distillation apsparatus. Extraction of artemisinin was carried out according to the method of Gupta et al. (5). Artemisinin was detected and quantified by thin layer chromatography (TLC) in combination with a densitometer.

RESULTS AND DISCUSSION

Plant Height. The effect of nitrogen doses on the plant height of *A. annua* L. was statistically significant ($p < 0.01$) only the first year. In 1997 the greatest plant height (250.1 cm) was obtained from N_{18} and in 1998 obtained from N_{12} with 235.5 cm (Table 1). N doses increased the plant height in both years of the study. N fertilization stimulates the vegetative development of the plant: the greater the application of N, the greater the height of the plant. Our findings were higher than those determined by Martinez and Staba (7).

The Number of Branches per Plant. The effects of nitrogen doses on the branch number per plant of *A. annua* L. were significant ($p < 0.05$) in both years of the experiment. The highest branch number was obtained from N_{18} with 90.93 in 1997 and from N_6 (82.67) and N_{12} (82.53) in 1998 (Table 1). In 1997, the number of branches were determined to increase in relation to increasing nitrogen doses. Our findings were higher than those of Elhag et al. (4).

Fresh Herb Yield. In terms of yield of fresh herb, the differences among the N applications in both years were significant ($p < 0.01$). The highest fresh herb yield was obtained from N_{18} with 4737 kg/da in 1997 and from N_{12} with 4211 kg/da in 1998 (Table 1). In 1997, fresh herb yield increased in relation to increasing N doses. In 1998, it increased with N_{12} application but with N_{18} it was determined to decrease. The fresh herb yield lower limit value (2300 kg/da-3300 kg/da) determined by Simon et al. (9) was lower than our findings, but the upper limit value was in the range of our findings.

Dry Herb Yield. In two experiment years the difference among the nitrogen applications in terms of dry herb yield was significant ($p < 0.01$). In 1997, the highest dry herb yield with 2504 kg/da was obtained from N_{18}. In 1998, the highest yield with 2377 kg/da was obtained from N_{12} (Table 2). In 1997, dry herb yield increased in relation to the increasing nitrogen doses, including the highest N dose (N_{18}). In 1998, it was determined to increase to application of N_{12} but decrease to application of N_{18}.

Dry Leaf Yield. In both experiment years the effect of nitrogen doses on dry leaf yield were statistically not significant. The highest dry leaf yield with 661.9 kg/da was obtained from N_{18} in 1997 and 673.3 kg/da was obtained from N_{12} in 1998 (Table 2). It was determined that N application did not increase dry leaf yield; however, previous research indicated that N application increases the dry leaf yield about two-fold (8).

TABLE 1. The effects of nitrogen doses on plant height, the number of branches per plant and fresh herb yield of *A. annua* L. in 1997 and 1998.

N Doses (kg/da)	Plant Height (cm)			Branch Number/Plant			Fresh Herb Yield (kg/da)		
	1997	1998	Mean	1997	1998	Mean	1997	1998	Mean
N_0	205.7 c	210.6	208.2	79.07 b	72.67 b	75.87	3099.4 c	3099.0 b	3099.2
N_6	221.2 bc	225.9	223.6	83.20 ab	82.67 a	82.94	4094.0 b	3511.0 b	3802.5
N_{12}	237.5 ab	235.5	236.5	83.13 ab	82.53 a	82.83	4210.5 b	4211.0 a	4210.8
N_{18}	250.1 a	214.0	232.1	90.93 a	72.93 b	81.93	4737.0 a	3002.0 b	3869.5
Mean	228.6	221.5		84.08	77.70		4035.2	3455.8	
CV (%)	4.01	5.49		4.72	4.33		6.22	9.89	
LSD (%)	18.32	NS		7.92	6.73		501.4	683.0	

* Different letters in columns indicate a significant difference at P < 0.05 level.

TABLE 2. The effects of nitrogen doses on dry herb yield, dry leaf yield and the essential oil content of *A. annua* L. in 1997 and 1998.

N Doses (kg/da)	Dry Herb Yield (kg/da)			Dry Leaf Yield (kg/da)			Essential Oil Content (%)		
	1997	1998	Mean	1997	1998	Mean	1997	1998	Mean
N_0	1596.3 c	1682.0 b	1639.2	493.3	535.7	514.5	0.80 b	0.82	0.81
N_6	2134.0 b	1688.1 b	1911.1	529.8	538.2	534.0	0.86 ab	0.85	0.86
N_{12}	2229.4 b	2377.0 a	2303.2	537.5	673.3	605.4	0.88 a	0.87	0.88
N_{18}	2504.2 a	1519.0 b	2011.6	661.9	465.2	563.6	0.86 ab	0.84	0.85
Mean	2116.0	1816.5		555.6	553.1		0.85	0.85	0.85
CV (%)	6.46	10.64		11.63	19.03		2.63	3.90	
LSD (%)	273.2	386.1		NS	NS		0.063	NS	

* Different letters in columns indicate a significant difference at P < 0.05 level.

The Essential Oil Content. The effect of nitrogen doses on the essential oil content of *A. annua* L. was statistically significant ($p < 0.05$) only the first year of the study. The highest essential oil content (0.88%) was obtained from N_{12} in 1997. Although the effect of nitrogen doses on essential oil content was nonsignificant in 1998, the highest value (0.87%) as in the first year was obtained from N_{12} (Table 2). In both experiment years the essential oil content in relation to nitrogen doses was determined to increase with N_{12} and then decrease with increasing applications. Our findings were lower than those essential oil contents reported by some researchers (10).

The Essential Oil Yield. The difference among nitrogen applications in terms of essential oil yield was significant ($p < 0.05$) only in 1997. The highest essential oil yield (5.69 l/da) was obtained from N_{18} in 1997. In 1998, the difference among nitrogen applications was not statistically significant; the highest yield (5.84 l/da) was obtained from N_{12} (Table 3). In 1997, essential oil yield increased in relation to the increasing nitrogen doses, and in 1998 it increased with the application of N_{12} then decreased.

Artemisinin Content. In both experiment years the effect of nitrogen doses on artemisinin content was not significant. The highest artemisinin content, with 0.017% in 1997 and 0.019% in 1998, was obtained from N_6 (Table 3). Although artemisinin contents were not significantly affected by nitrogen applications, in both experiment years the contents increased with application of N_6 and then decreased with increasing applications. Values obtained were lower than those of Nagalhaes et al. (8) who reported that artemisinin content in relation to the increasing nitrogen in *A. annua* L. decreased from 1.11% to 0.87%.

Artemisinin Yield. In both experiment years the effect of nitrogen applications on artemisinin yield of *A. annua* L. was not significant. The highest

TABLE 3. The effects of nitrogen doses on the essential oil yield, artemisinin content and artemisinin yield of *A. annua* L. in 1997 and 1998.

N Doses (kg/da)	Essential Oil Yield (l/da)			Artemisinin Content (%)			Artemisinin Yield (kg/da)		
	1997	1998	Mean	1997	1998	Mean	1997	1998	Mean
N_0	3.94 b	4.42	4.18	0.016	0.017	0.017	0.077	0.091	0.084
N_6	4.55 b	4.59	4.57	0.017	0.019	0.018	0.091	0.099	0.095
N_{12}	4.73 ab	5.84	5.29	0.014	0.015	0.015	0.076	0.101	0.089
N_{18}	5.69 a	3.91	4.80	0.012	0.013	0.013	0.071	0.060	0.066
Mean	4.73	4.69		0.015	0.016		0.079	0.088	
CV (%)	10.72	18.99		24.97	19.97		27.23	29.91	
LSD (%)	1.013	NS		NS	NS		NS	NS	

* Different letters in columns indicate a significant difference at P < 0.05 level.

artemisinin yield values were obtained in 1997 (0.091 kg/da) and 1998 (0.101 kg/da) from the applications of N_6 and N_{12}, respectively (Table 3). Our determined values were lower than those of Nagalhaes et al. (8) who reported that artemisinin yield with N applications increased from 4.04 kg/da to 4.13 kg/da.

REFERENCES

1. Baytop, T. 1984. Türkiye'de Bitkilerle Tedavi (Therapy by plants in Turkey). İ.Ü. Eczacilik Fakültesi Yayinlari No. 40, İstanbul.

2. Davis, P.H. 1975. Flora of Turkey and the East Aegean Islands, Vol. V. Edinburgh.

3. Delabays, N., A. Benakis, G. Collet 1993. Selection and breeding for high artemisinin (Qinghaosu) yielding strains of *Artemisia annua* L. Acta Horticultura, 330, 203-207.

4. Elhag, H.M., M.M. El-Domiaty, F.S. El-Feraly, J.S. Mossa, M.M. Olemy 1992. Selection and micropropagation of high artemisinin producing clones of *Artemisia annua* L. Phytotherapy Research, 6, 20-24.

5. Gupta, M.M., D.C. Jain, R.K Verma, A.P. Gupta 1996. A rapid analytical method for the estimation of artemisinin in *Artemisia annua* L. Journal of Medicinal and Aromatic Plant Sciences, 18, 5-6.

6. Laughlin, J.C. 1994. Agricultural production of artemisinin–A review. Transaction of The Royal Society of Tropical Medicine and Hygiene, 88 Supplement 1, 21-22.

7. Martinez, B.C., E.J. Staba 1988. The production of artemisinin in *Artemisia annua* L. tissues cultures. Advances in Cell Culture, 6, 69-87.

8. Nagalhaes, P., J. Raharinaivo, N. Delabays 1996. Influence de la dose et du type d'azote sur la production en Artemisine d' *Artemisia annua* L. Revue Suisse de Viticulture, d' Arboriculture et d' Horticulture, 28 (6), 349-353.

9. Simon, J.E., D. Charles, E. Cebert, L. Grant, J. Janick, A. Whipkey 1990. *Artemisia annua* L.: A promising aromatic and medicinal. Advances in New Crops, 522-526.

10. Woerdenbag, H.J., B.C. Lugt, N. Pras (1990). *Artemisia annua* L. A source of novel antimalarial drugs. Pharmaceutisch Weekbland Scientific, 12 (5), 169-181.

11. Woerdenbag, H.J., R. Bos, M.C. Salomons, H. Hendriks, N. Pras, T.M. Malingre 1992. Volatile constituents of *Artemisia annua* L. (Asteraceae). Flavour and Fragrance Journal, 8 (3), 131-137.

Propagation
of Some Native Grown Medicinal Plants
by Stem Cuttings

Filiz Ayanoğlu
Ahmet Mert
Cahit Erdoğan
Alpaslan Kaya

SUMMARY. *Salvia indica* L., *Helichrysum plicatum* subsp. *plicatum* D.C., and *Satureja thymbra* L. are some native grown plants which are used as folk medicine in Hatay province, Turkey. The propagation possibilities of these plants by stem cuttings were investigated. Different IBA (Indole 3-butiric acid) doses were applied to 10-12 cm cuttings as a rooting agent: 100, 250, 500, 1000 and 2000 ppm for *H. plicatum*; 250, 500, 1000, 2000 and 4000 ppm for *S. thymbra*; and 100, 250, 500, 1000, 2000 and 4000 ppm for *S. indica* L. The experimental design was a completely randomized block with four replications. Each replication had 25 cuttings. Cuttings were immersed in different IBA doses for 5 seconds. No IBA was applied to the control cuttings. Rooting ratio (%), root length (cm), root number (number/cutting) and rooting quality of cuttings were measured.

Filiz Ayanoğlu, Ahmet Mert, Cahit Erdoğan, and Alpaslan Kaya are affiliated with Mustafa Kemal University, Faculty of Agriculture, Field Crops Department, 31034, Antakya, Hatay, Turkey.

[Haworth co-indexing entry note]: "Propagation of Some Native Grown Medicinal Plants by Stem Cuttings." Ayanoğlu, Filiz et al. Co-published simultaneously in *Journal of Herbs, Spices & Medicinal Plants* (The Haworth Herbal Press, an imprint of The Haworth Press, Inc.) Vol. 9, No. 4, 2002, pp. 405-411; and: *Breeding Research on Aromatic and Medicinal Plants* (ed: Christopher B. Johnson, and Chlodwig Franz) The Haworth Herbal Press, an imprint of The Haworth Press, Inc., 2002, pp. 405-411. Single or multiple copies of this article are available for a fee from The Haworth Document Delivery Service [1-800-HAWORTH 9:00 a.m. - 5:00 p.m. (EST). E-mail address: getinfo@haworthpressinc.com].

The best rooting ratios were obtained from 250 and 1000 ppm IBA doses for *H. plicatum*, 1000 and 2000 ppm for *S. thymbra* and *S. indica*. *[Article copies available for a fee from The Haworth Document Delivery Service: 1-800-HAWORTH. E-mail address: <getinfo@haworthpressinc.com> Website: <http://www.HaworthPress.com> © 2002 by The Haworth Press, Inc. All rights reserved.]*

KEYWORDS. *Salvia*, sage, *Helichrysum, Satureja,* propagation, cutting, IBA

INTRODUCTION

The flora of Turkey is exceedingly rich as a temperate flora, comparable to the high species diversity found in tropical climates. Over 30% of the 8,800 plant species found in the country are endemic to Turkey. This rich biodiversity is also due to Turkey's location at the junction of several major floristic regions: Europe, the Mediterranean, and Central Asia. These rich plant genetic resources have provided the raw material for much temperate world agriculture. *Salvia indica, Helichrysum plicatum* subsp. *plicatum* D.C., and *Satureja thymbra* L. are some of native grown medicinal plants in Hatay province and have been used as folk medicine for a long time in the region (3,6)

Great differences exist among plants of different species and cultivars in the rooting ability of cuttings taken from them. Cuttings of some "difficult" cultivars can be rooted only if various influencing factors are taken into consideration and maintained at the optimum condition. IBA (indole 3-butiric acid) is probably the best material for general use, because it is generally nontoxic to plants over a wide concentration range and is effective in promoting rooting of a large number of plant species (5).

IBA is used as a quick immersion (5 seconds) rooting agent for Salvia (1), Jojoba (2), Thymbra, Origanum and Lavandula (4) and many other plants.

The objective of this research was to determine the propagation possibilities of *S. indica, H. plicatum* and *S. thymbra* by stem cuttings.

MATERIALS AND METHODS

S. indica, H. plicatum and *S. thymbra* cuttings were gathered from the flora of Hatay to determine the best rooting treatment in a two-year experiment (1999 and 2000). In the study, approximately 10-12 cm top cuttings of all three species were taken and only the basal leaves of cuttings were removed. IBA was used as a rooting hormone, applied to the basal 1 cm of the cuttings. Cuttings were dipped in IBA for 5 seconds (5) then inserted into the rooting medium. Six different IBA doses (100, 250, 500, 1000, 2000 and 4000 ppm) for *Salvia*, five doses for *Helichrysum* (100, 250, 500, 1000 and 2000 ppm) and five doses for *Satureja* (250, 500, 1000, 2000 and 4000 ppm) were the experimental treatments and cuttings with no IBA treatment were used as the control. The experimental design was a randomized complete block with four replications. Each replication had 25 cuttings. Rooting ratio (%), root length (cm), root number/cutting and rooting quality of cuttings were measured. Points (from 1 to 5) were given to each cutting to determine the rooting quality.

RESULTS

The rooting ratio of *S. indica* cuttings was not significantly affected by IBA doses in both years of the study but IBA doses significantly ($p < 0.01$ in 1999, $p < 0.05$ in 2000) affected the rooting quality of cuttings in both years. Generally the control gave the lowest rooting quality than any of the IBA treatments. IBA doses of 100, 250, and 500 ppm gave the best rooting quality in 1999 and similar results were obtained in 2000, except for the 4000 ppm IBA doses (Table 1). The number of roots per cutting was significantly ($p < 0.01$) affected by IBA doses in both years. The control gave the lowest root number. The IBA dose of 2000 ppm gave the best result in 1999 as did the 500 ppm IBA in 2000. Differences in root length were significant ($p < 0.01$) only in the first year of the study. Maximum root length (19.35 cm) was obtained from the 1000 ppm IBA application (Table 2).

TABLE 1. Rooting ratio (%) and rooting quality of *Salvia indica* L. cuttings.

IBA	Rooting ratio (%)		Rooting quality (1-5)	
(ppm)	1999	2000	1999	2000
Control	73 (59.54)*	86 (68.58)	2.33 c	4.01 b
100	85 (68.34)	97 (83.02)	3.92 a	4.48 ab
250	82 (65.81)	93 (76.73)	3.73 a	4.52 ab
500	81 (64.63)	92 (73.85)	3.64 a	4.35 ab
1000	87 (72.57)	94 (79.85)	3.34 ab	4.00 b
2000	90 (72.52)	89 (71.07)	2.86 bc	4.21 b
4000	72 (58.11)	93 (77.13)	2.85 bc	4.75 a
LSD 0.05	ns**	ns	0.6215	0.4791

*Values in parenthesis show Arcsin transformation; ** not significant

Rooting ratio of *H. plicatum* cuttings was not affected by IBA doses in both years of the study. However the rooting quality was affected significantly (p < 0.01) by IBA applications only in the first year of the study and the least rooting quality was obtained from the control (Table 3). The number of roots per plant was significantly (p < 0.01) affected by IBA doses in both years of the study. Maximum root number was obtained from the 2000 ppm IBA application. Root length was affected by IBA doses significantly (p < 0.01) only in the first year of the study. Maximum root length was obtained from 1000 and 2000 ppm IBA applications and the control gave the least results (Table 4).

TABLE 2. Root number per cutting and root length (cm) of *Salvia indica* L. cuttings.

IBA	Root number/cutting		Root length (cm)	
(ppm)	1999	2000	1999	2000
Control	4.54 c	6.19 d	14.94 c	18.67
100	10.82 ab	7.96 bcd	19.35 a	18.56
250	9.40 ab	10.02 ab	16.73 bc	19.05
500	8.74 ab	11.10 a	18.53 ab	17.92
1000	10.75 ab	7.77 cd	18.96 ab	20.53
2000	11.00 a	8.72 bc	18.11 ab	20.06
4000	7.72 bc	9.85 abc	15.28 c	22.35
LSD 0.05	3.201	1.932	2.612	ns*

* not significant

TABLE 3. Rooting ratio (%) and rooting quality of *H. plicatum* cuttings.

IBA	Rooting ratio (%)		Rooting quality (1-5)	
(ppm)	1999	2000	1999	2000
Control	78 (62.28)*	85 (67.35)	1.65 b	3.78
100	85 (68.38)	73.75 (61.13)	3.05 a	3.94
250	91 (73.01)	62.5 (52.50)	2.63 a	3.87
500	90 (74.19	56.25 (48.88)	2.92 a	3.36
1000	91 (77.47)	73.75 (59.75)	3.12 a	3.61
2000	88 (70.47)	55 (48.03)	3.12 a	3.79
LSD 0.05	ns**	ns	0.4884	ns

*Values in parenthesis show Arcsin transformation;** not significant

Rooting ratio of *S. thymbra* cuttings was affected significantly ($p < 0.01$) by IBA doses in both years of the study and the rooting ratios obtained from 1000, 2000 and 4000 ppm IBA doses were higher than the other IBA applications. The control gave the lowest results (Table 5). Differences in rooting quality were not significant in both years (Table 5). The number of roots per cutting were significantly ($p < 0.05$) affected by IBA doses only the second year of the study and low concentrations of IBA gave the lowest root numbers. Differences in root lengths were significant ($p < 0.01$) in both years. The application of 250 and 500 ppm IBA gave the best results (Table 6).

TABLE 4. Root number per cutting and root length (cm) of *H. plicatum* cuttings.

IBA	Root number/cutting		Root length (cm)	
(ppm)	1999	2000	1999	2000
Control	9.74 d	8.92 c	7.58 c	10.73
100	20.27 c	8.10 c	8.78 b	11.61
250	21.01 c	8.94 c	9.06 ab	12.56
500	25.72 bc	10.21 bc	9.48 ab	10.62
1000	30.49 ab	13.57 ab	10.01 a	11.11
2000	33.12 a	16.06 a	9.91 a	12.96
LSD 0.05	5.884	3.919	1.087	ns*

* not significant

TABLE 5. Rooting ratio (%) and rooting quality of *S. thymbra* cuttings.

IBA	Rooting ratio (%)		Rooting quality (1-5)	
(ppm)	1999	2000	1999	2000
Control	21 (26.91)* d	30.5 (32.83) c	2.12	3.51
250	27 (30.85) cd	45 (42.72) c	3.09	4.21
500	41 (39.72) bc	72 (58.25) b	3.37	4.10
1000	61 (51.62) a	82 (65.57) ab	2.97	3.74
2000	61 (51.47) a	92 (78.57) a	3.28	3.76
4000	50 (42.50) ab	90 (74.18) a	2.52	3.44
LSD 0.05	11.0	14.32	ns**	ns**

*Values in parenthesis show Arcsin transformation;** not significant

TABLE 6. Root number per cutting and root length (cm) of *S. thymbra* cuttings.

IBA (ppm)	Root no./cutting		Root length (cm)	
	1999	2000	1999	2000
Control	6.27	3.96 b	4.33 b	16.78 bc
250	9.19	5.58 ab	9.65 a	22.18 a
500	12.26	6.48 a	10.34 a	21.64 a
1000	11.33	6.75 a	8.49 a	18.57 b
2000	10.90	7.30 a	9.57 a	16.92 bc
4000	10.87	6.90 a	8.37 a	14.77 c
LSD 0.05	ns*	1.774	2.957	2.139

* not significant

DISCUSSION

Although a 100 ppm IBA concentration is recommended for rooting of *S. officinalis* L. cuttings (1), in this study no significant differences were determined among IBA applications for *S. indica* L. cuttings (Table 1). It was also determined that *S. indica* L. has high rooting ability. Therefore, there is no need to apply IBA for rooting of *S. indica* L. cuttings. However, IBA doses have some significant effects on rooting quality, the number of roots per cutting and root length, and better results were obtained than control. Therefore, 100 ppm IBA can be applied to obtain enough rooting quality.

There were no significant differences among the IBA concentrations applied to *Helichrysum* cuttings. Therefore, application of IBA to *Helichrysum* cuttings is not necessary or recommended. But IBA application gave better result than control (no application) for rooting quality, the number of roots per cutting and root length. So, low concentrations of IBA, such as 100 ppm, can be recommended for rooting quality.

The present findings suggested that IBA concentrations significantly influenced and increased rooting of *Satureja* cuttings. An IBA application of 2000 ppm gave the best results. IBA treatment also increased the root number and root length.

REFERENCES

1. Arslan, N., Gürbüz, B. and Yılmaz, G. (1993). Adaçayı (*Salvia officinalis* L.) 'ında Tohum Tutma Orani ve Çelik Alma Zamani ile Indol Butirik Asitin (IBA) Gövde Çeliklerinin Köklenmesine Etkileri Üzerine Araştirmalar. (Determination of seed set ratio and cutting time of *S. officinalis,* and the effects of IBA on the rooting of cuttings) *Turkish Journal of Agriculture and Forestry* 19: 83-87.

2. Ayanoğlu, F., Ayanoğlu, H. and Karagüzel, O. (1995). Jojoba (*Simmondsia chinensis*) Bitkisinin Çelikle Köklendirilmesi Üzerinde Araştirmalar (The propagation of Jojoba by stem cuttings), Türkiye II. Ulusal Bahçe Bitkileri Kongresi. Adana Cilt II. 645-649.

3. Ayanoğlu, F., Mert, A. and Kaya, D.A. (1999). Hatay yöresinde halk arasinda kullanilan bazi önemli tibbi ve kokulu bitkilerin tespiti ve toplanmasi. (Determination and collection of some important medicinal and aromatic plants used among people in the province of Hatay) *M.K.Ü. Ziraat Fakültesi Dergisi* (in print).

4. Ayanoğlu, F., Mert, A. and Kaya, A. (1999). Farkli IBA dozlarının doğal olarak yetişen bazı uçucu yağ bitkilerinin köklenmeleri üzerine etkileri (The effects of different IBA doses on the rooting of some native essential oil plants). 1st International Symposium on Protection of Natural Environment and Ehrami Karaçam. 23-25th Sept., Kütahya-Türkiye, 373-378

5. Hartmann, H.T., Kester, D.E. and Davies, F.T. (1990). Plant Propagation–Principles and Practices. Prentice Hall, Englewood Cliffs, New Jersey, 647 pp.

6. Zeybek N. and Zeybek U. (1994). *Farmasötik Botanik (Pharmaceutics Botany).* Ege Üniversitesi, Eczacilik Fakültesi Yayinlari, No. 2, 436.

The Effects of Different Plant Densities on Yield, Yield Components and Quality of *Artemisia annua* L. Ecotypes

Ahmet Mert
Saliha Kırıcı
Filiz Ayanoğlu

ABSTRACT. This study was conducted between 1997 and 1998 at Çukurova University, Faculty of Agriculture to determine the effects of different plant densities (5, 10, 15, 20, 25 and 30 plants/m^2) on yield and yield components and quality of *Artemisia annua* L. ecotypes (Adana, Samankaya and Serinyol from Turkey). In the study, the plant height (cm), fresh weight (kg/da), dry weight (kg/da), dry leaf weight (kg/da), essential oil content (%), essential oil yield (l/da), artemisinin content (%), and artemisinin yield (kg/da) of ecotypes were investigated. The results of this study showed that the highest essential oil content and the highest essential oil yield were found in the Samankaya ecotype when planted at 15 plants/m^2. However, the highest artemisinin content and highest artemisinin yield were found in the Serinyol ecotype when planted at 15 plants/m^2. *[Article copies available for a fee from The Haworth Document Delivery Service: 1-800-HAWORTH. E-mail address: <getinfo@haworthpressinc.com> Website: <http://www.HaworthPress.com> © 2002 by The Haworth Press, Inc. All rights reserved.]*

Ahmet Mert and Filiz Ayanoğlu are affiliated with the Department of Field Crops, Faculty of Agriculture, Mustafa Kemal University, 31034 Antakya, Hatay, Turkey.

Saliha Kırıcı is affiliated with the Department of Field Crops, Faculty of Agriculture, University of Çukurova, 01330 Adana, Turkey.

[Haworth co-indexing entry note]: "The Effects of Different Plant Densities on Yield, Yield Components and Quality of *Artemisia annua* L. Ecotypes." Mert, Ahmet, Saliha Kırıcı, and Filiz Ayanoğlu. Co-published simultaneously in *Journal of Herbs, Spices & Medicinal Plants* (The Haworth Herbal Press, an imprint of The Haworth Press, Inc.) Vol. 9, No. 4, 2002, pp. 413-418; and: *Breeding Research on Aromatic and Medicinal Plants* (ed: Christopher B. Johnson, and Chlodwig Franz) The Haworth Herbal Press, an imprint of The Haworth Press, Inc., 2002, pp. 413-418. Single or multiple copies of this article are available for a fee from The Haworth Document Delivery Service [1-800-HAWORTH 9:00 a.m. - 5:00 p.m. (EST). E-mail address: getinfo@haworthpressinc.com].

KEYWORDS. *Artemisia annua* L., annual wormwood, plant density, artemisinin, essential oil

INTRODUCTION

Artemisia annua L. (Asteraceae) is a highly aromatic annual herb with potential value as a source of artemisinin and essential oils (7). Annual wormwood *A. annua* L. has been used in Chinese traditional medicine for centuries to treat malaria and other fevers (5). Artemisinin (Qinghaosu), a sesquiterpene lactone endoperoxide isolated from the herb *A. annua* L., is a highly potent antimalarial compound (1). Artemisinin and related compounds have received considerable attention during the past decade and have opened broad perspectives for new antimalarials (8). As the compounding of artemisinin is complex and has a very low yield, only extraction from cultivated plant material is economically viable (1). In addition to artemisinin, *A. annua* L. also produces an essential oil in its leaves and flowers which can be used in perfumery and as an antimicrobial agent (4,8). The purpose of this study was to determine the effects of different plant densities on the plant parameters, yield, yield components and quality of *A. annua* L. ecotypes.

MATERIALS AND METHODS

In the study, *A. annua* L. ecotypes Adana, Samankaya and Serinyol from Turkey were used and densities of 5, 10, 15, 20, 25 and 30 plants/m^2 were applied over two years between 1997-1998 in Çukurova. *Artemisia* seeds were sown on the 15th of January and transplanted on the 15th of March. The experimental design was split plot with three replications. Plots were six rows of 3 m length with inter and intra row spacings of 30 and 15 cm, respectively The plots were harvested at full flowering period (second half of September). The essential oil content was obtained by a Neo-Clavenger distillation apparatus. Extraction of artemisinin was carried out according to the method of Gupta et al. (3). Artemisinin was detected and quantified by thin layer chromatography (TLC) in combination with a densitometer. The means were grouped in order to perform a Duncan's test.

RESULTS AND DISCUSSION

Dry Herb Yield. The effects of ecotypes ($p < 0.05$) and plant densities ($p < 0.01$) on dry herb yield of *A. annua* L. was significant in 1997. In 1998, the effects of ecotypes ($p < 0.01$) and plant densities ($p < 0.01$) and ecotypes $\times$ plant densities interaction (p < 0.01) on the dry herb yield was significant. The highest dry herb yield (3248 kg/da) among ecotypes was obtained from Serinyol and among plant densities with 2980 kg/da from 15 plant/m^2 in 1997. The highest value (3618 kg/da) was obtained from Serinyol × 15 plant/m^2 in 1998 (Table 1). The best ecotype was Seriyol and the most suitable densities were 15 and 20 plant/m^2.

Dry Leaf Yield. The effects of ecotypes (p < 0.05) and plant densities (p < 0.01) on the dry leaf yield was significant in 1997. The effects of ecotypes (p < 0.05), plant densities (p < 0.01) and ecotypes × plant densities interaction (p < 0.01) on the dry leaf yield was statistically significant in 1998. The highest value with 1099.2 kg/da was obtained from Serinyol × 15 plant/m^2 (Table 1). Our findings were higher than those reported by Delabays et al. (2).

Essential Oil Content. The effects of ecotypes (p < 0.01), plant densities (p < 0.01) and ecotypes × plant densities interaction (p < 0.01) on the essential oil content of *A. annua* L. were significant in both years of the experiment. The highest essential oil content with 1.40% was obtained from Samankaya × 20 plant/m^2 in 1997. In 1998, the highest essential oil content was obtained from Samankaya × 15 plant/m^2 with 1.38% (Table 2). Our findings were lower than

TABLE 1. The effects of plant densities on the dry herb yield and dry leaf yield of *A. annua* ecotypes (1997-1998).

Plant/m^2	Dry herb yield (kg/da)			Mean	Dry leaf yield (kg/da)			Mean
	Ecotypes				Ecotypes			
	Adana	Samankaya	Serinyol		Adana	Samankaya	Serinyol	
1997								
5	1797 h	2491 efg	3017 bcd	2435 b	566.0 f	780.4 cde	775.8 cde	707.4 c
10	2156 gh	2857 cde	3335 ab	2783 a	630.3 ef	814.5 cde	1034.6 ab	826.5 ab
15	2415 fg	2893 cde	3631 a	2980 a	747.3 c-f	868.7bcd	1107.4 a	907.8 a
20	2506 efg	2898 cde	3275 abc	2893 a	735.2 c-f	856.5bcd	911.0 bc	834.2 ab
25	2646 def	2990 bcd	3235 abc	2957 a	715.2 c-f	797.9 cd	880.8 bcd	798.0abc
30	2370 fg	2882 cde	2998 bcd	2750 a	694.9 def	760.6 c-f	800.3 cde	751.9 bc
Mean	2315 b	2835 a	3248 a		681.5 b	813.1 ab	918.3 a	
1998								
5	2169 g	2241 g	2877 b-e	2429 d	709.9 fgh	664.4 h	944.2 bc	772.9 cd
10	2404 fg	2616 ef	3142 b	2721 c	798.5 efg	800.9 efg	957.8 b	852.4 b
15	2639 def	2842 b-e	3618 a	3033 a	840.7 b-e	810.0 efg	1099.2 a	916.6 a
20	2842 b-e	2902 b-e	3478 a	3074 a	935.4 bcd	836.3 c-f	1070.3 a	947.3 a
25	2955 bcd	2826 b-e	2992 bc	2925 ab	835.7 c-f	785.7 e-h	815.0 def	812.1 bc
30	2696 c-f	2743 cde	2936 b-e	2792 bc	686.1 gh	714.4 e-h	757.6 e-h	719.4 d
Mean	2618 b	2695 b	3174 a		801.0 b	768.6 b	940.7 a	

Different letters in colums indicate a significant difference at p < 0.05 level.

TABLE 2. The effects of plant densities on the essential oil content and essential oil yield of *A. annua* ecotypes (1997-1998).

Plant/m^2	Essential oil content (%)				Essential oil yield (l/da)			
	Ecotypes			Mean	Ecotypes			Mean
	Adana	Samankaya	Serinyol		Adana	Samankaya	Serinyol	
1997								
5	0.74 g	1.34 b	0.82 def	0.97 b	4.18 h	10.45 abc	6.36 fg	7.00 c
10	0.78 fg	1.37 ab	0.86 de	1.00 a	4.91 gh	11.13 ab	8.93 cde	8.32 ab
15	0.83 def	1.37 ab	0.87 d	1.02 a	6.17 fgh	11.92 a	9.67 bc	9.25 a
20	0.82 def	1.40 a	0.81 ef	1.01 a	6.04 fgh	12.00 a	7.41 def	8.48 ab
25	0.81 ef	1.22 c	0.80 ef	0.94 bc	5.83 fgh	9.74 bc	7.07 ef	7.55 bc
30	0.79 fg	1.20 c	0.79 fg	0.93 c	5.51 fgh	9.13 cd	6.34 fg	6.99 c
Mean	0.80 c	1.32 a	0.83 b		5.44 c	10.73 a	7.63 b	
1998								
5	0.78 c	1.33 a	0.83 c	0.98 ab	5.54 h	8.83 cd	7.80 def	7.39 c
10	0.79 c	1.33 a	0.84 c	0.98 ab	6.28 fgh	10.59 ab	8.04 cde	8.30 b
15	0.80 c	1.38 a	0.86 c	1.01 a	6.73 e-h	11.23 a	9.45 bc	9.13 a
20	0.80 c	1.36 a	0.84 c	1.00 a	7.49 d-g	11.33 a	8.99 cd	9.27 a
25	0.80 c	1.13 b	0.83 c	0.92 bc	6.69 e-h	8.76 cd	6.80 e-h	7.42 c
30	0.78 c	1.10 b	0.82 c	0.90	5.35 h	7.84 def	6.21 g	6.47 d
Mean	0.79 b	1.27 a	0.84 b		6.35 c	9.76 a	7.88 b	

Different letters in colums indicate a significant difference at p < 0.05 level.

those reported by Woerdenbag et al. (8); lower limit values reported by Laughlin (4) were in the range of our findings. Although there are no significant differences among 5, 10, 15 and 20 plant/m^2 densities, higher essential oil content was obtained at 15 plant/m^2. Samankaya ecotype was also determined the best suitable ecotype for essential oil content.

Essential Oil Yield. The effects of ecotypes and plant densities on the essential oil yield of *A. annua* L. were significant (p < 0.01) in both years. The highest essential oil yield was obtained from Samankaya ecotype with 10.73 l/da in 1997 and 9.76 l/da in 1998. The highest essential oil yield was obtained in 1997 (9.25 l/da) and 1998 (9.27 l/da) from 15 plant/m^2 and 20 plant/ m^2, respectively (Table 2).

Simon et al. (7) reported that essential oil yield ranged from 5.8 to 8.1 l/da depending on plant density. While the lower values of their results were smaller than ours, the higher values were in the same range as our results. We concluded that the most suitable ecotype was Samakaya and plant density 15 or 20 plant/ m^2, for getting the highest essential oil yield.

Artemisinin Content. The effects of ecotypes (p < 0.01) and plant densities (p < 0.05) on the artemisinin content in *A. annua* L. was significant in 1997. In 1998, the effects of ecotypes (p < 0.01), plant densities (p < 0.01) and ecotypes × plant densities interaction (p < 0.01) on the artemisinin content in *A. annua* L. was statistically significant. The highest artemisinin content value (0.019%) among ecotypes in 1997 was obtained from Serinyol; the highest artemisinin contents (0.016%) among plant densities in 1997 were obtained from 15, 20 and 25 plant/m². The highest values in 1998 (0.020%) were obtained from Serinyol × 10 plant/m² and Serinyol × 15 plant/m² (Table 3). Our findings were in between those artemisinin contents reported by Laughlin (4).

Artemisinin Yield. The effects of ecotypes (p < 0.01) and plant densities (p < 0.05) on artemisinin yield in *A. annua* L. was statistically significant in 1997. In 1998, the effects of ecotypes (p < 0.01), plant densities (p < 0.01) and ecotypes × plant densities interaction (p < 0.05) on artemisinin yield in *A. annua* L. was statistically significant. The highest dry artemisinin yield (0.173 kg/da) among ecotypes in 1997 was obtained from Serinyol; the highest dry artemisinin yield (0.151 kg/da) among plant densities in 1997 was obtained from 15 plant/m². The highest value (0.219 kg/da) was obtained from Serinyol × 15 plant/m² in 1998 (Table 3). Our findings were lower than those artemisinin yields reported by Woerdenbag et al. (8) and Nagalhaes et al. (6). It is suggested that the most suitable ecotype for artemisinin yield is Serinyol at a plant density of 15 plant/m².

TABLE 3. The effects of plant densities on the artemisinin content and artemisinin yield of *A. annua* ecotypes (1997-1998).

Plant/m²	Artemisinin content (%)				Artemisinin yield (kg/da)			
	Ecotypes			Mean	Ecotypes			Mean
	Adana	Samankaya	Serinyol		Adana	Samankaya	Serinyol	
1997								
5	0.006 e	0.010 de	0.018 ab	0.012 b	0.035 f	0.079 def	0.141bcd	0.085 b
10	0.008 de	0.016 abc	0.019 ab	0.014 ab	0.051 ef	0.132 b-e	0.192 ab	0.125 ab
15	0.009 de	0.017 abc	0.021 a	0.016 a	0.070 def	0.148bcd	0.234 a	0.151 a
20	0.013bcd	0.017 abc	0.019 ab	0.016 a	0.097 c-f	0.143bcd	0.173 abc	0.138 a
25	0.013bcd	0.017 abc	0.019 ab	0.016 a	0.093 c-f	0.135 b-e	0.169 abc	0.132 a
30	0.011 cde	0.017 abc	0.016abc	0.015 a	0.077 def	0.132 b-e	0.101 b-e	0.113 ab
Mean	0.010 b	0.016 a	0.019 a		0.070 c	0.128 b	0.173 a	
1998								
5	0.007 g	0.011 def	0.017 b	0.012 b	0.050 i	0.075 ghi	0.157 c	0.094 c
10	0.010 f	0.017 b	0.020 a	0.015 a	0.080 f-i	0.134 cde	0.188 b	0.134 ab
15	0.011 ef	0.018 ab	0.020 a	0.016 a	0.092 fgh	0.142 cd	0.219 a	0.151 a
20	0.011 ef	0.018 ab	0.018 ab	0.016 a	0.103 efg	0.151 c	0.193 ab	0.149 a
25	0.013 de	0.016 bc	0.016 b	0.015 a	0.106 efg	0.127 cde	0.133 cde	0.122 b
30	0.010 f	0.015 bc	0.014 cd	0.013 b	0.066 hi	0.111 def	0.104 efg	0.094 c
Mean	0.010 c	0.016 b	0.017 a		0.083 c	0.123 b	0.166 a	

Different letters in colums indicate a significant difference at p < 0.05 level.

REFERENCES

1. Delabays, N., Benakis, A. and Collet, G. 1993. Selection and breeding for high Artemisinin (Qinghaosu) yielding strains of *Artemisia annua* L. Acta Horticulturae, 330, 203-207.

2. Delabays, N., Jenelten, U., Paris, M., Pivot, D. and Galland, N. 1994. Aspects agronomiques et genetiques de la production d' artemisine a partir d' *Artemisia annua* L. Vitic. Arboric. Hortic., 26 (5), 291-296.

3. Gupta, M.M., Jain, D.C., Verma, R.K. and Gupta, A.P. 1996. A rapid analytical method for the estimation of artemisinin in *Artemisia annua* L. Journal of Medicinal and Aromatic Plant Sciences, 18, 5-6.

4. Laughlin, J.C. 1994. Agricultural production of artemisinin–a review. Transaction of The Royal Society of Tropical Medicine and Hygiene, 88 Supplement 1, 21-22.

5. Laughlin, J.C. 1995. The influence of distribution of antimalarial constituents in *Artemisia annua* L. on time and method of harvest. Acta Horticulturae, 390, 67-73.

6. Nagalhaes, P., Raharinaivo, J. and Delabays, N. 1996. Influence de la dose et du type d'azote sur la production en Artemisine d' *Artemisia annua* L. Revue Suisse de Viticulture, d' Arboriculture et d' Horticulture, 28 (6), 349–353.

7. Simon, J.F., Charles, D., Cebert, E., Grant, L., Janick, J. and Whipkey, A. 1990. *Artemisia annua* L.: A promising aromatic and medicinal. Advances in New Crops, 522-526. Timber Press, Portland.

8. Woerdenbag, H.J., Lugt, B.C. and Pras, N. 1990. *Artemisia annua* L.: A Source of novel antimalarial drugs. Pharmaceutisch Weekbland Scientific, 12 (5), 169-181.

Index

Page numbers followed by "f" indicate figures; page numbers followed by "t" indicate tables; page numbers followed by "b" indicate boxed material.

9 780789 019738